T0214623

Applied and Numerical Harmonic Analysis

John J. Benedetto
University of Maryland
College Park, MD, USA

Editorial Advisory Board

Akram Aldroubi
Vanderbilt University
Nashville, TN, USA

Douglas Cochran
Arizona State University
Phoenix, AZ, USA

Hans G. Feichtinger
University of Vienna
Vienna, Austria

Christopher Heil
Georgia Institute of Technology
Atlanta, GA, USA

Stéphane Jaffard
University of Paris XII
Paris, France

Jelena Kovačević
Carnegie Mellon University
Pittsburgh, PA, USA

Gitta Kutyniok
Technische Universität Berlin
Berlin, Germany

Mauro Maggioni
Duke University
Durham, NC, USA

Zuowei Shen
National University of Singapore
Singapore, Singapore

Thomas Strohmer
University of California
Davis, CA, USA

Yang Wang
Michigan State University
East Lansing, MI, USA

Birkhäuser

Applied and Numerical Harmonic Analysis

More information about this series at http://www.springer.com/series/4968

Alexander I. Saichev · Wojbor Woyczynski

Distributions in the Physical and Engineering Sciences, Volume 1

Distributional and Fractal Calculus, Integral Transforms and Wavelets

 Birkhäuser

Alexander I. Saichev
Mathematical Department
State University of Nizhny Novgorod
Nizhny Novgorod, Russia

Wojbor Woyczynski
Department of Mathematics, Applied
  Mathematics and Statistics
Case Western Reserve University
Cleveland, OH, USA

ISSN 2296-5009   ISSN 2296-5017   (electronic)
Applied and Numerical Harmonic Analysis
ISBN 978-3-030-07427-2   ISBN 978-3-319-97958-8   (eBook)
https://doi.org/10.1007/978-3-319-97958-8

This book is published under the imprint Birkhäuser, www.birkhauser-science.com by the registered
company Springer Nature Switzerland AG part of Springer Nature
The registered company address is: Gewerbestrasse 11, 6330 Cham, Switzerland

Foreword

This volume is being reissued in a new format to coincide with the publication of Volume 3 of our textbook series *Distributions in the Physical and Engineering Sciences*. The latest volume is devoted to random and anomalous fractional dynamics in continuous media, while Volume 2 concentrated on linear and nonlinear dynamics in continuous media.

Our original intent for these volumes was to fill a gap in the mathematical coursework of future scientists and engineers, as well as to help improve communication between applied scientists and mathematicians. Given the increasingly interdisciplinary nature of modern research, this aim is just as, if not more, relevant today than when this volume first appeared in 1997. It is our hope that reissuing Volume 1 to coincide with the publication of Volume 3 will not only enhance the cohesiveness of the three volumes, but also make this material more accessible to all interested students and researchers.

—Wojbor Woczyński

Alexander I. SAICHEV
University of Nizhniy Novgorod
and
Wojbor A. WOYCZYŃSKI
Case Western Reserve University

DISTRIBUTIONS IN THE PHYSICAL AND ENGINEERING SCIENCES

Volume 1
Distributional and Fractal Calculus, Integral Transforms and Wavelets

BIRKHÄUSER
Boston   Basel   Berlin

Alexander I. Saichev
Radio Physics Department
University of Nizhniy Novgorod
Nizhniy Novgorod, 603022
Russia

Wojbor A. Woyczyński
Department of Statistics and Center
for Stochastic and Chaotic Processes
in Science and Technology
Case Western Reserve University
Cleveland, Ohio 44106
U.S.A.

Library of Congress Cataloging In-Publication Data
Woyczyński, W. A. (Wojbor Andrzej), 1943-
    Distributions in the physical and engineering sciences / Wojbor A.
Woyczyński, Alexander I. Saichev.
      p. cm. -- (Applied and numerical harmonic analysis)
    Includes bibliographical references and index.
    Contents: V. 1. Distributional and fractal calculus, integral
transforms, and wavelets.
    ISBN-13: 978-1-4612-8679-0   e-ISBN-13: 978-1-4612-4158-4
    DOI: 10.1007978-1-4612-4158-4

    1. Theory of distributions (Functional analysis) I. Saichev, A.
I.   II. Title.   III. Series.
QA324.W69 1996
515'.782 ' 0245--dc20

                          96-39028
                            CIP

Printed on acid-free paper

© 1997 Birkhäuser Boston   *Birkhäuser* ®

ISBN-13: 978-1-4612-8679-0

Camera-ready text prepared in LATEX by T & T TechWorks Inc., Coral Springs, FL.

9 8 7 6 5 4 3 2 1

Alexander I. Suchev
Radio Physics Department
University of Nizhny Novgorod
Nizhniy Novgorod, 60??
Russia

Wojbor A. Woyczyński
Department of Statistics and Center
for Stochastic and Chaotic Processes
in Science and Technology
Case Western Reserve University
Cleveland, Ohio 44106
U.S.A.

Library of Congress Cataloging-in-Publication Data
Woyczyński, W. A. (Wojbor Andrzej), 19??-
Distributions in the physical and engineering sciences / Wojbor A.
Woyczyński, Alexander I. Saichev.
p.   cm. — (Applied and numerical harmonic analysis)
Includes bibliographical references and index.
Contents: 1. Distributional and fractal calculus, integral transforms
and wavelets.
ISBN 0-8176-??-X (v. 1) (alk. paper) ISBN 3-7643-???-X
Boston. ISBN ? (Berlin)
1. Theory of distributions (Functional analysis) I. Saichev, A.
II. Title.   III. Series.
QA324.W65   1996
515.?82—dc??
CIP

Printed on acid-free paper

ISBN 0-8176-3924-1

Typeset by the authors in LaTeX. X 64, T & T Productions Ltd., London S.W.19, U.K.

9 8 7 6 5 4 3 2 1

Contents

Introduction

Goals and audience

The usual calculus/differential equations sequence taken by the physical sciences and engineering majors is too crowded to include an in-depth study of many widely applicable mathematical tools which should be a part of the intellectual arsenal of any well educated scientist and engineer. So it is common for the calculus sequence to be followed by elective undergraduate courses in linear algebra, probability and statistics, and by a graduate course that is often labeled *Advanced Mathematics for Engineers and Scientists*. Traditionally, it contains such core topics as equations of mathematical physics, special functions, and integral transforms. This book is designed as a text for a modern version of such a graduate course and as a reference for theoretical researchers in the physical sciences and engineering. Nevertheless, inasmuch as it contains basic definitions and detailed explanations of a number of traditional and modern mathematical notions, it can be comfortably and profitably taken by advanced undergraduate students.

It is written from the unifying viewpoint of distribution theory and enriched by such modern topics as wavelets, nonlinear phenomena and white noise theory, which became very important in the practice of physical scientists. The aim of this text is to give the readers a major modern analytic tool in their research. Students will be able to independently attack problems where distribution theory is of importance.

Prerequisites include a typical science or engineering 3-4 semester calculus sequence (including elementary differential equations, Fourier series, complex variables and linear algebra—we review the basic definitions and facts as needed). No probability background is necessary as all the concepts are explained from scratch. In solving some problems, familiarity with basic computer programming methods is necessary although using a symbolic manipulation language such as *Mathematica*, *MATLAB* or *Maple* would suffice. These skills should be acquired during freshman and sophomore years.

The book can also form the basis of a special one/two semester course on the theory of distributions and its physical and engineering applications, and serve as a supplementary text in a number of standard mathematics, physics and engineering courses such as *Signals and Systems, Transport Phenomena, Fluid Mechanics, Equations of Mathematical Physics, Theory of Wave Propagation, Electrodynamics, Partial Differential Equations, Probability Theory*, and so on, where, regrettably, the distribution-theoretic side of the material is often superficially treated, dismissed with the generic statement "... and this can be made rigorous within the distribution theory..." or omitted altogether.

Finally, we should make it clear that the book is **not** addressed to pure mathematicians who plan to pursue research in distributions theory. They do have many other excellent sources; some of them are listed in the Bibliographical Notes.

Typically, a course based on this text would be taught in a Mathematics/Applied Mathematics Department. However, in many schools, some non-mathematical sciences departments (such as Physics and Astronomy, Electrical, Systems, Mechanical and Chemical Engineering) could assume responsibility.

Philosophy

The book covers distributions theory from the applied view point; abstract functional-theoretic constructions are reduced to a minimum. The unifying theme is the Dirac delta and related one- and multidimensional distributions. To be sure, these are the distributions that appear in the vast majority of problems encountered in practice.

Our choice was based on the long experience in teaching mathematics graduate courses to physical scientists and engineers which indicated that distributions, although commonly used in their faculty's professional work, are very seldom learned by students in a systematic fashion; there is simply not enough room in the engineering curricula. This induced us to weave distributions into an exposition of integral transforms (including wavelets and fractal calculus), equations of mathematical physics and random fields and signals, where they enhance the presentation and permit achieving both, an additional insight into the subject matter and a computational efficiency.

Distribution theory in its full scope is quite a complex, subtle and difficult branch of mathematical analysis requiring a sophisticated mathematical background. Our goal was to restrict exposition to parts that are obviously effective tools in the above mentioned areas of applied mathematics. Thus many arcane subjects such as the nuclear structure of locally convex linear topological spaces of distributions are not included.

We made an effort to be reasonably rigorous and general in our exposition: results are proved and assumptions are formulated explicitly, and in such a way that the resulting proofs are as simple as possible. Since in realistic situations similar sophisticated assumptions may not be valid, we often discuss ways to expand the area of applicability of the results under discussion. Throughout we endeavor to favor constructive methods and to derive concrete relations that permit us to arrive at numerical solutions. Ultimately, this is the essence of most of problems in applied sciences.

As a by-product, the book should help in improving communication between applied scientists on the one hand, and mathematicians on the other. The first group is often only vaguely aware of the variety of modern mathematical tools that can be applied to physical problems, while the second is often innocent of how physicists and engineers reason about their problems and how they adapt pure mathematical theories to become effective tools. Experts in one narrow area often do not see the vast chasm between mathematical and physical mentalities. For instance, a mathematician rigorously proves that

$$\lim_{x \to \infty} \big(\log(\log x)\big) = \infty,$$

while a physicist, usually, would not be disposed to follow the same logic. He might say:

—Wait a second, let's check the number 10^{100}, which is bigger than most physical quantities—I know that the number of atoms in our Galaxy is less than 10^{70}. The iterated logarithm of 10^{100} is only 2, and this seems to be pretty far from infinity.

This little story illustrates psychological difficulties which one encounters in writing a book such as this one.

Finally, it is worth mentioning that some portions of material, especially the parts dealing with the basic distributional formalism, can be treated within the context of symbolic manipulation languages such as *Maple* or *Mathematica* where the package DiracDelta.m is available. Their use in student projects can enhance the exposition of the material contained in this book, both in terms of symbolic computation and visualization. We used them successfully with our students.

Organization

Major topics included in the book are split between two parts:

Part 1. *Distributions and their basic physical applications*, containing the basic formalism and generic examples, and

Part 2. *Integral transforms and divergent series* which contains chapters on Fourier, Hilbert and wavelet transforms and an analysis of the uncertainty principle, divergent series and singular integrals.

A related volume (*Distributions in the Physical and Engineering Sciences, Volume 2: Partial Differential Equations, Random Signals and Fields*, to appear in 1997) is also divided into two parts:

Part 1. *Partial differential equations*, with chapters on elliptic, parabolic, hyperbolic and nonlinear problems, and

Part 2. *Random signals and fields*, including an exposition of the probability theory, white noise, stochastic differential equation and generalized random fields along with more applied problems such as statistics of a turbulent fluid.

The needs of the applied sciences audience are addressed by a careful and rich selection of examples arising in real-life industrial and scientific labs. They form a background for our discussions as we proceed through the material. Numerous illustrations (62) help better understanding of the core concepts discussed in the text. A large number (125) of exercises (with answers and solutions provided in a separate chapter) expands on themes developed in the main text.

A word about notations and the numbering system for formulas. The list of notation is provided following this introduction. The formulas are numbered separately in each section to reduce clutter, but, outside the section in which they appear, referred to by three numbers. For example, formula (4) in section 3 of chapter 1 will be referred to as formula (1.3.4) outside Section 1.3. Sections and chapters can be easily located via the running heads.

Acknowledgments

The authors would like to thank Dario Gasparini (Civil Engineering Department), David Gurarie (Mathematics Department), Dov Hazony (Electrical Engineering and Applied Physics Department), Philip L. Taylor (Physics Department) of the Case Western Reserve University, Valery I.Klyatskin of the Institute for Atmospheric Physics, Russian Academy of Sciences, Askold Malakhov and Gennady Utkin of the Radiophysics Faculty of the Nizhny Novgorod University, George Zaslavsky of the Courant Institute at New York University, and Kathi Selig of the Fachbereich Mathematik, Universität Rostock, who read parts of the book and offered their valuable comments. A CWRU graduate student Rick Rarick also took upon himself to read carefully parts of the book from a student viewpoint and his observations were helpful in focusing our exposition. Finally, the anonymous referees issued reports on the original version of the book that we found extremely helpful and that led to a complete revision of our initial plan. Birkhäuser editors Ann Kostant and Wayne Yuhasz took the book under their wings and we are grateful to them for their encouragement and help in producing the final copy.

The second named author also acknowledges the early distribution-theoretic influences of his teachers; as a graduate student at Wrocław University he learned some of the finer points of the subject (such as Gevrey classes theory and hypoelliptic convolution equations) from Zbigniew Zieleźny (now at SUNY at Buffalo) who earlier also happened to be his first college calculus teacher at the Wrocław Polytechnic. Working with Kazimierz Urbanik (who in the 50s, simultaneously with Gelfand, created the framework for generalized random processes) as a thesis advisor also kept the functional perspective in constant view. Those interest were kept alive with the early 70s visits to Séminaire Laurent Schwartz at Paris École Polytechnique.

Authors

Alexander I. SAICHEV, received his B.S. in the Radio Physics Faculty at Gorky State University, Gorky, Russia, in 1969, a Ph.D. from the same faculty in 1975 for a thesis on *Kinetic equations of nonlinear random waves*, and his D.Sc. from the Gorky Radiophysical Research Institute in 1983 for a thesis on *Propagation and backscattering of waves in nonlinear and random media*. Since 1980 he has held a number of faculty positions at Gorky State University (now Nizhniy Novgorod University) including the senior lecturer in statistical radio physics, professor of mathematics and chairman of the mathematics department. Since 1990 he has visited a number of universities in the West including the Case Western Reserve University, University of Minnesota, etc. He is a co-author of a monograph *Nonlinear Random Waves and Turbulence in Nondispersive Media: Waves, Rays and Particles* and served on editorial boards of *Waves in Random Media* and *Radiophysics and Quantum Electronics*. His research interests include mathematical physics, applied mathematics, waves in random media, nonlinear random waves and the theory of turbulence. He is currently Professor of Mathematics at the Radio Physics Faculty of the Nizhniy Novgorod University.

Wojbor A. WOYCZYŃSKI received his B.S./M.Sc. in Electrical and Computer Engineering from Wrocław Polytechnic in 1966 and a Ph.D. in Mathematics in 1968 from Wrocław University, Poland. He has moved to the U.S. in 1970, and since 1982, has been Professor of Mathematics and Statistics at Case Western Reserve University in Cleveland, and served as chairman of the department there from 1982 to 1991. Before, he has held tenured faculty positions at Wrocław University, Poland, and at Cleveland State University, and visiting appointments at Carnegie-Mellon University, Northwestern University, University of North Carolina, University of South Carolina, University of Paris, Gottingen University,

Aarhus University, Nagoya University, University of Minnesota and the University of New South Wales in Sydney. He is also (co-)author and/or editor of seven books on probability theory, harmonic and functional analysis, and applied mathematics, and serves as a member of editorial boards of the *Annals of Applied Probability, Probability Theory and Mathematical Statistics,* and the *Stochastic Processes and Their Applications.* His research interests include probability theory, stochastic models, functional analysis and partial differential equations and their applications in statistics, statistical physics, surface chemistry and hydrodynamics. He is currently Director of the CWRU Center for Stochastic and Chaotic Processes in Science and Technology.

Notation

$\lceil \alpha \rceil$	—	least integer greater than or equal to α
$\lfloor \alpha \rfloor$	—	greatest integer less than or equal to α
C	—	concentration
$\mathbf{C}$	—	complex numbers
$C(x)$	$=$	$\int_0^x \cos(\pi t^2/2)\, dt$, Fresnel integral
C^∞	—	space of smooth (infinitely differentiable) functions
$\mathcal{D}$	$=$	C_0^∞, space of smooth functions with compact support
$\mathcal{D}'$	—	dual space to $\mathcal{D}$, space of distributions
$\bar{D}$	—	the closure of domain D
D/Dt	$=$	$\partial/\partial t + v \cdot \nabla$, substantial derivative
$\delta(x)$	—	Dirac delta centered at 0
$\delta(x-a)$	—	Dirac delta centered at a
Δ	—	Laplace operator
$\mathcal{E}$	$=$	C^∞-space of smooth functions
$\mathcal{E}'$	—	dual to $\mathcal{E}$, space of distributions with compact support
$\mathrm{erf}\,(x)$	$=$	$(2/\sqrt{\pi}) \int_0^x \exp(-s^2)\, ds$, the error function
$\tilde{f}(\omega)$	—	Fourier transform of $f(t)$
$\{f(x)\}$	—	smooth part of function f, see page 104
$\lfloor f(x) \rceil$	—	jump of function f at x
ϕ, ψ	—	test functions
$\gamma(x)$	—	canonical Gaussian density
$\gamma_\epsilon(x)$	—	Gaussian density with variance ϵ
$\Gamma(s)$	$=$	$\int\limits_0^\infty e^{-t} t^{s-1} dt$, gamma function
(h, g)	$=$	$\int h(x) g(x) dx$, the Hilbert space inner product
$\chi(x)$	—	canonical Heaviside function, unit step function
$\hat{H}$	—	the Hilbert transform operator
j, J	—	Jacobians
$I_A(x)$	—	the indicator function of set A ($=1$ on A, $=0$ off A)
$\mathrm{Im}\, z$	—	the imaginary part of z
$\lambda_\epsilon(x)$	$=$	$\pi^{-1}\epsilon(x^2 + \epsilon^2)^{-1}$, Cauchy density

$L^p(A)$	—	Lebesgue space of functions f with $\int_A |f(x)|^p \, dx < \infty$
$\mathbf{N}$	—	nonnegative integers
$\phi = O(\psi)$	—	ϕ is of the order not greater than ψ
$\phi = o(\psi)$	—	ϕ is of the order smaller than ψ
$\mathcal{PV}$	—	principal value of the integral
$\mathbf{R}$	—	real numbers
$\mathbf{R}^d$	—	d-dimensional Euclidean space
Re z	—	the real part of z
ρ	—	density
sign (x)	=	1 if $x > 0$, -1 if $x < 0$, and 0 if $x = 0$
sinc ω	=	$\sin \pi \omega / \pi \omega$
$\mathcal{S}$	—	space of rapidly decreasing smooth functions
$\mathcal{S}'$	—	dual to $\mathcal{S}$, space of tempered distributions
$S(x)$	=	$\int_0^x \sin(\pi t^2/2) \, dt$, Fresnel sine integral
T, S	—	distributions
$T[\phi]$	—	action of T on test function ϕ
T_f	—	distribution generated by function f
$\widehat{T}$	—	generalized Fourier transform of T
z^*	—	complex conjugate of number z
$\mathbf{Z}$	—	integers
∇	—	gradient operator
$\mapsto$	—	Fourier map
$\rightarrow$	—	converges to
$\Rightarrow$	—	uniformly converges to
$*$	—	convolution
$[\![\, . \,]\!]$	—	physical dimensionality of a quantity
$\emptyset$	—	empty set
$\blacksquare$	—	end of proof, example

Part I

DISTRIBUTIONS AND THEIR BASIC APPLICATIONS

Chapter 1

Basic Definitions and Operations

1.1 The "delta function" as viewed by a physicist and an engineer

The notion of a *distribution* (or a *generalized function*—the term often used in other languages) is a comparatively recent invention, although the concept is one of the most important in mathematical areas with physical applications. By the middle of the 20th century, the theory took final shape, and distributions are commonly used by physicists and engineers today.

This book presents an exposition of the theory of distributions, their range of applicability, and their advantages over familiar smooth functions. The *Dirac delta function*—more often called the *delta function*—is the most fundamental distribution, introduced by the physicists as a convenient "automation" tool for handling unwieldy calculations. Its introduction was preceded by the practical use of another standard discontinuous function, the so-called *Heaviside function*, which was applied in the analysis of electrical circuits. However, as is the case of many mathematical techniques that are heuristically applied by physicists and engineers, such as the nabla operator or operational calculus, intuitive use can sometimes lead to false conclusions, which explains the need for a rigorous mathematical theory.

Let us begin with describing the way in which distributions and, in particular, the "delta function", are usually introduced in physical sciences.

Typically, the delta function is defined as a limit, as $\varepsilon \to 0$, of certain rectangular functions (see Fig.1.1.1),

$$f_\varepsilon(x) = \begin{cases} 1/2\varepsilon, & \text{for } |x| < \varepsilon; \\ 0, & \text{for } |x| > \varepsilon. \end{cases} \tag{1}$$

As $\varepsilon \to 0$, the rectangles become narrower, but taller. However, their areas always

© Springer Nature Switzerland AG 2018
A. I. Saichev and W. Woyczynski, *Distributions in the Physical and Engineering Sciences, Volume 1*, Applied and Numerical Harmonic Analysis, https://doi.org/10.1007/978-3-319-97958-8_1

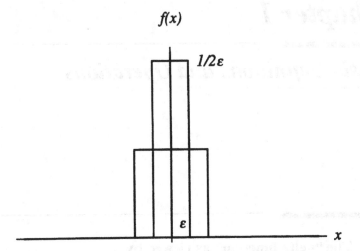

FIGURE 1.1.1
A naive representation of the delta function as a limit of rectangular functions.

remains constant, since for any ε,

$$\int f_\varepsilon(x)dx = 1. \tag{2}$$

In other words, the delta-function is being defined as a pointwise limit

$$\delta(x) = \lim_{\varepsilon \to 0} f_\varepsilon(x). \tag{3}$$

This pointwise limit, as can be easily seen, is zero everywhere except at the point $x = 0$ where it is infinity. Therefore, the common in the applied literature definition of the delta function is

$$\delta(x) = \begin{cases} \infty, & \text{for } x = 0; \\ 0, & \text{for } |x| > 0, \end{cases} \tag{4}$$

under the additional condition that the area beneath it is equal to one. This, in particular, yields the well-known *probing* property of the delta function when convolved with any continuous function:

$$\int \delta(x - a)\phi(x)dx = \phi(a). \tag{5}$$

In other words, integrating ϕ against a delta function we recover the value of ϕ at the (only) point where the delta function is not equal to zero. Here, and throughout the remainder of the book, an integral written without the limits will indicate integration over the entire infinite line (plane, space, etc.), that is, from $-\infty$ to $+\infty$.

At this point we would also like to bring up the question of dimensionality which is always of utmost importance to physicists and engineers, but usually neglected by mathematicians. The delta function is one of the few *self-similar functions* whose argument can be a dimensional variable, for example a spatial coordinate x or time t, and depending on the dimension of its argument, the delta function itself has a nonzero dimension. For instance, the dimension of the delta function of time is equal to the inverse time,

$$[\![\delta(t)]\!] = 1/T,$$

i.e., the dimension of frequency since, by definition, the integral of the delta function of time with respect to time is equal to one—a dimensionless quantity.

Notice that $\delta(t)$ has the same dimension as the inverse power function $1/t$. In what follows (see Section 6.2) we will derive formulas important for physical applications formulas which provide a deeper inner connection between such seemingly unrelated functions.

1.2 A rigorous definition of distributions

The physical definition of the delta function introduced in the previous section is not mathematically correct. Even if we skip over the question of whether functions can take ∞ as a value, the integral of the delta function given by equality (1.1.4) is either not well defined if understood as a Riemann integral, or equals zero if understood as a Lebesgue integral. Observe, however, that for each $\varepsilon > 0$, the integral

$$T_\varepsilon[\phi] = \int f_\varepsilon(x)\phi(x)dx \tag{1}$$

exists for any fixed continuous *test function* ϕ, and as $\varepsilon \to 0_+$, it converges to the value of the test function at zero:

$$\lim_{\varepsilon \to 0_+} T_\varepsilon[\phi] = T[\phi] = \phi(0). \tag{2}$$

As we show below, one of the possible mathematically correct definitions of the delta function can be based on integral equalities of type (2) and their interpretation as limits of integrals of type (1) rather than on the pointwise limits of ordinary functions. Recall that the integral (1) represents what in mathematics is called a *linear functional* on test functions $\phi(x)$, generated by the function $f_\varepsilon(x)$ which determines all the functional's properties and is called the *kernel* of a functional.

The notion of a functional is more general than that of a function of a real variable. A functional depends on a variable which is a function itself but its values are real numbers. This modern mathematical notion will help us develop a rigorous definition of the delta function. It should be noted that Paul DIRAC, "father" of the delta function and one of the creators of quantum mechanics, recognized the necessity of a functional approach to distributions earlier than most of his fellow physicists. For this reason the rigorous version of the intuitive delta function will be called henceforth the *Dirac delta distribution*, or simply the *Dirac delta*. We will use that term to emphasize that the delta function is not a function.

Let us consider a linear functional $T[\phi]$ on test functions ϕ, generated by an integral

$$T[\phi] = \int f(x)\phi(x)dx \tag{3}$$

with kernel $f(x)$. Test functions ϕ will come from a certain set D of test functions which will be selected later. Once this set of test functions is chosen, the set of linear functionals on D, called the *dual space* of D and denoted D', will be automatically determined. It is these functionals that will be identified later with distributions. The functional which assigns to each test function its value at 0 will correspond to the Dirac delta distribution. It is worthwhile to observe that the narrower the set of the test functions, the broader the set of linear functionals defined on the latter, and vice versa. Therefore, as a rule, to obtain a large set of distributions, we have to impose rather strict constraints on the set of test functions. At the same time the set of test functions should not be too small. This would restrict the range of problems where the distribution theoretic tools can be used.

There are a few natural demands on the set D of test functions. In particular, it has to be broad enough to identify usual continuous kernels f via the integral functional (3). In other words, once the values of functional $T[\phi]$ are known for all $\phi \in D$, kernel f has to be uniquely determined. Paraphrasing, we can say that we require that the set D' of distributions be rich enough to include all continuous functions.

It turns out that the *family of all infinitely differentiable functions with compact support* is a good candidate for the space D of test functions. From now on, we shall reserve D for this particular space. Recall that a function is said to be of *compact support* if it is equal to zero outside a certain bounded set on the x-axis. The support of f itself, denoted supp f, is by definition the closure of the set of x's such that $f(x) \neq 0$.

Let us show that *the value of a continuous function f at any point x is determined by the values of functional (3) on all test functions $\phi \in D$.*

Consider the function

$$\omega(x) = \begin{cases} C \exp\{-(1-x^2)^{-1}\}, & \text{for } |x| < 1; \\ 0, & \text{for } |x| \geq 1, \end{cases} \tag{4}$$

where the constant C is selected in such a way that the normalization condition

$$\int \omega(x)dx = 1$$

is satisfied. It turns out that $C \approx 2.25$, and the bell-shaped function is pictured in Fig. 1.2.1. It can be easily shown that the function ω is an infinitely differentiable function with compact support. Indeed, it vanishes outside the $[-1, 1]$ interval which is bounded, and it has derivatives of arbitrary order everywhere, including the two delicate points $+1$, and -1, where at $+1$ one checks that all the left derivatives are zero and the right derivatives are obviously identically zero, and one proceeds similarly at -1.

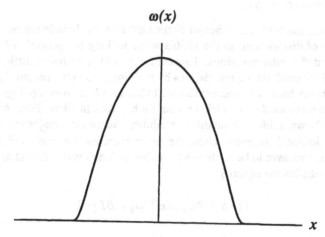

$$\omega(x)$$

FIGURE 1.2.1
Graph of a bell-shaped function which is both very smooth and has compact support.

Rescaling the function ω, for each $\varepsilon > 0$, we can produce a new function

$$\omega_\varepsilon(x) = \frac{1}{\varepsilon}\omega\left(\frac{x}{\varepsilon}\right).$$

Clearly, it has compact support as it vanishes outside the interval $[-\varepsilon, \varepsilon]$, and it is also infinitely differentiable, as can be checked by an application of the chain rule. Moreover, changing the variables one can check that

$$\int \omega_\varepsilon(x)dx = 1.$$

It follows from the generalized mean value theorem for integrals, and from the continuity of the function $f(x)$ that, as $\varepsilon \to 0$, the value of the functional

$$T_\varepsilon[f] = \int f(x)\omega_\varepsilon(x)dx \to f(0).$$

Thus the value of f at 0 can be recovered by evaluating functionals T_ε at f. Values of f at other y's can be recovered by evaluating the integral functionals on test functions ω_ε shifted by y. This gives a proof of our statement.   ∎

By definition, *any linear functional $T[\phi]$ which is continuous on the set D of infinitely differentiable functions with compact support is called a distribution. The set of all distributions on D, that is the dual space D', is often called the Sobolev-Schwartz space.*

A Russian mathematician Sergei SOBOLEV laid the foundation of the rigorous theory of distributions in the 1930's while looking for generalized solutions for partial differential equations. Laurent SCHWARTZ, a French mathematician, completed the work on the foundations by building a precise structure for the distribution theory based on concepts that are called locally convex topological vector spaces. He was awarded the Fields Medal for his work in 1950. Thus, for the first time since Newton, ideas about differentiability underwent a major revision.

A few additional comments about the above definition are warranted, and some of its statements have to be made more precise. A functional T is said to be *linear* on D if it satisfies the equality

$$T[\alpha\phi + \beta\psi] = \alpha T[\phi] + \beta T[\psi]$$

for any test functions ϕ and ψ in D and arbitrary real (or complex) numbers α and β. A functional T on D is called *continuous* if for any sequence of functions $\phi_k(x)$ from D which converge to a test function $\phi(x)$, the numbers $T[\phi_k]$, representing the values of the functional T on ϕ_k's, converge to the number $T[\phi]$. The convergence of the sequence ϕ_k of test functions in D to ϕ is understood in this case as meaning

(1) The supports of the ϕ_k's, that is the (closures of) sets of x's where $\phi_k \neq 0$, are all contained in a fixed bounded set on the x-axis, and

(2) As $k \to \infty$, the functions ϕ_k themselves, and all their derivatives $\phi_k^{(n)}(x), n = 1, 2, \ldots$, converge *uniformly* to the corresponding derivatives of the limit test function $\phi(x)$, that is, for each $n = 0, 1, 2, \ldots$,

$$\phi_k^{(n)} \Longrightarrow \phi^{(n)}. \tag{5}$$

Let us give several examples of such functionals.

Example 1. Let the function $f(x)$ appearing on the right hand side of (3) be *locally integrable*, that is, integrable over any finite interval of the x-axis. A continuous function on the whole axis is an example of such a function as well as a function which is simply integrable. Then the right-hand side of (3) is well defined for any $\phi \in \mathcal{D}$ and clearly defines a linear functional on it. Its continuity in the sense defined above is immediately verifiable. A distribution defined in such a way, with the help of a standard "good" function f, will be called a *regular distribution* and denoted T_f. In this context, we can say that locally integrable functions can be identified with certain distributions in the distribution space $\mathcal{D}'$. ∎

However, some linear continuous functionals (distributions) on $\mathcal{D}$ cannot be identified with locally integrable kernels, and are then called *singular distributions*.

Example 2. The simplest example of a singular distribution is a functional that assigns to each test function ϕ in $\mathcal{D}$ its value at $x = 0$. This distribution is traditionally denoted by δ, and thus by definition,

$$\delta[\phi] = \phi(0). \tag{6}$$

It is not a regular distribution generated by a locally integrable function, and it is called the *Dirac delta*. The above defining equation is sometimes written heuristically in the integral form

$$\int \delta(x)\phi(x)dx = \phi(0),$$

although formally the integral on the left-hand side does not make any sense. However, the above equation can serve as an intuitive mnemotechnic rule that, if used judiciously, will greatly facilitate actual calculations involving Dirac delta distributions. By now, it should be also clear to the reader that the name "delta function" is a misnomer. The delta function is not a function but a singular distribution functional, and one should not talk lightly about its value at x. Writing the argument x is however convenient since in the future it will permit us to talk about the distribution functional $\delta(x - a)$ defined by the formula

$$\int \delta(x - a)\phi(x)dx = \phi(a); \tag{7}$$

it could be thought of as the Dirac delta shifted by a. Another more rigorous possibility would be to denote by δ_a the Dirac delta centered at a, but this notation becomes unwieldy if a has to be replaced in the subscript by a more complex expression. ∎

In what follows, in addition to routine (ab)use of the integrals (3) and (7), to denote the action of the distribution functional on test functions, we will utilize another convenient compact notation

$$T_f[\phi] = \int f(x)\phi(x)dx. \tag{8}$$

Remark 1. When discussing both regular functions and singular distributions, a vital role was played by the notion of the support of a function. Recall that support of a regular function $f(x)$ was defined as closure of the set of x's where $f(x)$ was different from 0. Thus the support of the bell-shaped function from (4) was the segment $[-1, 1]$. Similarly, one can define the notion of support of a distribution functional. The distribution T is considered to be equal to zero in the open region B on the x-axis if $T[\phi] = 0$ for all the test functions ϕ with supports contained in B. The complement of the largest open region in which distribution T is equal to zero will be called the *support of distribution* T and denoted supp T. It immediately follows from the above definition that the support of the delta function consists of a single point $x = 0$, that is,

$$\text{supp } \delta = \{0\}.$$

1.3 Singular distributions as limits of regular functions

Although the Dirac delta distribution itself cannot be represented in the form of an integral functional, it can be obtained as a limit of a sequence of integral functionals

$$T_k[\phi] = \int f_k(x)\phi(x)dx \tag{1}$$

with respect to kernels that are regular functions, for example, the rectangular functions introduced in Section 1.1. In a sense the distribution δ can then be understood as being represented by such an approximating sequence $\{f_k(x)\}$, and many properties of δ can be derived from the properties of the sequence f_k. The approximation of δ by f_k in the sense that, for each test function ϕ in $\mathcal{D}$

$$T_k[\phi] \to \delta[\phi]$$

as $k \to \infty$, is called the *weak approximation* and the corresponding convergence —*the weak convergence.*

The choice of a weakly convergent sequence of distributions $\{T_k\}$ represented by regular functions $\{f_k\}$ is clearly not unique, and instead of rectangular functions from Section 1.1, it is always possible and often more convenient to select them in such a way that functions $f_k(x)$ are infinitely differentiable (although not necessarily with compact support).

Example 1. Consider the family of Gaussian functions (see Fig. 1.3.1)

$$\gamma_\varepsilon(x) = \frac{1}{\sqrt{2\pi\varepsilon}} \exp\left(-\frac{x^2}{2\varepsilon}\right) \tag{2}$$

parametrized by a parameter $\varepsilon > 0$, and take as a weakly approximating sequence $f_k(x) = \gamma_{1/k}(x), k = 1, 2, \ldots$. Notice that the constant in front of the exponential function in (2) has been selected in such a way that

$$\int \gamma_\varepsilon(x)dx = 1.$$

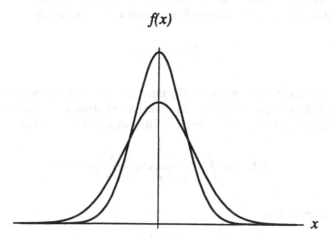

$f(x)$

FIGURE 1.3.1
Graphs of the first two elements of the sequence of Gaussian functions weakly convergent to Dirac delta.

As $k \to \infty$, we have that $\varepsilon \to 0$, and the approximating Gaussian functions become higher and higher peaks, more and more concentrated around $x = 0$, while preserving the total area underneath them. This satisfies the above normalization condition. ∎

Example 2. Let us consider another sequence of regular functionals converging weakly to the Dirac delta, determined by the kernels $f_k = \lambda_{1/k}$, where

$$\lambda_\varepsilon(x) = \frac{1}{\pi} \frac{\varepsilon}{x^2 + \varepsilon^2}. \tag{3}$$

Physicists often call these functions *Lorentz curves*, and mathematicians call them *Cauchy densities*.   ∎

Although at first sight Lorentz functions look somewhat like Gaussian functions (see Fig. 1.3.1) there are some significant differences. Both are infinitely differentiable and integrable functions (satisfying the above normalization condition) on the entire x-axis, since the indefinite integral

$$\int \frac{1}{x^2 + 1} dx = \arctan x$$

has a finite limit in ∞ and $-\infty$, and both have values at zero that blow up to $+\infty$ as $\varepsilon \to 0$, since

$$\gamma_\varepsilon(0) = \frac{1}{\sqrt{2\pi\varepsilon}}, \quad \text{and} \quad \lambda_\varepsilon(0) = \frac{1}{\pi\varepsilon}.$$

But whereas a Gaussian function decays exponentially to 0 as $x \to \pm\infty$, the asymptotic behavior of the Lorentz functions at $x \to \pm\infty$ is only

$$\lambda_\varepsilon(x) \sim \frac{\varepsilon}{\pi x^2}$$

so that they decay to 0 much less rapidly than the Gaussian functions, and the areas underneath their graphs are much less concentrated around the origin $x = 0$ than those of Gaussian functions. Hence, in particular, if $f(x) = x^2$ then

$$T_f[\gamma_\varepsilon] = \int x^2 \frac{1}{\sqrt{2\pi\varepsilon}} \exp\left(-\frac{x^2}{2\varepsilon}\right) dx = \varepsilon \qquad (4)$$

is well defined, while

$$T_f[\lambda_\varepsilon] = \int x^2 \frac{1}{\pi} \frac{\varepsilon}{x^2 + \varepsilon^2} dx$$

is not, since the integral on the right diverges. However, for all the test function $\phi \in \mathcal{D}$, and for $\varepsilon \to 0$,

$$T_\phi[\lambda_\varepsilon] \to \phi(0)$$

in view of the compact support of test functions.

Here a note of caution is in order lest the reader get the impression that regular functions weakly approximating the Dirac delta must concentrate their nonzero values in the neighborhood of $x = 0$. This is not the case once we abandon the restriction (which appeared without mentioning it in the above two examples) that the weakly approximating regular functions be positive or even real-valued.

Example 3. Consider complex-valued oscillating functions

$$f_\varepsilon(x) = \sqrt{\frac{i}{2\pi\varepsilon}} \exp\left(-\frac{ix^2}{2\varepsilon}\right), \tag{5}$$

parametrized by $\varepsilon > 0$. Their real parts are pictured in Fig. 1.3.2. They are frequently encountered in quantum mechanics and quasi-optics. In quasi-optics they appear as the Green's functions of a monochromatic wave in the Fresnel approximation. The modulus of these function is constant

$$|f_\varepsilon(x)| = \frac{1}{\sqrt{2\pi\varepsilon}},$$

for any x, which diverges to ∞ as $\varepsilon \to 0$.

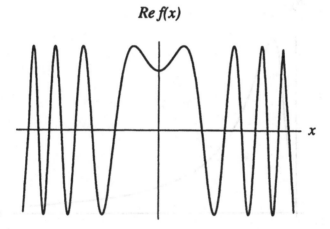

Re f(x)

FIGURE 1.3.2
Graph of an element of a sequence of functions $f_\varepsilon(x)$ (5) which do not converge to 0 for $x \neq 0$ as $\varepsilon \to 0$, and still weakly converge to the Dirac delta.

Nevertheless, as $\varepsilon \to 0$, these functions converge weakly to the Dirac delta. In physical terms it can be explained by the fact that function $f_\varepsilon(x)$ defined by (5) oscillates at a higher and higher rate the smaller ε becomes. As a result, the integrals of their products with any test function $\phi(x)$ supported by a region which excludes point $x = 0$ converge to $\phi(0) = 0$ as $\varepsilon \to 0$. ∎

Notice that all of the above examples of weakly approximating families f_ε for the Dirac delta have been constructed with the help of a single function f, be it Gaussian, Lorentz or oscillating complex-valued, which was later rescaled following the same rule:

$$f_\varepsilon(x) = \frac{1}{\varepsilon} f\left(\frac{x}{\varepsilon}\right).$$

The properties of the limiting Dirac delta really do not depend on the particular analytic form of the original regular function f. Practically, any smooth enough function satisfying the normalization condition $\int f(x)\,dx = 1$ will do. In particular, function f need not be symmetric (even). The sequence produced by rescaling function

$$f(x) = \begin{cases} x^{-2} \exp(-1/x), & \text{for } x > 0; \\ 0, & \text{for } x \leq 0, \end{cases} \tag{6}$$

whose plot is represented in Fig. 1.3.3 will also weakly approximate the delta function. However, in what follows, we shall see that the fine structure of function f should not be always ignored by the physicists and that it can affect the final physical result.

$f(x)$

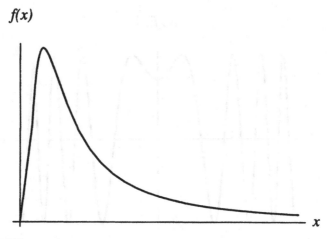

FIGURE 1.3.3
An example of a function f which is not even but such that $f(x/\varepsilon)/\varepsilon$, weakly converge to δ as $\varepsilon \to 0$.

1.4 Derivatives; linear operations

The infinite differentiability of the chosen set $\mathcal{D}$ of test functions $\phi(x)$ allows us to define, for any distribution $T \in \mathcal{D}'$, a derivative of arbitrary order, thus freeing us from a constant worry about differentiability within the class of regular functions.

It is one of the main advantages the theory of distributions has over the classical calculus of regular functions. Before we provide a general definition, let us observe that the familiar integration-by-parts formula in the integral calculus applied to a differentiable function $f(x)$ and a test function $\phi(x) \in \mathcal{D}$ reduces to

$$\int f'(x)\phi(x)dx = -\int f(x)\phi'(x)dx, \tag{1}$$

since the boundary term

$$f(x)\phi(x)\Big|_{-\infty}^{\infty} = 0,$$

because the test function ϕ is zero outside a certain bounded set on the x-axis. If we think about the regular function f as representing a distribution T_f, then equation (1) can be rewritten as

$$(T_f)'[\phi] = -T_f[\phi'], \tag{2}$$

which is valid for any test function ϕ. We can take the above equality as a definition of the functional on the left-hand side and call it the derivative of the distribution T_f —notice that the right hand side does not depend on the differentiability of f. This idea can be extended to any distribution.

If T is a distribution in $\mathcal{D}'$ then its derivative T' is defined as a distribution in $\mathcal{D}'$ which is determined by its values (as a functional) on test functions $\phi \in \mathcal{D}$ by the equality

$$T'[\phi] = -T[\phi'].$$

It is always well defined, since it is a linear and continuous functional on $\mathcal{D}$. Derivatives of higher order are defined by consecutive application of the operation of the first derivative. Hence, by definition, if T is a distribution on $\mathcal{D}$ then its n-th derivative $T^{(n)}$ is again a distribution in $\mathcal{D}$ determined by its values on test functions $\phi \in \mathcal{D}$ by

$$T^{(n)}[\phi] = (-1)^n T[\phi^{(n)}].$$

So, distributions *always* have derivatives of all orders. It is a very nice universe, indeed. Let us illustrate the concept of the distributional derivative on the Dirac delta distribution.

Example 1. Consider the distribution $\delta(x - a)$ which is defined by the probing property at the point $x = a$:

$$\delta(x - a)[\phi] = \phi(a),$$

or, writing informally, by the condition

$$\int \delta(x - a)\phi(x) = \phi(a).$$

Hence its nth derivative $(\delta(x-a))^{(n)}$ is defined by the equality

$$(\delta(x-a))^{(n)}[\phi] = (-1)^n \phi^{(n)}(a).$$

In particular, the first derivative δ' of the Dirac delta is the functional on $\mathcal{D}$ defined by the equality

$$\delta'[\phi] = -\phi'(0).$$

The weak approximation of δ' by regular functions can be accomplished, for example, by taking a sequence of derivatives γ'_ε of Gaussian functions from formula (1.3.2) (see Fig. 1.4.1).

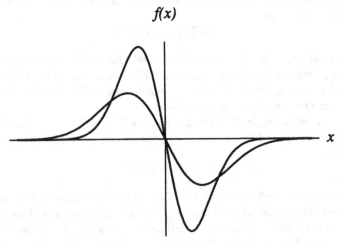

f(x)

FIGURE 1.4.1
Approximating functions of the first derivative of the Dirac delta.

(or of any other smooth weak approximants) since

$$\int \gamma'_\varepsilon(x)\phi(x)dx = -\int \gamma_\varepsilon(x)\phi'(x)dx \to -\delta[\phi'] = \delta'[\phi]$$

as $\varepsilon \to 0$.   ∎

Notice that the *operation of differentiation is a linear operation on the space of distributions* in the sense that if we define the *linear combination* of two distributions T and S from $\mathcal{D}'$ by the equality

$$(\alpha T + \beta S)[\phi] = \alpha T[\phi] + \beta S[\phi],$$

where α and β are numbers, then

$$(\alpha T + \beta S)' = \alpha T' + \beta S'.$$

The proof of this fact is immediate from the above basic definitions.

1.5 Multiplication by a smooth function; Leibniz formula

Another linear operation, which produces a new distribution from a distribution T and an infinitely differentiable function g, is the multiplication of T by g. Denote the set of all infinitely differentiable functions (but not necessarily with compact support) on the x-axis by C^∞.

By definition, the product gT of a function $g \in C^\infty$ by a distribution $T \in \mathcal{D}'$ is a distribution in $\mathcal{D}'$ determined by

$$(gT)[\phi] = T[g\phi], \quad \phi \in \mathcal{D}. \tag{1}$$

The right-hand side is well defined since the product of an infinitely differentiable $g(x)$ by a test function from $\mathcal{D}$ is again a function from $\mathcal{D}$, and in particular it has compact support. The above formula obviously corresponds to a formula for regular functions:

$$\int \Big(g(x)f(x)\Big)\phi(x)dx = \int f(x)\Big(g(x)\phi(x)\Big)dx.$$

The above definition, in the particular case of a constant function $g(x) = c$, which certainly is infinitely differentiable, provides a definition of cT—a product of the number c by the distribution $T \in \mathcal{D}'$:

$$(cT)[\phi] = T[c\phi].$$

Example 1. Let us calculate the product of an arbitrary infinitely differentiable g with the delta function $\delta(x - a)$. By definition

$$(g\delta(x - a))[\phi] = \delta(x - a)[g\phi] = g(a)\phi(a) = g(a)\delta(x - a)[\phi]$$

and we have demonstrated that the distribution

$$g\delta(x - a) = g(a)\delta(x - a). \qquad \blacksquare$$

Observe that our definition does not allow multiplication of distributions by functions that are not infinitely differentiable. The product $g\phi$ on the right-hand side of the defining formula (1) has to be infinitely differentiable if we are to apply the functional T to it, and that cannot be guaranteed unless g itself is infinitely differentiable. This is an essential restriction that has to be kept in mind.

The differentiation of distributions and their multiplication by a smooth function are tied together by an analogue of the classical *Leibniz formula* for the derivative of a product of two functions.

If g is a function from C^∞ and T is a distribution from $\mathcal{D}'$ then

$$(gT)' = g'T + gT'. \tag{2}$$

Indeed, by (1.4.2), applying the left-hand side to a test function ϕ we get that

$$(gT)'[\phi] = -(gT)[\phi'] = -T[g\phi'] = -T[(g\phi)' - g'\phi]$$

$$= -T[(g\phi)'] + T[g'\phi] = T'[g\phi] + (g'T)[\phi]$$

$$= (gT')[\phi] + (g'T)[\phi] = (g'T + gT')[\phi]. \qquad ∎$$

Similarly, one can prove a *general Leibniz formula*

$$(gT)^{(n)} = \sum_{m=0}^{n} \binom{n}{m} g^{(m)} T^{(n-m)}, \tag{2a}$$

for distributions. It is well known for smooth functions from the standard calculus courses.

Formulas (2) and (2a) may look nice and elegant, but in practice, different portions of the above chain of equalities may turn out to be more useful in the evaluation of the derivative of a product gT.

Example 2. Applying the Leibniz formula to the product of g and a distribution $\delta(x - a)$, we immediately get from the second equality in the above chain that

$$(g\delta(x - a))'[\phi] = -g(a)\phi'(a).$$

Thus, in particular, for a constant c

$$(c\delta(x - a))'[\phi] = -c\phi'(a).$$

The above formula can be obtained in a more straightforward manner by observing that

$$(g\delta(x-a))'[\phi] = -(g\delta(x-a))[\phi'],$$

and then using the calculation of $g\delta(x-a) = g(a)\delta(x-a)$ from Example 1. In a similar fashion, by the repeated use of the above argument one can show that

$$(g\delta(x-a))^{(n)} = g(a)\delta^{(n)}(x-a). \tag{3}$$

This equality expresses again the remarkable *multiplier probing property* of the Dirac delta distribution, complementing the equality $\delta[\phi] = \phi(0)$ discussed before. It can be also expressed as follows: *a function multiplier of the Dirac delta can be viewed as a constant which can be factored outside the test functional.* In the future we will often refer to the multiplier probing property analyzing various applied problems. ∎

Example 3. Let us find a distribution equal to the product of a function $g \in C^\infty$ and the derivative of the Dirac delta $\delta'(x-a)$. Following the above rules of differentiation of distributions and their multiplication by smooth functions, we get

$$(g\delta')[\phi] = \delta'[g\phi] = -\delta[(g\phi)'] = -\delta[g'\phi + g\phi']$$
$$= -g'(a)\phi(a) - g(a)\phi'(a).$$

Thus, the derivative of the Dirac delta loses the multiplier probing property of the Dirac delta itself—it is a linear combination of values of both g and g' at the point $x = a$. In the particular case of $g(x) = x$ and $a = 0$ the above calculation gives

$$x\delta'(x) = -\delta(x). \tag{4}$$

∎

A by-product of the above example is that the Dirac delta is a generalized distributional solution of the differential equation

$$xT' = -T.$$

This fact, as we shall see later, will have useful consequences for solving real-life physical and engineering problems. The elegant equation (4), as well as the more general formula

$$x^n \delta^{(n)}(x) = (-1)^n n! \delta(x),$$

cannot be derived if one sticks to the intuitive understanding of the delta function described in Section 1.1, and it shows the power of mathematical tools introduced on the last few pages.

A word of warning is in order here. Under no circumstances can both sides of formula (4) be divided by x since, obviously,

$$\delta'(x) \neq -\frac{\delta(x)}{x},$$

not to mention the fact that the right hand side is not well defined because the function $1/x$ does not belong to C^∞. This illuminates difficulties with the operation of division of a distribution by functions that vanish at a certain point. We shall return to this problem later.

On the other hand, the above properties of the Dirac delta distributions allow us to sometimes solve the different division problem of finding a distribution T from $\mathcal{D}'$ which satisfies equation $gT = 0$, where g is a known smooth function. If T represents a regular function f, such an equation obviously has a multitude of solutions as it implies only that $f(x)$ and $g(x)$ cannot be different from 0 at the same point x. In other words the intersection of supports of f and g has to be empty:

$$f(x)g(x) = 0 \quad \Longleftrightarrow \quad \operatorname{supp} f \cap \operatorname{supp} g = \emptyset.$$

In the generalized sense, however, such equations may appear in different applications (for example, in the analysis of the propagation of waves in dispersive media) and may have nontrivial solutions. As an exercise one can check that the distribution of the form

$$T = c_0\delta(x) + \ldots + c_{n-1}\delta^{(n-1)}(x), \tag{5}$$

where $c_0, c_1, \ldots, c_{n-1}$, is a solution of equation

$$x^n T = 0. \tag{6}$$

Hence, in the above sense, it solves the problem of division of zero.

1.6   Integrals of distributions; the Heaviside function

By analogy with classical calculus, one could define an (indefinite) integral of a distribution T as a distribution S such that $S' = T$. Without searching for the general solution of this problem, let us observe that its solution for the Dirac delta is easy. Consider the so-called *Heaviside* or *unit step* function

$$\chi(x) = \begin{cases} 1, & \text{for } x \geq 0; \\ 0, & \text{for } x < 0, \end{cases} \tag{1}$$

often encountered in physical applications and pictured in Fig. 1.6.1.

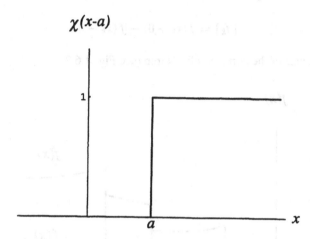

FIGURE 1.6.1
The graph of the shifted Heaviside function $\chi(x - a)$.

In the sense of classical analysis it has no derivative at $x = 0$, but its distributional derivative is well defined, and it is easy to see that

$$\chi' = T_\chi' = \delta. \tag{2}$$

Indeed, checking the values of the left hand side as a functional on test functions, we get that

$$T_\chi'[\phi] = -T_\chi[\phi'] = -\int \chi(x)\phi'(x)dx = -\int_0^\infty \phi'(x)dx = \phi(0) = \delta[\phi].$$

Having found the derivative of the Heaviside function, one can compute easily the distributional derivative of any piecewise-smooth function $f(x)$ which has jump discontinuities at points $x_k, k = 1, 2, \ldots, n$. Such a function can be always represented as a sum of its continuous piecewise-smooth part f_s without jumps, and pure jump part in the following form

$$f(x) = f_s(x) + \sum_{k=1}^n \Big(f(x_k + 0) - f(x_k - 0) \Big) \chi(x - x_k),$$

or in a more compact form

$$f(x) = f_s(x) + \sum_{k=1}^{n} \lfloor f_k \rfloor \chi(x - x_k),$$

where

$$\lfloor f_k \rfloor = f(x_k + 0) - f(x_k - 0)$$

denotes the size of the corresponding jump (see Fig. 1.6.2).

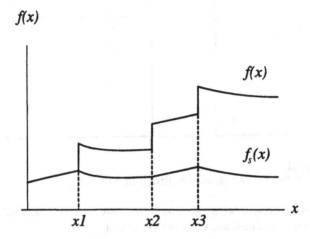

FIGURE 1.6.2
Graphs of a function $f(x)$ with jumps and the corresponding continuous function $f_s(x)$ which has been obtained from f by the removal of its jumps.

Since the derivative is a linear operation we immediately see that

$$f' = \{f_s'\} + \sum_{k=1}^{n} \lfloor f_k \rfloor \delta(x - x_k),$$

which, read in the reverse order, gives a formula for an indefinite integral of any distribution of the following form: a locally integrable function plus a linear combination of Dirac deltas centered at points of jumps.

It can happen that a function has first $n - 1$ derivatives in the classical sense and only the derivative of order $n - 1$ displays some discontinuities and its derivative has to be considered in the distributional sense. Before providing an example, let us define the function sign (x) which is another of those special discontinuous functions that we will encounter often in what follows. By definition

$$\text{sign}\,(x) = \begin{cases} +1, & \text{for } x > 0; \\ 0, & \text{for } x = 0; \\ -1, & \text{for } x < 0. \end{cases} \tag{3}$$

Its graph is presented on Fig. 1.6.3. By a computation similar to that above, one can check that $\text{sign}'\,(x) = 2\delta(x)$.

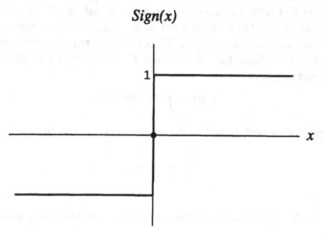

FIGURE 1.6.3
Graph of the function sign (x).

Example 1. Consider the function $f(x) = x^2 \,\text{sign}\,(x)$. It is differentiable in the classical sense and

$$f'(x) = 2|x|$$

for any point x. The derivative, however, is not differentiable at $x = 0$, but in the distributional sense it is easy to check that

$$f''(x) = 2\,\text{sign}\,(x),$$

so that,

$$f'''(x) = 2\,\text{sign}'\,(x) = 4\delta(x). \qquad\blacksquare$$

At this point it should be observed that there is some flexibility in computing a function whose distributional derivative is equal to the Dirac delta. The distributional derivative is determined by its functional action on test functions. So if we change the value of the Heaviside function at a single point, we also obtain an indefinite integral of the Dirac delta. As a consequence function

$$f(x) = \frac{\text{sign}(x)}{2}$$

also satisfies equality $f' = \delta$.

As far as the *definite integral of a distribution* on the entire real line is concerned, it is clear that some additional assumptions are necessary. One such possible restriction is that the distribution T has compact support. Any such distribution can be identified with a continuous linear functional on the space C^∞, in the sense that its value $T[\phi]$ is defined not only on any infinitely differentiable test function with compact support $\phi \in \mathcal{D}$, but also on any infinitely differentiable function $\phi \in C^\infty$. Since for a distribution T_f representing a regular function f with compact support

$$T_f[\phi] = \int f(x)\phi(x)dx,$$

it is natural, for a distribution T with compact support, to define

$$\int T = T[1],$$

where 1 on the right hand side stands for a function identically equal to 1. In particular,

$$\int \delta = \delta[1] = 1.$$

This line of thinking can be extended to introduce another linear operation on distributions, namely, their convolution with a smooth function. This will be done in Section 1.8.

1.7 Distributions of composite arguments

The reader should have already noticed that the only singular distribution explicitly defined so far was the Dirac delta distribution and whatever we could obtain from it by the linear operations of differentiation and multiplication by a smooth function from C^∞. In this section we continue using this method of producing new distributions from the ones already constructed by introducing new linear operations on general distributions. As usual our guide will be how the analogous operation on regular functions can be expressed in terms of the integral functional.

Let us begin with *distributions of a composite argument*, that is, a composition of a distribution with a function of the x-variable. In the case of a regular function $f(x)$, the interplay between the integration and composite arguments is expressed by the usual change-of-variable formula

$$\int f(\alpha(x))\phi(x)dx = \int f(y)\phi(\beta(y))|\beta'(y)|dy,$$

where $y = \alpha(x)$, and $x = \beta(y)$ represents the function inverse to $\alpha(x)$, such that $\beta(\alpha(x)) = x$. An assumption guaranteeing validity of the above formula is that the function $\alpha(x)$ is strictly monotone and that it maps the x-axis onto the entire y-axis.

If we want to use this equality in the functional setting, it is clear that further restrictions on the composite argument $\alpha(x)$ are necessary. Namely, to assure that the factor $\phi(\beta(y))|\beta'(y)|$ on the right-hand side is a function in $\mathcal{D}$, it is not sufficient to assume that function $\alpha(x)$ is strictly monotone and that it maps the x-axis onto the entire y-axis. We also need $\beta(y)$ to be an infinitely differentiable function $\mathbf{R} \to \mathbf{R}$.[1]

So, under the above restrictions on the composite argument $\alpha(x)$, it is clear how we should proceed in the case of distributions.

By definition, *the formula*

$$T(\alpha(x))[\phi(x)] = T[\phi(\beta(y))|\beta'(y)|] \tag{1}$$

determines the composition of the distribution T with the function $\alpha(x)$.

Example 1. Consider a shift function $\alpha(x) = x - a$. Composition of this function with the Dirac delta clearly gives

$$\delta(\alpha(x)) = \delta(x - a),$$

where $\delta(x - a)$ was introduced earlier. ∎

Example 2. Consider the distribution $\delta(\alpha(x) - a)$ defined by the equality

$$\delta(\alpha(x) - a)[\phi(x)] = \delta(y - a)[\phi(\beta(y))|\beta'(y)|] = \phi(\beta(a))|\beta'(a)|,$$

which can be symbolically written as

$$\delta(\alpha(x) - a) = \frac{\delta(x - \beta(a))}{|\alpha'(\beta(a))|}, \tag{2}$$

---

[1] What to do if $\alpha(x)$ is not one-to-one (for example, $\alpha(x) = x^2$) will be discussed elsewhere in this book.

where we have taken into account the fact that $\beta'(a) = 1/\alpha'(\beta(a))$. This formula is most frequently applied to a linear composition function $\alpha(x) = cx$, where $c \neq 0$. In this case, we get that

$$\delta(cx) = \frac{\delta(x)}{|c|}.$$

The above equality expresses the previously mentioned *self-similarity* property of the Dirac delta distribution.

As any other distribution, the distribution of a composite argument can be differentiated and, in general, standard formulas from classical analysis can be applied. Let us demonstrate the above statement by rewriting the relation (2) in another equivalent form. Assuming, for definiteness, that $\alpha(x)$ is a strictly increasing function, the absolute value signs can be dropped in (2), and it can be written, using the multiplier probing property of the Dirac delta, as

$$\delta(\alpha(x) - y) = \frac{\delta(x - \beta(y))}{\alpha'(\beta(y))} = \frac{\delta(x - \beta(y))}{\alpha'(x)},$$

which gives

$$\alpha'(x)\delta(\alpha(x) - y) = \delta(x - \beta(y)). \tag{3}$$

If we differentiate both sides of the above equality with respect to y we get

$$\alpha'(x)\frac{\partial}{\partial y}\delta(\alpha(x) - y) = \frac{\partial}{\partial y}\delta(\beta(y) - x).$$

On the other hand, by the classical rules of calculus,

$$\frac{\partial}{\partial x}\delta(\alpha(x) - y) = -\alpha'(x)\frac{\partial}{\partial y}\delta(\alpha(x) - y).$$

Hence we arrive at useful relation

$$\frac{\partial}{\partial x}\delta(\alpha(x) - y) + \frac{\partial}{\partial y}\delta(\beta(y) - x) = 0. \tag{4}$$

Bear in mind that both variables x and y above have the same status, and that the above distributional equality can be tested with test functions $\phi(x)$ and $\phi(y)$. Once this is done, we recover the familiar chain rules of differential calculus:

$$\frac{d}{dy}\phi(\beta(y)) = \beta'(y)\frac{d}{d\beta}\phi(\beta),$$

and

$$\frac{d}{dx}\phi(\alpha(x)) = \alpha'(y)\frac{d}{d\alpha}\phi(\alpha).$$

1.8 Convolution

A combination of the shift transformation and integration gives rise to another important linear operation on distributions: the *convolution* with a function $\phi \in \mathcal{D}$.

By definition, *the convolution $T * \phi$ is a regular C^∞ function defined by the formula*

$$(T * \phi)(x) = T_t[\phi(x - t)].$$

Notice that it is defined point-wise for every x separately, and that t is the running argument of the distribution T and the test function on the right-hand side. In particular

$$(T * \phi)(0) = T[\tilde{\phi}]$$

where $\tilde{\phi}(t) = \phi(-t)$, so that

$$\delta * \phi = \phi.$$

In other words, the Dirac delta behaves as a unity for convolution "multiplication".

Remark 1. The convolution operation can be similarly defined for the distribution T with compact support, and an arbitrary infinitely differentiable function ϕ.

If we want to extend the above operation to permit convolution of two distributions, a "weak" approach is necessary.

If T is a distribution with compact support and S is an arbitrary distribution in $\mathcal{D}'$, then their convolution $T * S$ is a distribution in $\mathcal{D}'$ acting on test functions $\phi \in \mathcal{D}$ as follows:

$$(T * S)[\phi] = T_x[S_y[\phi(x + y)]]$$

or equivalently

$$(T * S)[\phi] = \Big(T * (S * \tilde{\phi})\Big)(0).$$

One easily checks that the Dirac delta is the unit element for this more general operation of convolution multiplication as well, that is

$$\delta * S = S.$$

If one differentiates the convolution of two distributions one gets that

$$(S * T)^{(k)} = S^{(k)} * T = T^{(k)} * S.$$

The convolution is a linear operation since

$$(\alpha_1 S_1 + \alpha_2 S_2) * T = \alpha_1 S_1 * T + \alpha_2 S_2 * T.$$

It is also commutative since
$$S * T = T * S,$$

and associative, i.e.,
$$S * (T * U) = (S * T) * U.$$

Finally, another important property of the convolution is that

$$\text{supp}\,(S * T) \subset \text{supp}\,S + \text{supp}\,T,$$

where for two sets A, B, by definition $A + B = \{x + y, x \in A, y \in B\}$. The same relationship is true for singular supports.

1.9   The Dirac delta on $\mathbf{R}^n$, lines and surfaces

By analogy with distributions on $\mathbf{R}$, *distributions on $\mathbf{R}^n$ are defined as linear continuous functionals on the space $\mathcal{D}(\mathbf{R}^n)$ of infinitely differentiable test functions $\phi(x)$ of compact support in $\mathbf{R}^n$.*

Again, if a function $f(x)$ of an n-dimensional variable $x = (x_1, \ldots, x_n)$ is locally integrable, then it defines a distribution on $\mathbf{R}^n$ by the formula

$$T_f[\phi] = \int \ldots \int f(x)\phi(x)d^n x,$$

where the integral is an n-tuple integral with respect to the differential $d^n x = dx_1 \ldots dx_n$. In the future, to avoid unwieldy formulas, we will denote the multiple integral $\int \ldots \int$ by a single integral sign $\int$ without any risk of confusion. The dimension will be clear from what appears under the integral sign.

It turns out that all the conclusions about distributions in $\mathcal{D} = \mathcal{D}(\mathbf{R})$ can be extended, with obvious adjustments, to distributions on multidimensional spaces. In particular, we will define a Dirac delta distribution $\delta(x - a)$ by

$$\delta(x - a)[\phi] = \phi(a).$$

With the help of the above Dirac delta we can, for example, define the *singular dipole function* as
$$-\Big(n \cdot \nabla \delta(x - a)\Big)p,$$

where $n = p/p$ is the unit vector in the direction of the dipole and the operator of directional derivative $n \cdot \nabla$ acts on the delta function via the equality

$$-\left(n \cdot \nabla \delta(x - a)\right)[\phi] = n \cdot \nabla \phi(a).$$

In view of these general similarities, we will not go through detailed introduction of multidimensional distributions, and concentrate instead on a few issues reflecting the special nature of the multidimensional spaces. We also restrict our attention to the Dirac delta distribution on the 3-D space.

As in the 1-D case, the Dirac delta $\delta(x)$ can be obtained as a weak limit of distributions represented by regular functions $f_k(x)$ on $\mathbf{R}^3$. For example, it is convenient to take

$$f_k(x) = g_k(x_1)g_k(x_2)g_k(x_3),$$

where $g_k(x_i)$ are regular functions of one variable approximating the one-dimensional Dirac delta $\delta(x_i)$. In this context, the 3-D Dirac delta can be intuitively viewed as a simple product of 1-D Dirac deltas

$$\delta(x) = \delta(x_1)\delta(x_2)\delta(x_3),$$

although that operation was never formally defined. The above picture, however, hides the important property of isotropy of the 3-D Dirac delta, which can be expressed as an invariance with respect to the group of rotations of $\mathbf{R}^3$. This isotropy becomes more transparent if we take the Gaussian function

$$\gamma_\varepsilon(x) = \frac{1}{\sqrt{2\pi}\,\varepsilon} \exp\left(\frac{-x^2}{2\varepsilon^2}\right)$$

with $\varepsilon = 1/k$ as the approximating one-dimensional regular function of δ. Its coordinatewise product

$$f_\varepsilon(x) = \left(\frac{1}{\sqrt{2\pi}\,\varepsilon}\right)^3 \exp\left(\frac{-r^2}{2\varepsilon^2}\right)$$

depends only on the magnitude (norm)

$$r = |x| = \sqrt{x_1^2 + x_2^2 + x_3^2}$$

of vector x and not on its orientation in space.

Let us also observe that, in a similar fashion, we can think of the Dirac delta in space-time as the product

$$\delta(x, t) = \delta(x)\delta(t).$$

As the next step, compute the Dirac delta $\delta(\alpha(x) - a)$ on $\mathbf{R}^3$ of the composite argument $\alpha(x)$, which corresponds to finding out how the Dirac delta is transformed under a change of the coordinate system $y = \alpha(x)$, where coordinate-wise

$$y_i = \alpha_i(x).$$

If we assume that $\alpha(x)$ is a one-to-one function which satisfies required differentiability conditions, then the equality

$$\delta(\alpha(x) - a) = \frac{\delta(\beta(a) - x)}{|J(x)|} \tag{1}$$

is valid with $x = \beta(y)$ representing coordinate transformation inverse to $\alpha(x)$, and J standing for the Jacobian

$$J(x) = \left| \frac{\partial \alpha_i(x)}{\partial x_j} \right|$$

of the transformation from y-coordinates to x-coordinates.

Equation (1) extrapolates to the Dirac delta, the classical change of variables formula for integrals of functions of several variables:

$$\int f(y - a)\phi(y)d^3y = \int f(\alpha(x) - a)\phi(\alpha(x))|J(x)|d^3x.$$

The absolute value of the Jacobian determinant describes the compression ($|J| < 1$) and stretching ($|J| > 1$) of the elementary volume d^3y in comparison with the original elementary volume d^3x. Symbolically, we can write this fact as a heuristic equation

$$|J(x)| = \frac{d^3y}{d^3x}.$$

The above discussion of the Dirac delta distribution on $\mathbf{R}^3$ with a single point support can be extended to introduce Dirac delta distributions whose singular supports are lines or surfaces in $\mathbf{R}^3$.

So, if σ is a surface in $\mathbf{R}^3$, then the *surface Dirac delta* δ_σ is defined by

$$\delta_\sigma[\phi] = \int_\sigma \phi(x)d\sigma$$

where $\phi \in \mathcal{D}(\mathbf{R}^3)$, and the integral on the right-hand side is the surface integral. In the same fashion, one defines a *line Dirac delta* for a curve ℓ in $\mathbf{R}^3$, by the condition

$$\delta_\ell[\phi] = \int_\ell \phi(x)d\ell$$

with the line integral on the right-hand side.

These distributions are often applied in physics, for example, as a mathematical model of electrically charged surfaces and strings. Standard operations with distributions can be extended to the above Dirac deltas in a natural manner, e.g., operation of multiplication by an infinitely differentiable function (surface or linear charge density), differentiation, etc. Thus, in electrodynamics, the Dirac delta

$$-\frac{\partial}{\partial n}\Big(f(x)\delta_\sigma\Big)$$

of a double layer is often encountered. It is functionally defined by the condition

$$-\int \frac{\partial}{\partial n}\Big(f(x)\delta_\sigma\Big)\phi(x)dx = \int_\sigma f(n \cdot \nabla\phi)d\sigma,$$

where n is the normal unit vector to the surface σ which describes, for example, a dipole surface.

In particular cases, the notation introduced above for surface and line Dirac deltas is not always used since the latter can be sometimes constructed from the usual one-dimensional Dirac delta. Thus, a surface Dirac delta corresponding to surface $x_1 = 0$ can be more readily interpreted as the usual Dirac delta $\delta(x_1)$, and the line Dirac delta concentrated on the x_3-axis can be written in the form of a product of Dirac deltas $\delta(x_1)\delta(x_2)$. In the same manner, the field of a spherical wave, propagating away from the origin with velocity c, can expressed with the help of the one-dimensional Dirac delta as follows:

$$U(x, t) = \frac{1}{|x|}\delta(|x| - ct).$$

1.10 Linear topological space of distributions

As we mentioned in Section 1.2, the set $\mathcal{D}$ of test functions forms a linear space, i.e., for any complex numbers a, b, and any test functions ϕ, ψ from $\mathcal{D}$, their linear combination

$$a\phi + b\psi \tag{1}$$

is also a test function in $\mathcal{D}$. A function identically equal to 0 plays the role of a *neutral element for addition*. Moreover, we defined in $\mathcal{D}$ a notion of convergence of sequences of test functions (in other words, a *topology* on $\mathcal{D}^2$), with respect to which the above linear combinations are continuous, thus determining what is called the structure of a *linear topological space* for $\mathcal{D}$.

A similar structure can be established for the set $\mathcal{D}'$ of distributions. Hence, for any complex numbers a, b and any distributions T, S, the linear combination

$$aT + bS$$

is again a distribution in $\mathcal{D}'$ defined by its action on test functions from $\mathcal{D}$ by

$$(aT + bS)[\phi] = aT[\phi] + bS[\phi].$$

The zero distribution, defined by condition $T[\phi] = 0$ for any $\phi \in \mathcal{D}$, plays the role of a neutral element for addition of distributions. The topology of $\mathcal{D}'$ is determined by the following definition of the convergence of a sequence T_k of distributions.

We shall say that, as $k \to \infty$,

$$T_k \to T$$

in $\mathcal{D}'$ if, for each test function $\phi \in \mathcal{D}$, (complex) numbers

$$T_k[\phi] \to T[\phi].$$

It is immediate to check that linear combinations of distributions are continuous in this topology, or in other words,

$$\lim_{k \to \infty} (aT_k + bS_k) = a \lim_{k \to \infty} T_k + b \lim_{k \to \infty} S_k.$$

The above topology is called the *weak topology*, or the *dual topology* to the topology of $\mathcal{D}$, and it will be the only convergence considered on $\mathcal{D}'$. The reader will recognize that an approximation of the Dirac delta by regular functions considered in Section 1.3 was conducted in the spirit of weak convergence.

Example 1. Consider distributions

$$T_\varepsilon = \frac{\varepsilon}{2}|x|^{\varepsilon-1}, \qquad \varepsilon > 0.$$

---

[2]To be more precise, the topology is defined by convergence of all, not necessarily countable, "sequences" but we will not dwell on that in this book.

Indeed, these functions are locally integrable, and they represent distributions in $\mathcal{D}'$, as long as $\varepsilon > 0$. Notice, however, that $|x|^{-1}$ is not a locally integrable function around $x = 0$, so it does not represent a distribution from $\mathcal{D}'$. Let us compute the limit

$$\lim_{\varepsilon \to 0+} T_\varepsilon.$$

In view of the above, it is obvious that if the above limit exists, it cannot be related to $|x|^{-1}$. The way to proceed is to check values of T_ε on test functions from $\mathcal{D}$. Let ϕ be a fixed test function from $\mathcal{D}$. Then, for some positive number M,

$$T_\varepsilon[\phi] = \int \frac{\varepsilon}{2}|x|^{\varepsilon-1}\phi(x)dx = \frac{\varepsilon}{2}\int_0^M x^{\varepsilon-1}\Big(\phi(x) + \phi(-x)\Big)dx,$$

because ϕ has compact support. On the other hand, by the mean value theorem, for each x,

$$\phi(x) = \phi(0) + x\phi'(\theta_x)$$

for a certain θ_x in the interval $[0, M]$. Hence,

$$T_\varepsilon[\phi] = \phi(0) \cdot \varepsilon \int_0^M x^{\varepsilon-1}dx + \frac{\varepsilon}{2}\int_0^M x^\varepsilon \Big(\phi'(\theta_x) + \phi'(\theta_{-x})\Big)dx. \qquad (2)$$

Since

$$\int_0^M |x|^{\varepsilon-1}dx = \frac{1}{\varepsilon}M^\varepsilon,$$

the first term on the right-hand side of (2) converges to $\phi(0)$ as $\varepsilon \to 0$. The second term converges to 0 since ϕ' remains bounded on $[-M, M]$, and since

$$\int_0^M |x|^\varepsilon dx = \frac{1}{\varepsilon+1}M^{\varepsilon+1}$$

is bounded as well. Therefore, we get that $T_\varepsilon[\phi] \to \phi(0)$, which gives

$$\lim_{\varepsilon \to 0} T_\varepsilon = \delta. \qquad \blacksquare$$

Mixing the weak limits and linear combinations, one can produce additional nontrivial distributions.

Example 2. For an arbitrary sequence $\{a_n\}$ of complex numbers, the series

$$\sum_{-\infty}^{\infty} a_n\delta(x - n)$$

defines a distribution in $\mathcal{D}$ even when the numerical series $\sum_{-\infty}^{\infty} a_n$ does not converge. Indeed, the series always converges weakly in $\mathcal{D}'$ since action on a fixed test function will always cut out all but finitely many terms of the series. ∎

In addition to the space $\mathcal{D}'$ of distributions, we will also consider some other spaces and equip them with the structure of a linear topological space. One such space, the space $\mathcal{E}'$ of distribution of compact support, has been introduced before. Its weak topology is determined by convergence on test functions from C^∞, not just on C^∞ functions with compact support, as is the case for weak convergence in $\mathcal{D}'$. Hence, it is more difficult to achieve.

Remark 1. The Dirac deltas and their derivatives are dense in the space of all distributions in the sense that any $T \in \mathcal{D}'$ is a weak limit of distributions of the form

$$\sum_k \sum_n a_k \delta(x - b_n).$$

Remark 2. If a distribution T has a support equal to $\{0\}$ then it is a finite sum of derivatives of the Dirac delta, that is

$$T = \sum_{k=1}^{n} a_k \delta^{(k)}(x)$$

for some constants a_k.

Remark 3. Nevertheless, one can find a family of test functions ϕ such that weakly, with respect to that family,

$$\lim_{n \to \infty} \sum_{k=0}^{n} \frac{a^k \delta^{(k)}(x)}{k!} = \delta(x + a).$$

1.11 Exercises

1. Find the weak limits, as $k \to \infty$, of the following sequences of functions:
(a) $-k^3 x \exp(-k^2 x^2)$.
(b) $k^3 x / ((kx)^2 + 1)^2$.
(c) $1/(1 + k^2 x^2)$.
(d) $\exp(-e^{-kx})$.

2. What conditions have to be imposed on a function $f(x)$ to guarantee that the sequence $\{k^2 f(kx)\}$ weakly converges to $\delta'(x)$?

3. Let $a > 0$. Find the derivative of the Heavisde function of the following composite arguments:

(a) $\chi(ax)$.

(b) $\chi(e^{\lambda x} \sin ax)$.

4. Calculate the distributional derivative $f'(x)$ of function $f(x) = \chi(x^4 - 1)$, and find the weak limit $f'(x/\varepsilon)/\varepsilon^2$, as $\varepsilon \to 0_+$.

5. Find distributional solutions of equation $(x^3 + 2x^2 + x)y(x) = 0$.

6. Find the nth distributional derivative of function $e^{\lambda x} \chi(x)$.

7. Prove the identity

$$x^m \delta^{(n)}(x) = (-1)^m \frac{n!}{(n-m)!} \delta^{(n-m)}(x), \qquad m \le n.$$

8. Using the standard one dimensional Dirac delta distribution, construct a surface Dirac delta corresponding to the level surface $\sigma : g(x) = a \in \mathbf{R}, x \in \mathbf{R}^3$ of a smooth function g on $\mathbf{R}^3$.

9. Using the standard one dimensional Dirac delta distribution, construct a line Dirac delta corresponding to the curve ℓ of intersection of two level surfaces $g_1(x) = a_1, g_2(x) = a_2, x \in \mathbf{R}^3$.

10. Find the length of the level curve $\ell : \psi(x) = c, x \in \mathbf{R}^2$, without finding a parametric description of it.

11. Define a *vector-valued distribution*

$$\boldsymbol{P} = \delta(\psi(x) - c) \boldsymbol{\nabla} \psi(x), \qquad x \in \mathbf{R}^3,$$

by its action on an arbitrary vector-valued test function $\boldsymbol{\phi}$

$$P[\phi] = \int \delta(\psi(x) - c) \Big(\boldsymbol{\nabla} \psi(x) \cdot \boldsymbol{\phi}(x) \Big) d^3x.$$

What is the physical meaning of the functional action $P[\phi]$?

Chapter 2

Basic Applications: Rigorous and Pragmatic

2.1 Two generic physical examples

In this section we give a couple of seemingly naive physical examples. Keeping them in mind, however, reinforces appropriate intuitive images of distributions that help solidify a formal mathematical understanding of the theory, and make it easier to grasp the automation of computations that can be achieved with help of distribution theory.

Example 1. Beads on a string. This rather elementary example illustrates possibilities to simplify many calculations if one uses the notion of the Dirac delta distribution. Let us try to describe the linear mass density of beads with masses $m_k, k = 1, \ldots, n$ strung along a tight string which coincides with the x-axis. Recall that the linear density $\rho(x)$ of the string itself is defined as the ratio $\Delta m / \Delta l$, where Δm is the mass of an infinitesimal string segment Δl. If the size of the beads is small relative to other scales: string length, distances between beads, etc., then the inner structure of the beads is insignificant for most calculations— beads can be assumed to be material points, and the linear mass density $\rho(x)$ can be accurately described with the help of a sum of Dirac deltas:

$$\rho(x) = \rho_0(x) + \sum_{k=1}^{n} m_k \delta(x - x_k), \tag{1}$$

where $x_1, \ldots, x_n$ are coordinates of the locations of the beads.

The generalized string density plot is displayed in Fig. 2.1.1, where arrows symbolically indicate the delta-shaped bead densities. Their different heights reflect variations among the masses of the beads. This generalized density is extremely convenient in calculations of various physical quantities, such as the string's mass center, moment of inertia, and many others. If, for instance, the string of beads

© Springer Nature Switzerland AG 2018
A. I. Saichev and W. Woyczynski, *Distributions in the Physical and Engineering Sciences, Volume 1*, Applied and Numerical Harmonic Analysis, https://doi.org/10.1007/978-3-319-97958-8_2

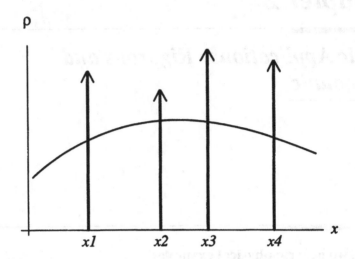

FIGURE 2.1.1
A schematic graph of the density of mass for beads on a string.

is placed in the force field $f(x)$, then the total force acting on the system can be calculated by means of

$$F = \int \rho(x)f(x)dx = \int \rho_0(x)f(x)dx + \sum_{k=1}^{n} m_k f(x_k). \qquad \blacksquare$$

Example 2. Dipole in an electrostatic field. Let us discuss the behavior of a molecule in an electrostatic field. It is often sufficient to consider it as an infinitesimal dipole with a given dipole moment p and not to worry about its internal microscopic structure. For simplicity, we shall again assume that the dipole is located at position a on the x-axis. The charge distribution of a single dipole can then be described with the help of the Dirac delta's derivative:

$$\rho(x) = -p\delta'(x - a). \qquad (2)$$

Hence, if a dipole is placed in an electric field $E(x)$ directed along the x-axis, then the force acting on the dipole is equal to

$$F = \int \rho(x)E(x)dx = pE'(a).$$

The formula makes it clear that, in an inhomogeneous (space-dependent) field, the dipole is moving in the direction of a stronger field. ∎

2.2 Systems governed by ordinary differential equations

We are now adequately prepared to consider one of the most fundamental areas of applying distribution theory, namely, integration of linear ordinary differential equations. The latter are the main mathematical tool in studying a variety of physical problems. In particular, equations describing the signal transfer through a linear system are typically inhomogeneous linear differential equations of the form

$$L_n\left(\frac{d}{dt}\right) x(t) = g(t), \tag{1}$$

where

$$L_n(p) = a_0 + a_1 p + a_2 p^2 + \ldots + a_n p^n \tag{2}$$

is a given polynomial of degree n ($a_n \neq 0$), $p^k = (d/dt)^k$ is taken to mean d^k/dt^k, and $g(t)$ is a known function of time t called the *input signal*. Notice that if the input signal $g(t)$ belongs to the space of test functions then the identity

$$g(t) = \int \delta(t - \tau)g(\tau)d\tau \tag{3}$$

is satisfied. It turns out that the solution of (1) which, in the systems engineering terminology, will be called the *output signal*, can be written in the form of the convolution integral

$$x(t) = \int H(t - \tau)g(\tau)d\tau, \tag{4}$$

which expresses the time invariance of the system's properties. The distribution H is called the *transfer function* of the system. Substituting (3) and (4) into (1), we get that

$$L_n\left(\frac{d}{dt}\right) H(t) = \delta(t). \tag{5}$$

In other words, the transfer function $H(t)$ is the response of the system to the Dirac delta input signal. In this context, $H(t)$ is also often called the *fundamental solution* or the *Green's function* of equation (1).

Physically, it is also clear that the solution of equation (5) should satisfy the *causality principle* according to which the system's response cannot occur prior to the appearance of the input signal, i.e.,

$$H(t) \equiv 0 \quad \text{for} \quad t < 0. \tag{6}$$

So, let us try to construct a solution to equation (5) satisfying the causality condition (6). To this end, consider an auxiliary homogeneous differential equation

$$L_n \left(\frac{d}{dt} \right) y(t) = 0, \tag{7}$$

with initial conditions

$$y(0) = y'(0) = \ldots = y^{(n-2)}(0) = 0, \quad y^{(n-1)}(0) = 1/a_n, \tag{8}$$

where a_n are the coefficients in (2). Function

$$H(t) = y(t)\chi(t) \tag{9}$$

obviously satisfies the causality condition (6). We shall check that H also satisfies equation (5). Indeed, following the differentiation rules for distributions

$$H'(t) = y'(t)\chi(t) + y(t)\delta(t).$$

However, the first initial condition and the Dirac delta's multiplier probing property imply that $y(t)\delta(t) = y(0)\delta(t) \equiv 0$ so that, actually, $H'(t) = y'(t)\chi(t)$. Repeating this argument we get that

$$H^{(k)}(t) = y^{(k)}(t)\chi(t), \quad k = 0, 1, 2, \ldots, n-1. \tag{10}$$

Using again the Dirac delta's probing property and the last initial condition, we also obtain the formula for the highest derivative:

$$H^{(n)}(t) = y^{(n)}(t)\chi(t) + \delta(t)/a_n. \tag{11}$$

Substituting (10) and (11) into (5), we obtain

$$\chi(t)L_n \left(\frac{d}{dt} \right) y(t) + \delta(t) = \delta(t).$$

This demonstrates that H is the desired fundamental solution since y is a solution of the homogeneous equation (7).

Substituting the above formula for H in the convolution (4), we arrive at an explicit expression for the output signal (solution of equation (1)), in the form

$$x(t) = \int_{-\infty}^{t} y(t - \tau)g(\tau)d\tau, \qquad (12)$$

which is often called the *Duhamel integral*. Equivalently we can write

$$x(t) = \int_{0}^{\infty} y(\tau)g(t - \tau)d\tau.$$

Example 1. Harmonic oscillator with damping. Let us apply the above general scheme to the equation

$$\ddot{x} + 2\alpha\dot{x} + \omega^2 x = g(t), \qquad \omega^2 > \alpha^2.$$

In this case the transfer function is completely determined by the solution of the corresponding homogeneous equation

$$\ddot{y} + 2\alpha\dot{y} + \omega^2 y = 0,$$

satisfying initial conditions $y(0) = 0$, $\dot{y}(0) = 1$. The solution is well known to be of the form

$$y(t) = \frac{1}{\omega_1}e^{-\alpha t}\sin(\omega_1 t),$$

where $\omega_1 = \sqrt{\omega^2 - \alpha^2}$. One can check by direct differentiation that function $y \in C^\infty$. Therefore, in this case, the output signal is described by the Duhamel integral

$$x(t) = \frac{1}{\omega_1}\int_{0}^{\infty} e^{-\alpha\tau}\sin(\omega_1\tau)g(t - \tau)d\tau.$$

The corresponding fundamental solution, that is, a response of the system to the Dirac delta input, is plotted in Fig. 2.2.1. ∎

The coefficients in equation (1) were constant. However, it is worth mentioning that similar distributional arguments apply also to equations with time-dependent coefficients. Without exposition of the full theory, let us illustrate this fact in a simple example.

Example 2. Time-dependent coefficients. Consider a first order equation

$$a(t)\dot{x} + b(t)x = g(t). \qquad (13)$$

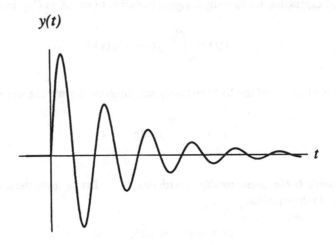

$y(t)$

t

FIGURE 2.2.1
The fundamental solution of the harmonic oscillator equation with damping.

Let $a(t), b(t) \in C^\infty(\mathbf{R})$, and assume additionally that $a(t)$ never vanishes. Then, analogous with (4), we can look for a special solution of the form

$$x(t) = \int H(t, \tau) g(\tau) d\tau, \qquad (14)$$

where the Green's function $H(t, \tau)$ satisfies equation

$$a(t)\dot{H} + b(t)H = \delta(t - \tau).$$

Utilizing properties of the Dirac delta, it is easy to check that the above equation has a solution which satisfies the causality principle and is of the form

$$H(t, \tau) = \chi(t - \tau) y(t, \tau),$$

where $y(t, \tau)$ is a solution of the homogeneous Cauchy problem

$$a(t)\dot{y} + b(t)y = 0, \quad y(t = \tau, \tau) = 1/a(\tau).$$

Hence, substituting

$$H(t, \tau) = \frac{\chi(t - \tau)}{a(\tau)} \exp\left[-\int_\tau^t \frac{b(t')}{a(t')} dt'\right]$$

into (14) we obtain the desired solution of equation (13):

$$x(t) = \int_{-\infty}^t \frac{g(\tau)}{a(\tau)} \exp\left[-\int_\tau^t \frac{b(t')}{a(t')} dt'\right] d\tau. \qquad \blacksquare$$

2.3 One-dimensional waves

The emergence of distribution theory greatly extended boundaries of rigorous theoretical analysis of mathematical physics problems. Let us consider a simple example which illustrates how distribution theory helps in dealing with typical physical situations.

Consider a field $u(x, t)$ satisfying the 1-D *wave equation*

$$\frac{\partial^2 u}{\partial t^2} = c^2 \frac{\partial^2 u}{\partial x^2} \qquad (1)$$

with initial conditions

$$u(x, t = 0) = g(x), \qquad \left.\frac{\partial u(x, t)}{\partial t}\right|_{t=0} = h(x). \qquad (2)$$

In order to obtain a solution to this *initial value problem* in the classical sense, functions g and h appearing in the initial conditions must be continuously differentiable at least two times. However, it is very often interesting to know how the initial rectangular pulse

$$g(x) = \chi(x + a) - \chi(x - a) \qquad (3)$$

propagates (if $h \equiv 0$). The initial condition (3) has no classical derivatives at points $x = \pm a$. So ordinary calculus is not helpful here. However, it is well known that the sum of two pulses

$$u(x, t) = \frac{1}{2}\Big(g(x - ct) + g(x + ct)\Big), \qquad (4)$$

traveling in opposite directions at speed c provides a solution to the above problem.

One way to arrive at such a solution would be to smooth out the initial pulse's edges in the ε-neighborhood of the jumps to get differentiable initial conditions, solve the equation, and then find the limit of the solution as $\varepsilon \to 0$.

A distributional approach permits us to avoid this unwieldy technique altogether and we shall check, by direct substitution, that function (4) satisfies the equation (1). Actually, in view of the linearity of equation (1), it suffices to check that one of the components of traveling waves in (4), say $\chi(x - ct + a)$, satisfies the wave equation (1). Taking the distributional chain rule into account, the second derivative of the above function with respect to t turns out to be

$$\frac{\partial^2}{\partial t^2}\chi(x - ct + a) = c^2\delta'(x - ct + a),$$

and, on the other hand,

$$c^2\frac{\partial^2}{\partial x^2}\chi(x - ct + a) = c^2\delta'(x - ct + a)$$

as well, so that the verification is complete.

Let us remark that the general solution of the initial value problem (1-2) is the well known *D'Alembert solution*

$$\overset{.}{u}(x, t) = \frac{1}{2}\Big(g(x - ct) + g(x + ct)\Big) + \frac{1}{2c}\int_{x-ct}^{x+ct} h(y)dy. \qquad (5)$$

Even in this well known case, distribution theory is useful in extending the class of admissible functions $g(x)$ and $h(x)$, and in facilitating a rigorous interpretation of (5) as a "generalized D'Alembert solution" of the wave equation.

2.4 Continuity equation

2.4.1. Continuity equation for the density of a single particle. In this section we discuss an—intuitively unexpected but important for physicists—example of the Dirac delta's application including differentiation of the Dirac delta of a composite argument and utilizing its multiplier probing property.

Consider a gas of moving particles. Denote the velocity of a particle which is located at point $x \in \mathbf{R}^3$ by $v(x, t)$. Then the motion of that particle satisfies a vector nonlinear ordinary differential equation

$$\frac{db(t)}{dt} = v(b(t), t),\tag{1}$$

where $b(t)$ is the position vector of a particle at a given instant t. Leaving aside the question of the inner structure of the particle, assume that its size is infinitesimal, and define its density by

$$\rho(x, t) = m\,\delta(b(t) - x),\tag{2}$$

where m is the particle mass. To derive an equation for the particle density, let us differentiate (2) with respect to t to get

$$\frac{\partial\rho}{\partial t} = m\frac{\partial}{\partial t}\delta(b(t) - x).$$

By the chain rule,

$$\frac{\partial}{\partial t}\delta(b(t) - x) = -\left(\frac{db(t)}{dt}\cdot\nabla_x\right)\delta(b(t) - x);$$

so, taking into account equation (1) of the particle motion, we obtain

$$\frac{\partial}{\partial t}\delta(b(t) - x) + \left(v(b(t), t)\cdot\nabla_x\right)\delta(b(t) - x) = 0.\tag{3}$$

Since the term $v(b(t), t)$ is independent of x, we take the velocity vector inside the ∇_x operator:

$$\left(v(b(t), t)\cdot\nabla_x\right)\delta(b(t) - x) = \left(\nabla_x\cdot v(b(t), t)\right)\delta(b(t) - x).$$

Now, in view of the Dirac delta's multiplier probing property,

$$v(b(t), t)\delta(b(t) - x) = v(x, t)\delta(b(t) - x).$$

As a result, equality (3) can be rewritten in the form

$$\frac{\partial}{\partial t}\delta(b(t) - x) + \left(\nabla_x\cdot v(x, t)\delta(b(t) - x)\right) = 0.$$

Multiplying both sides by the particle mass, and recalling the definition (2) of particle density, we arrive at

$$\frac{\partial \rho}{\partial t} + \left(\nabla_x \cdot v\rho \right) = 0, \tag{4}$$

which is the traditional *continuity equation*, and is often written in a more transparent physically, divergence notation:

$$\frac{\partial \rho}{\partial t} + \text{div } (v\rho) = 0. \tag{5}$$

This simple equation occupies one of the central places in physics and deserves additional commentaries. The above derivation strikes many physicists as unexpected because the continuity equation is derived typically as a corollary to the mass conservation law in an ideal continuous medium, thus ignoring its particle nature. This is how the name "continuity equation" originated. The traditional derivation required not only continuity, but also sufficient smoothness of fields $\rho(x, t)$ and $v(x, t)$ describing the motion of an ideal continuous medium. Our derivation shows that the continuity equation remains valid for a singular density of each separate particle in the medium as well.

Hence, it even makes sense to talk about the continuity equation for a gas consisting of just one particle. Such a degenerate gas should not elicit skepticism among physicists. Mathematicians study similar objects quite seriously, having in mind not material particles but a gas of points in *phase space*. The density of a point gas in phase space, as any real gas, satisfies the continuity equation. Notice that equation (5) can be viewed as an equation for a density of a point in phase space, since the mathematical phase space of solutions of equation (1) coincides, in that case, with the physical space $\mathbf{R}^3$.

2.4.2. Mass conservation law. A smooth mass density $\rho(x, t)$ of an ideal continuous medium satisfies the mass conservation law, and so does the singular density (2) as

$$\int \rho(x, t)d^3x = m \int \delta(b(t) - x)d^3x = m.$$

This follows from Dirac delta's basic properties. One says that the mass conservation law is guaranteed by the *divergence form* of the continuity equation (5). We shall show this by integrating all summands of equation (5) over a certain fixed (time-independent) space region Ω. As a result, we get that

$$\frac{d}{dt}m_\Omega + \int_\Omega \text{div } (v\rho)d^3x = 0, \tag{6}$$

where

$$m_\Omega = \int_\Omega \rho(x, t) d^3 x$$

is mass of the medium contained in region Ω. Now recall that the *Gauss divergence theorem* states that if σ is a smooth surface enclosing a bounded three dimensional region Ω, and n is the external normal vector to σ then, for any smooth function f on $\mathbf{R}^3$,

$$\oint_\sigma (f \cdot n) d\sigma = \int_\Omega \operatorname{div} f d^3 x. \tag{7}$$

This theorem can be readily extended to the case where function f is replaced by a vector distribution. Now, if we transform the second term in (6) using this generalized Gauss divergence theorem, we get

$$\int_\Omega \operatorname{div} (v\rho) \, d^3 x = \oint_\sigma \rho(v \cdot n) d\sigma. \tag{8}$$

If, during time interval $[t_1, t_2]$, no particle crosses the surface σ which bounds region Ω, then the surface integral in (8) is automatically equal to 0, and equation (6) gives the condition

$$\frac{d}{dt} m_\Omega = 0,$$

which implies the mass conservation law

$$m_\Omega = \text{const} \qquad (t \in [t_1, t_2]).$$

2.4.3. Continuity equation for continuous media.

Continuity equation (5) was derived for the density of a single particle. However, in view of its linearity, the continuity equation is also satisfied by a superposition of densities of different particles and thus remains valid for the full microscopic density of the medium

$$\rho(x, t) = \sum_i m_i \delta(b_i(t) - x), \tag{9}$$

where the summation extends over all the medium particles, and m_i and $b_i(t)$ denote, respectively, mass and position of the ith particle.

A physicist, who has read the above material and until now assumed $\rho(x, t)$ to be a smooth function may be tempted to swing to the other extreme and start suspecting that all physically realizable solutions of the continuity equation are singular and have a structure similar to (9). This is not necessarily so since we will show that, in the analysis of the motions of a macroscopic medium, at scales very large compared to the distances between adjacent particles, it is still natural to assume $\rho(x, t)$ to be a smooth function.

Let us begin by rewriting density (2) of a single particle in the following form

$$\rho(x, t; y) = m\delta\big(b(y, t) - x\big), \tag{10}$$

which explicitly involves the initial particle position

$$b(y, t = 0) = y.$$

In practice, the particle's initial position is never known accurately. This fact can be modeled mathematically by taking the convolution

$$\bar{\rho}(x, t) = \frac{1}{l^3} \int g\left(\frac{y-z}{l}\right)\rho(x, t; z)d^3z, \tag{11}$$

of density (10) with an "initial position uncertainty function" $g(y/l)/l^3$. Here $g(z)$ is a normalized ($\int g(z)d^3z = 1$) function of a dimensionless variable z, while l can be viewed as a measure of uncertainty about the initial particle position. It is clear that the "averaged" density $\bar{\rho}(x, t)$ obtained in such a way also satisfies the continuity equation

$$\frac{\partial \bar{\rho}}{\partial t} + \text{div }(v\bar{\rho}) = 0. \tag{12}$$

However, in contrast to the singular density (2), $\bar{\rho}$ also satisfies smooth initial condition

$$\bar{\rho}(x, t = 0) = \frac{1}{l^3}g\left(\frac{y-x}{l}\right).$$

Similarly, the averaged density (9) satisfies equation (12) and a smooth initial condition

$$\bar{\rho}(x, t = 0) = \frac{1}{l^3}\sum_i m_i g\left(\frac{y_i - x}{l}\right). \tag{13}$$

From the physicist's viewpoint, function g describes an uncertain initial particle position, and l accounts for a macroscopic maximum accuracy of the measuring instrument. We should add that the concrete shape of the function $g(z)$ is not significant. One can rigorously prove that, under fairly general assumptions, and for l's large in comparison to the distances between adjacent particles, in the limit of infinitely particles, the sum in (13) approximates a "generic" macroscopic initial density $\rho_0(x)$, whose form is independent of the form of the individual summands.

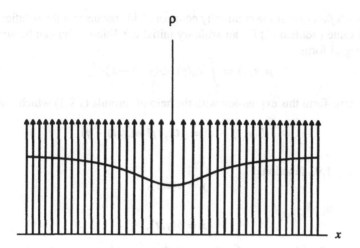

FIGURE 2.4.1
The figure shows symbolically, in the 1-D case, the initial singular density
$\rho_0(x)$ of particles with identical mass m, located at points $y_i = \beta(is\Delta)$, where
$\beta(y) = 2y(|y|+s)/(2|y|+s)$, $\Delta = s/10$, and the corresponding averaged density $\bar{\rho}_0(x)$ (continuous line) obtained with the help of the uncertainty function
$g(y/l)/l = \exp(-y^2/l^2)/(\sqrt{\pi}l)$, $l = 4\Delta$. It is clear from the graphs that the
averaged density has a minimum at the origin where the density of particles
is smaller.

2.5 Green's function of the continuity equation and Lagrangian coordinates

The preceding section makes it clear why it is important for physicists and
engineers to solve the initial value problem

$$\frac{\partial \rho}{\partial t} + \operatorname{div}(\rho v) = 0, \tag{1}$$

$$\rho(x, t = 0) = \rho_0(x),$$

with smooth initial data. However, as we will see in this section, the physical singular density of microparticles (2.4.10), while remaining very important, acquires
a new, mathematical, interpretation. It turns out that

$$G(x, t, y) = \delta(b(y, t) - x) \tag{2}$$

is *Green's function* of the continuity equation. This means that the solution of the initial value problem (1) for an arbitrary initial condition $\rho_0(x)$ can be written in the integral form

$$\rho(x, t) = \int \rho_0(y)\delta(b(y, t) - x)d^3y. \tag{3}$$

Let us transform this expression with the help of formula (1.9.1) which gives that

$$\delta(b(y, t) - x) = j(x, t)\delta(a(x, t) - y). \tag{4}$$

In Section 1.9., functions

$$y = a(x, t), \tag{5}$$

and

$$x = b(y, t), \tag{6}$$

are inverse functions of each other, and they provide a connection between the initial position y and the current position x of the same particle of the medium.

This seems to be the right place to recall traditional hydrodynamics terminology. Current coordinates x of a particle in a certain fixed coordinate system are traditionally called the *Eulerian coordinates*. Initial coordinates y of the particle are then called its *Lagrangian coordinates*. Both the Eulerian and the Lagrangian coordinates determine spatial position of the particle. But whereas an external observer may find it preferable to use the Eulerian coordinates, the observer traveling with the particle itself may find the Lagrangian coordinates more convenient. The situation is similar to that of a person who prefers to identify himself by his birthplace rather than by the listing of his current address. Formulas (5) and (6) define laws of transformation of Eulerian coordinates of a particle into its Lagrangian coordinates, and vice versa. Function

$$j(x, t) = \left|\frac{\partial a(x, t)}{\partial x}\right| \tag{7}$$

appearing in (4) is the Jacobian of the transformation of Lagrangian into Eulerian coordinates.

Let us return now to expression (3) for the density field. Substituting into (3) the right-hand side of equality (4), and applying the Dirac delta's multiplier probing property, we find that density

$$\rho(x, t) = \rho_0(a(x, t))j(x, t), \tag{8}$$

is proportional to the initial density $\rho_0(y)$ in the neighborhood of a particle which is at x at time t, and to the Jacobian $j(x, t)$ which takes into account changes in density caused by infinitesimal deformations of the medium's volume in the neighborhood of a particle.

2.6 Method of characteristics

In this section we will present the *method of characteristics*—the standard general method of solving first-order partial differential equations. Our exposition will be restricted to the continuity equation. In this case the physical appeal of the method is that the characteristics are physically observable paths of particles.

Often, an effective mathematical approach reduces the problem at hand to a simpler problem which has been solved earlier (one can say that by "moving backwards" mathematicians manage to move forward and obtain new results). The method of characteristics follows a similar path by reducing partial differential equations to more familiar ordinary differential equations. To see how this works, let us rewrite the continuity equation (2.5.4) in the form

$$\frac{\partial \rho}{\partial t} + (v \cdot \nabla)\rho = -u\rho, \tag{1}$$

where

$$u(x, t) = (\nabla \cdot v) = \operatorname{div} v(x, t) \tag{2}$$

is an auxiliary scalar field. We will solve equation (1) under the above mentioned initial condition $\rho(x, t = 0) = \rho_0(x)$.

A solution to any partial differential equation is, by definition, a function of several independent variables. In the case of density field $\rho(x, t)$, there are three spatial coordinates and the time variable. The method of characteristics makes an assumption that all of these variables depend on a single common parameter, and dependence is selected in such a way that the partial differential equation is transformed into an ordinary differential equation. For the continuity equation, it is convenient to select the time t as that common parameter and assume that the spatial coordinates x are functions of time, that is $x = b(t)$. In this case, by the chain rule, as long as function $b(t)$ satisfies

$$\frac{db}{dt} = v(b, t), \tag{3}$$

the left-hand side of (1) turns out to be an ordinary derivative, and the continuity equation itself becomes an ordinary differential equation

$$\frac{d\rho}{dt} = -u\rho. \tag{4}$$

The system (3-4) of ordinary differential equations is called the *characteristic equations*.

In our particular case, the system splits into two groups of equations for two different types of functions. The first group consists of equations (3), whose solutions represent *characteristics*, i.e., paths $x = b(t)$ in phase space. The last group consists of a single equation (4) which describes the density evolution along the characteristics. Observe that the characteristic equations (3) coincide with equations of motion (2.4.1) for the medium's particle. For that reason, in hydrodynamics, the ordinary derivative with respect to time in equation (4) is called the *substantial derivative* and the special notation

$$\frac{D}{Dt} = \frac{\partial}{\partial t} + v \cdot \nabla$$

is used.

By complementing equations (3) with the initial condition

$$b(t = 0) = y, \tag{5}$$

we fix the point in space where the characteristic originates, i.e., the point from

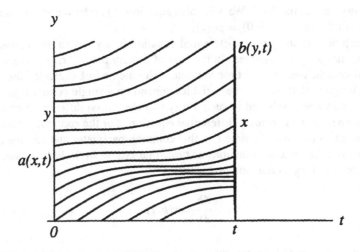

FIGURE 2.6.1
1-D case. Set of characteristic curves of the continuity equation mapping Lagrangian coordinates into Eulerian coordinates and vice versa.

which the particle starts its motion. Therefore, the corresponding initial condition in equation (4) should be the initial density $\rho_0(x)$ at point y, that is

$$\rho(t = 0) = \rho_0(y). \tag{6}$$

From the hydrodynamic viewpoint, initial conditions (5) fix the Lagrangian coordinate of a particle. Consequently, the equation (4) determines evolution of the density in the neighborhood of a particle with a given Lagrangian coordinate, and *is the continuity equation in the Lagrangian coordinate system*. Thus, solutions

$$\rho = R(y, t), \quad b(y, t), \tag{7}$$

of the initial value problem (3-6) describe the density field in the Lagrangian coordinate system.

However, if we observe the evolving state of a hydrodynamic field by sensors in fixed spatial positions, then we are interested in a description of the density field $\rho(x, t)$ in the Eulerian coordinate system. The problem is that, for a given y, field $R(y, t)$ determines density at a point with coordinates $b(y, t)$ which may not coincide with Eulerian coordinates x at the point of observation. In the 1-D case, the above situation is pictured in Fig. 2.6.1.

The situation is saved by the fact that, by varying y, we obtain a *family of characteristics* with, hopefully, one of them hitting point x at time t. Mathematically speaking, for a given mapping $y \in \mathbf{R}^3 \mapsto x \in \mathbf{R}^3$ given by the formula

$$x = b(y, t), \tag{8}$$

we have to find an inverse transformation

$$y = a(x, t), \tag{9}$$

determining the Lagrangian coordinate of the particle which at time t is located at point x. Substituting that inverse transformation into $R(y, t)$, we obtain the required density field

$$\rho(x, t) = R(a(x, t), t) \tag{10}$$

in the Eulerian coordinate system.

It is easy to see that the solution of continuity equation (4) in the Lagrangian coordinate system can be written in the form

$$R(y, t) = \rho_0(y) \exp\left[-\int_0^t u(b(y, \tau), \tau)d\tau\right]. \tag{11}$$

As a result the Eulerian density field is given by the expression

$$\rho(x, t) = \rho_0(a(x, t))j(x, t), \tag{12}$$

where, as is clear from (11),

$$j(x, t) = \exp\left[-\int_0^t u(b(y, \tau), \tau)d\tau\right]\Bigg|_{y=a(x,t)},\tag{13}$$

is the Jacobian of transformation (2.5.7) of Lagrangian into Eulerian coordinates.

Until now, to simplify the exposition, we left out an important mathematical question of the existence and uniqueness of an inverse transformation (9) for all $x \in \mathbf{R}^3$. Now is a good time to address this question.

As is well known, transformation (8) is an *isomorphism* (i.e., a continuously differentiable, and one-to-one mapping with a continuously differentiable inverse) if and only if the Jacobian j in (2.5.7) is bounded and positive for any $x \in \mathbf{R}^3$. It is clear from (13) that a sufficient condition for (8) to be an isomorphism is that the field $u(x, t)$ of (2) be bounded. Actually this assumption was made implicitly above when we omitted the sign of absolute value around j in (2.5.4) (which the general formula requires).

Let us also remark that the method of characteristics just described, is embedded, as if in the genetic code, in Green's function (2.5.2) and in its probing properties. As a consequence, solution (12)- (13) can be obtained without resorting to the method of characteristics. But the time and space used to describe that alternative method has not been wasted as it sheds light on the problem from a different viewpoint.

2.7   Density and concentration of the passive tracer

The continuity equation plays an important role in the study of ecological problems related to dispersion of pollutants in the environment. Obviously, in addition to the medium density, also the density of any *passive tracer* suspended in and carried by the medium, satisfies the continuity equation. A suitable example here is the density of smoke particles released by the smokestack of a power station. The continuity equation describes evolution of the passive tracer density if $\rho_0(x)$ in (2.5.1) is replaced by the initial density of the passive tracer.

Sometimes for instance in the analysis of chemical reactions, the density itself is not as important as the *concentration* $C(x, t)$. It measures not the absolute, as the density does, but the relative proportion of the tracer in a physically infinitesimal unit of the medium's volume. That's why, while the density increases when the medium is compressed and decreases when the medium expands, the concentration preserves its value in the neighborhood of an arbitrary fixed particle. In other words, in the Lagrangian coordinate system, the concentration does not depend on time, i.e.,

$$C(y, t) = C_0(y) = const,\tag{1}$$

where $C_0(x)$ is the initial concentration field, and it satisfies equation

$$\frac{DC}{Dt} = 0, \tag{2}$$

where, in Eulerian coordinates, $D/Dt = \partial/\partial t + v \cdot \nabla$ is the substantial derivative introduced in Section 2.6. Thus, the Eulerian field of concentration satisfies
equation

$$\frac{\partial C}{\partial t} + (v \cdot \nabla)C = 0. \tag{3}$$

Its solution can be obtained by expressing the Lagrangian coordinates in (1) through
the Eulerian coordinates. As a result we get that

$$C(x, t) = C_0\Big(a(x, t)\Big). \tag{4}$$

Let us also observe that the equation (3) is an obvious consequence of a mathematically more important initial value problem

$$\frac{\partial y}{\partial t} + (v \cdot \nabla)y = 0, \quad y(x, t = 0) = x, \tag{5}$$

whose solution $y = a(x, t)$ describes a transformation of the Eulerian coordinates
into the Lagrangian coordinates.

2.8   Incompressible medium

Both the Green function for concentration equation (2.7.3) and for the Lagrangian coordinates (2.7.5) are equal to the Dirac delta which appears on the
right-hand side of equality (2.5.4). Hence, we can write

$$C(x, t) = \int C_0(y)\delta(a(x, t) - y)d^3y. \tag{1}$$

If a medium is such that

$$j(x, t) \equiv 1, \tag{2}$$

then, as is clear from (2.5.4), the Green's function of the continuity equation coincides with the Green's functions of equations (2.7.3) and (2.7.5), and the continuity

equation itself takes the form

$$\frac{\partial \rho}{\partial t} + (v \cdot \nabla)\rho = 0. \tag{3}$$

A medium where the identity (2) is satisfied is an *incompressible medium*. Condition (2) implies that the volume of a region "frozen" into the medium, does not change in time. It is clear from (2.6.2) and (2.6.13) that the condition

$$\nabla \cdot v = \text{div } v(x, t) \equiv 0 \tag{4}$$

is necessary and sufficient for the medium to be incompressible. The above identity means that the velocity field of an incompressible medium is purely rotational. Indeed, by the *fundamental theorem of vector calculus*, any smooth vector field can be represented as a sum of a potential and a rotational components, that is

$$v = v_p + v_r, \tag{5}$$

where

$$v_p = \nabla \varphi = \text{grad } \varphi(x, t), \tag{6}$$

for a certain φ called the *scalar potential* of the vector field v, and

$$v_r = \nabla \times \psi = \text{rot } \psi(x, t) \tag{7}$$

for a certain ψ which is called the *vector potential* of the same field v.

The divergence of the rotational part

$$\nabla \cdot v_r = \nabla \cdot (\nabla \times \psi) = (\nabla \times \nabla) \cdot \psi \equiv 0, \tag{8}$$

a condition that coincides with the incompressibility condition (4). The divergence of the potential component is equal to

$$\nabla \cdot v_p = \nabla \cdot \nabla \varphi = (\nabla \cdot \nabla)\varphi = \Delta \varphi, \tag{9}$$

where the letter Δ is used to denote the *Laplace operator* which, in the 3-D Cartesian coordinate system, has the form

$$\Delta = \nabla \cdot \nabla = \frac{\partial^2}{\partial x_1^2} + \frac{\partial^2}{\partial x_2^2} + \frac{\partial^2}{\partial x_3^2}. \tag{10}$$

If the Laplacian of a function φ is identically zero then function φ is affine everywhere, that is

$$\varphi(x) = a + v \cdot x.$$

Here the first term a is just a constant—a potential is determined by condition (6) only up to an arbitrary constant. The second term describes a parallel uniform motion which is not affected by compressibility. This means that the velocity field in an incompressible medium is purely rotational.

Finally, let us observe that in the two dimensional medium ($x \in \mathbf{R}^2$), which is used as a model of surface, or oceanographic phenomena, the vector potential of a purely rotational velocity field degenerates into a scalar function $\psi(x, t)$ called the *stream function*, and components of the velocity field in the Cartesian coordinate system can then be expressed through the stream function by formulas

$$v_1 = \frac{\partial \psi}{\partial x_2}, \quad v_2 = -\frac{\partial \psi}{\partial x_1}. \tag{11}$$

2.9 Pragmatic applications: beyond the rigorous theory of distributions

The rigorous distribution theory discussed up to now relied on narrow spaces of test functions like $\mathcal{D}$, $\mathcal{E}$ (or the space $\mathcal{S}$ of rapidly decreasing functions to be introduced in later chapters). However, in various physical and engineering applied problems one is often tempted to go beyond those well defined spaces and apply results of the rigorous theory in a formal fashion. This sometimes leads to correct results but it has to be done carefully lest erroneous conclusions are arrived at. Basically there are two ways to assure a correct outcome.

The first method is to build from scratch a distribution theory suitable for the problem under investigation by selecting a specialized space of test functions, and then to follow the general scheme developed in Chapter 1. Such an approach is worth the time and effort required only if the potential application area is large enough. We will pursue this path ourselves by introducing the space $\mathcal{S}$ of rapidly decreasing smooth functions in connection with the study of Fourier integrals.

However this approach is not practical in every situation—construction of a plethora of different distribution spaces adapted to each new situation is best left to theoretical mathematicians. For an applied scientist such a "micromanagment" would often be confusing.

Thus, the second, more pragmatic approach is often used. One borrows and formally applies relations and formulas from the rigorous distribution theory, and then one checks their validity in each particular case. If this is done with care, no harm will result and the desired calculations can be completed expeditiously. The general rule is that any extension of the test function space results in the narrower class of linear functionals on them (distributions). This can entail a loss of some of the crucial properties of distributions, such as their infinite differentiability. To avoid unexpected pitfalls, it is also essential to develop an intuition based on experience in pragmatic applications.

In this section we analyze five typical examples of the second approach:

- distributions on a finite interval;

- differential equations with singular coefficients;

- nonmonotonic composite arguments of the Dirac delta;

- nonlinear functions of distributions;

- supersingular distributions.

All five present delicate mathematical questions when framed within the rigorous distribution theory. Yet, in physical and engineering practice, we see them successfully handled in a nonrigorous fashion. So, it is only reasonable to discuss them openly without pretending that they do not exist, see how it is being done, and how to avoid related potential dangers.

2.9.1. Distributions on a finite interval. Many applied problems require calculating integrals with finite limits such as

$$\int_a^b f(x)\phi(x)dx. \tag{1}$$

Often, one is tempted to operate with analogous functionals but with function $f(x)$ replaced by a distribution and, in particular, by the Dirac delta. We have already met such a situation in Example 2.1.1 of beads threaded on a finite length string. The above integral is not well defined within distribution theory even if $\phi(x)$ is a test function in $\mathcal{D}$. Indeed, finite integral limits actually mean that we use as a test function the truncated function

$$\tilde{\phi}(x) = \phi(x)\Big(\chi(x-a) - \chi(x-b)\Big),$$

where $\chi(x)$ is the Heaviside function. Nevertheless, the functional seems to be well defined if restricted to distributions with support contained in the interval of integration. Such is the case of $\delta(x - c)$ with $a < c < b$ or its derivatives. It is relatively easy to extend the standard distribution theory on the entire real line to the case of distributions with supports inside a given open interval.

The rigorous theory developed in Chapter 1 is not, however, able to suggest what should be expected if the support of the Dirac delta is one of the endpoints of the integration interval. Let us consider this situation in more detail in a typical example of the functional

$$J = \int \delta(x)\chi(x)dx. \tag{2}$$

Having absolutely no idea of what the meaning of the above integral is, one could start with its evaluation by a formal change of variables $y = \chi(x)$, notice that $\delta(x)dx = dy$, and conclude that

$$J = \int_0^1 y\,dy = \frac{1}{2}.$$

This heuristic answer, which is extensively used in applications, can be justified more formally.

To get a deeper insight into the situation, let us follow a more rigorous path of evaluating the functional (2) as a limit $J = \lim_{k\to\infty} J_k$ of ordinary integrals

$$J_k = \int \delta_k(x)\chi_k(x)dx, \tag{3}$$

where $\delta_k(x)$ and $\chi_k(x)$ are regular functions, weakly converging to distributions $\delta(x)$ and $\chi(x)$, respectively. As examples of such weakly converging sequences let us take

$$\delta_k(x) = k\lambda(kx - \alpha), \quad \text{and} \quad \chi_k(x) = \Lambda(kx), \tag{4}$$

where

$$\Lambda(x) = \int_{-\infty}^x \lambda(y)dy,$$

with

$$\lambda(x) = \frac{1}{\pi}\frac{1}{x^2 + 1}$$

being the Lorentz kernel, so that

$$\Lambda(x) = \frac{1}{\pi}\left(\arctan(x) + \frac{\pi}{2}\right). \tag{5}$$

It can be easily seen that the values of integrals $J_k = J(\alpha)$ do not depend on k, and range over the interval $(0, 1)$ as α varies from $-\infty$ to $+\infty$. Therefore, unlike the values of functionals on the test function space $\mathcal{D}$, the value of functional (2) depends on the "inner structure" of a distribution. Functions $\chi_k(x) - 1/2$ are clearly odd functions and, in physical applications one often "imagines" the Dirac delta $\delta(x)$ as an "even function". Under such an assumption, we would always have $J = 1/2$. Because of this, physicists are accustomed to using the following working formula: For any continuous function $\varphi(x)$,

$$\int_a^b \delta(x - c)\phi(x)dx = \begin{cases} \varphi(c) & \text{if } c \in (a, b); \\ 0 & \text{if } c \notin [a, b]; \\ \varphi(c)/2 & \text{if } c = a \text{ or } b. \end{cases} \quad (6)$$

Of course, the above physical assumption of the Dirac delta's evenness has to be taken with a grain of salt, and if used recklessly it can result in an incorrect solution of the particular physical problem. Thus, in the case of beads on a string occupying interval $(0, l)$, the bead at the left endpoint of the interval with density $\rho = m\delta(x)$ either is assumed to be a part of the string or not. Therefore, the contribution of that bead to the force acting on the string has to be taken either as $f(0)m$ or zero, depending on the actual physical situation. No middle ground is sensible here.

In a more general case of weakly converging sequences (4), parameter α is used to describe an infinitesimal quantity, comparable to the "thickness" of the Dirac delta, displacing it to the left ($\alpha < 0$) or to the right ($\alpha > 0$) from its basic location at $x = 0$. Under zero displacement, we obtain from (4) that $J = 1/2$, and for $\alpha \to -\infty$, when the support of the Dirac delta becomes completely separated from the region of nonzero values of the Heaviside function, we have $J \to 0$.

2.9.2. Differential equations with singular coefficients.

Let us discuss another interesting and instructive example demonstrating a limited nature of the range of applications of formula (6). Consider the following differential equation for function $f(x)$:

$$f' = p\delta(x - c)f, \quad f(0) = f_0. \quad (7)$$

Here p is a known constant and $c > 0$. A formal solution of this equation

$$f(x) = f_0 \exp\left(p\chi(x - c)\right),$$

obtained by the separation of variables has a discontinuity at $x = c$ with the size of the jump

$$\lfloor f \rfloor = f_0\left(e^p - 1\right). \quad (8)$$

Hence, distribution theory indicates that the solution of equation (7) can be written in the form

$$f(x) = f_0 + \lfloor f \rfloor \chi(x - c),$$

and that its derivative is

$$f'(x) = \lfloor f \rfloor \delta(x - c).$$

Substituting the above formulas in equation (7), rearranging the terms, and integrating over all x's, we get that the value of the integral

$$J = \int \delta(x - c)\chi(x - c)dx = \frac{\lfloor f \rfloor - pf_0}{p\lfloor f \rfloor} = \frac{e^p - 1 - p}{p(e^p - 1)}.$$

The graph of J as a function of p is presented in Fig. 2.9.1.

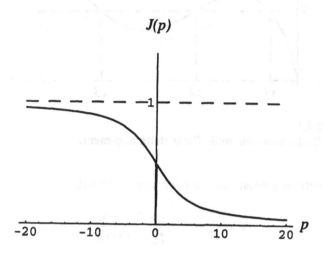

FIGURE 2.9.1
Possible values of integral J as a function of p.

The above "calculation" makes it clear that as p varies from $-\infty$ to $+\infty$, J decreases from 1 to 0, and as $p \to 0$, the functional J converges to $1/2$. The latter situation corresponds to what the physicists call the solution of (7) in the *first perturbation approximation* (or, the *Born approximation*). It is obtained by replacing f by f_0 on the right-hand side.

Let us also remark that expressions like (8) for the jump of a solution at the point of coefficient's singularity are used in complex problems which require gluing two classical solutions on either side of a singularity point.

2.9.3. Nonmonotonic composite arguments of delta functions. Another example of a situation where we are forced to abandon the framework of the rigorous distribution theory is related to the necessity of operating with Dirac delta $\delta(\alpha(x) - y)$ of a nonmonotone composite arguments $\alpha(x)$. In this case, equation $y = \alpha(x)$ may either have no roots at all, or may have multiple roots (see Fig. 2.9.2) $x_k = \beta_k(y), k = 1, 2, \ldots, n(y)$.

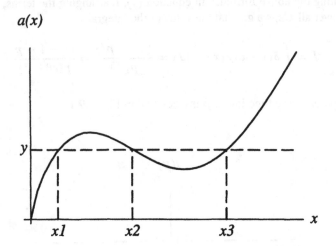

FIGURE 2.9.2
An example of a nonmonotonic Dirac delta argument.

Accordingly, in applications, one often uses the formula

$$\delta(\alpha(x) - y) = \sum_{k=1}^{n} \frac{\delta(x - x_k)}{|\alpha'(x_k)|}, \tag{9}$$

or, in view of the Dirac delta's multiplier probing property,

$$\delta(\alpha(x) - y) = \frac{1}{|\alpha'(x)|} \sum_{k=1}^{n} \delta(x - x_k),$$

which turns out to also function well in the case of nonmonotonic functions $\alpha(x)$ as long as all its terms are well defined, i.e., provided function $\alpha(x)$ has continuous nonzero derivatives in the vicinity of points x_k. In the example that goes back to Dirac himself, the relationship

$$\delta(x^2 - a^2) = \frac{1}{2|a|}\Big(\delta(x + a) + \delta(x - a)\Big) \tag{10}$$

remains valid for all $|a| > 0$. Sometimes, one can even remove the above mentioned requirement that $|\alpha'(x_k)| > 0$. Indeed, the relationship

$$2x\delta(x^2 - a^2) = \delta(x - |a|) - \delta(x + |a|)$$

is satisfied for any a since the multiplier $2x$ removes the singularity of the corresponding functional at point $x = 0$.

Another typical situation where relations like (9) are useful arises when one tries to count the number of intersections of the level y by the graph of function $\alpha(x)$ over the interval $(0, z)$. Indeed, by (9), the integral,

$$N(y, z) = \int_0^z |\alpha'(x)| \delta(\alpha(x) - y) dx \tag{11}$$

gives the desired crossing number. Similarly,

$$N_+(y, z) = \int_0^z \alpha'(x) \chi(\alpha'(x)) \delta(\alpha(x) - y) dx \tag{12}$$

counts the number of upcrossings of the level y by function $\alpha(x)$, and

$$N_-(y, z) = -\int_0^z \alpha'(x) \chi(-\alpha'(x)) \delta(\alpha(x) - y) dx \tag{13}$$

counts the number of downcrossings of the level y by the same function.

Counters of this type are used extensively in processing observational data.

2.9.4. Nonlinear transformations of distributions. One of the most obvious and essential difficulties encountered in distribution theory is the problem of nonlinear transformations. One has to remember that the ability to analyze nonlinear functions such as $\exp(f(x))$ or $f^2(x)$ is a major appeal of classical analysis. But if we replace function f by the Dirac delta or its derivatives, the above expressions do not make sense. That's why distribution theory is most effective when applied to linear problems.

Nonlinear transformations of distributions are, however, not always meaningless and, even if applied heuristically, lead sometimes to correct physical results. In those cases, the rule of thumb is: the less singular the distributions, the more freedom one has in handling their nonlinear transformations. For example, equalities

$$g(\chi(x)) = \begin{cases} g(1), & \text{for } x \geq 0; \\ g(0), & \text{for } x < 0. \end{cases}$$

and, in particular,

$$\chi^n(x) = \chi(x),$$

are completely rigorous and can be established by pointwise inspection. When dealing with more singular functions, e.g., the Dirac delta, more attention should be paid, as was emphasized earlier, to the symmetry and the inner structure of distributions. Hence, when facing the integral

$$\int \delta(x)\chi^n(x)dx \tag{14}$$

one tries to make sense out of it by making certain assumptions (resulting usually from the physics of the problem) about the relationship of smooth approximants to the singular distributions $\delta(x)$ and the Heaviside function $\chi(x)$. If, for example, sequences converging weakly to these distributions satisfy the following consistency condition

$$\chi_k(x) = \int_{-\infty}^{x} \delta_k(y)dy,$$

then one can use the above mentioned change of variables $y = \chi(x)$ to obtain

$$\int \delta(x)\chi^n(x)dx = \int_0^1 y^n dy = \frac{1}{n+1}.$$

Also, it is relatively easy to rigorously define a product of two distributions if their singular supports are disjoint. However, the corresponding relationships turn out to be, as a rule, rather trivial and are of the type of the formula

$$\delta(x-a)\delta(x-b) \equiv 0, \quad a \neq b.$$

There exists, however, a possibility to rather rigorously define nontrivial nonlinear combinations of singular distributions. It is related to *parameter-dependent distributions*. Let $T(x, y)$ be a distribution in variable x depending on parameter y. Distribution $\delta(x - y)$ can serve here as an example. A functional

$$T(x, y)[\phi(x)] = \psi(y) \tag{15}$$

maps the set $\mathcal{D}$ of test functions $\phi(x)$ onto a set $\mathbf{L}$ of functions $\psi(y)$. Provided $\mathbf{L} \subset \mathcal{D}$, function (15) can be used itself as a test function for another distribution S, that is we can evaluate $S(y)[\psi(y)]$. The last expression can be used to define

$$\int S(y)T(x, y)\, dy$$

as a distribution determined by the following functional action:

$$\int S(y)T(x, y)\, dy[\phi(x)] = S[\psi(y)].$$

A convolution of two distributions, which was introduced rigorously earlier, is a special example of such a quadratic form often encountered in applications.

Another nontrivial and fruitful type of multiplication of singular distributions is the *direct product of distributions* which permits a construction of distributions on multidimensional spaces. For example, the direct product $\delta(x)\delta(y)$ can be defined rigorously by

$$\int \delta(x)\delta(y)\phi(x, y)dx\, dy = \phi(0, 0).$$

Distributions on d-dimensional spaces are, however, a separate and interesting matter discussed in Section 1.9.

2.9.5. Supersingular distributions. It can happen that the function (15) does not have any meaning within the framework of the standard distribution theory, while the functional

$$T(x, y)[\psi(y)] = g(x), \tag{16}$$

is well defined for test functions $\psi(y) \in \mathcal{D}$, for each value of parameter x. Then it makes sense to say that the functional (15) determines a new distribution R:

$$T(x, y)[\phi(x)] = R(y). \tag{17}$$

Its algorithmic action postulates an extension to the functional case of the *Fubini Theorem* on preservation of the value of the double integral under change of the integration order:

$$T(x, y)[\psi(y)][\phi(x)] = R(y)[\psi(y)],$$

and, consequently,

$$R(y)[\psi(y)] = \int g(x)\phi(x)\, dx.$$

Let us illustrate the above concept in the typical example of

$$T(x, y) = \delta(\chi(x) - y). \tag{18}$$

Since the Heaviside function $\chi(x)$, appearing as an argument of the Dirac delta (18) clearly violates assumptions on thus far allowable arguments of distributions, the formula (18) does not define a distribution in a rigorous sense. Moreover, applying to (18) the pragmatic equality (9) also leads to a nonsensical result. Nevertheless, equation (18) rigorously defines a distribution (depending on parameter x) on the set of test functions $\{\psi(y)\}$ such that

$$\delta(\chi(x) - y)[\psi(y)] = \psi(\chi(x)). \tag{19}$$

Multiplying this equality by $\phi(x)$ and integrating it over all x's, we find the algorithm of action of the distribution $R(y)$ (17) in our special case (18):

$$R(y)[\psi(y)] = \psi(0) \int_{-\infty}^{0} \phi(x)\,dx + \psi(1) \int_{0}^{\infty} \phi(x)\,dx.$$

Hence, the sought distribution is

$$R(y) = \delta(\chi(x) - y)[\phi(x)] = \delta(y) \int_{-\infty}^{0} \phi(x)\,dx + \delta(y - 1) \int_{0}^{\infty} \phi(x)\,dx. \quad (20)$$

This equality displays the algorithm of action of distribution (18)on the family
of test functions $\{\phi(x)\}$. It is clear from (20), that this distribution can be called
neither regular nor singular. Indeed, the right-hand side of (20) is not a continuous
operation on the set of test functions $\{\phi(x)\}$. For this reason a new term is needed
and we will call distributions of type (18) the *supersingular distributions*. Notice
that the supersingular Dirac delta (18) no longer enjoys the usual probing property
since it depends on the values of the test function $\phi(x)$ on the entire x-axis.

In the latter part of this book we will discover that equalities like (20) find an
application in solving stochastic problems and provide a clear-cut probabilistic
interpretation. In this section we will illustrate the situation similar to (20) in an
example of gas dynamics discussed in Section 2.5.

Example 1. Density of the gas of sticky particles. Particles of the
1-D gas, initially (for $t = 0$) distributed on the x-axis with density $\rho_0(x)$, move
with constant velocity $v(x)$. In this case, the relation between the Lagrangian
coordinates y and the Eulerian coordinates x of the particles is given by the obvious
equality

$$x = b(y, t) = y + v(y)t. \quad (21)$$

Assume that $v(x)$ is a continuously differentiable function such that

$$\min v'(x) = -u < 0, \qquad u > 0.$$

The flow's evolution can be split into two qualitatively distinct stages.
Up to time

$$t^* = 1/u$$

the particles preserve their spatial order and move in the *single-stream regime*.
The function $x = b(y, t)$ (21), and its inverse function $y = a(x, t)$ are strictly
monotone and continuously differentiable, while the particle density is given by
the rigorous formula

$$\rho(x, t) = \int \rho_0(y)\delta(b(y, t) - x)\,dy, \quad (22)$$

analogous to the equation (2.5.3).

For $t > t^*$, there appears a new, *multi-stream*, regime of motion when some particles catch up with other particles, and there are points on the x-axis where simultaneously several particles are located moving with different velocities. At that stage, the functional (22) has to be calculated by means of the pragmatic formula (2.9.9) which leads to

$$\rho(x, t) = \sum_{n=1}^{N} \rho_0(a_n(x, t))|j_n(x, t)|, \tag{23}$$

where the summation is carried over all N ($N \geq 1$) roots $y = a_n(x, t)$ of the equation

$$b(y, t) = x,$$

solved for y for fixed x and t, and

$$j_n(x, t) = \frac{\partial}{\partial x} a_n(x, t).$$

Expression (23) has a clear-cut physical meaning: the density of particles at the point x is equal to the sum of densities (2.5.4) of all streams arriving at this point. The graph of the function $x = b(y, t)$ in the single-stream and multi-stream regimes is shown in Fig. 2.9.3(a).

Now let us consider the situations where the particles are sticky and are forbidden to overtake each other. After a "collision" they move together. Mathematically, the phenomenon of particles sticking together corresponds to the passage from a nonmonotone (for $t > t^*$) function $b(y, t)$ to a monotone function $\bar{b}(y, t)$, where the nonmonotone piece of the former was replaced by the horizontal piece of the latter. The position $x^*(t)$, the coordinate of adhesion of sticky particles, is determined, for example, by the momentum conservation law. The typical graph of function $\bar{b}(y, t)$ is shown in Fig. 2.9.3(b).

The corresponding particle density is described by

$$\rho(x, t) = \int \rho_0(y)\delta(\bar{b}(y, t) - x)\, dy \tag{24}$$

containing a supersingular Dirac delta. To find its action on the test function $\rho_0(y)$ note that, as in (19),

$$\delta(\bar{b}(y, t) - x)[\phi(x)] = \phi(\bar{b}(y, t)).$$

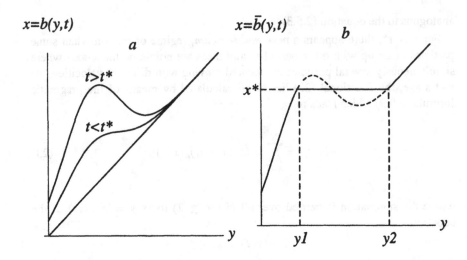

FIGURE 2.9.3
The graphs of functions representing the Lagrangian coordinates y and the
Eulerian coordinates x in the two cases of: (a) noninteracting particles with
resolved multi-stream motion for $t > t^*$; (b) sticky particles.

Multiplying this equality by $\rho_0(y)$ and integrating it over all y we obtain, in view of the Fubini postulate (see (17) and the following comments), that

$$\int \rho(x,t)\phi(x)\,dx = \phi(x^*(t)) \int_{y_1(t)}^{y_2(t)} \rho_0(y)\,dy + \left[\int_{-\infty}^{y_1(t)} + \int_{y_2(t)}^{\infty}\right] \phi(\bar{b}(y,t))\rho_0(y)\,dy.$$

Here, $y_1(t) < y_2(t)$ are edges of the function $b(y,t)$'s plateau, where $b = x^*(t)$ (see Fig. 2.9.3(b)). Now, let $y = \bar{a}(x,t)$ be the inverse function to the monotone function $x = \bar{b}(y,t)$. Choosing a new integration variable x, connected with y via the equality $y = \bar{a}(x,t)$, we merge the last two integrals into a single integral:

$$\int \rho(x,t)\phi(x)\,dx = m(t)\phi(x^*) + \int \phi(x)\rho_0(\bar{a}(x,t))j(x,t)\,dx, \qquad (25)$$

where

$$m(t) = \int_{y_1(t)}^{y_2(t)} \rho_0(y)\,dy$$

is the total mass of particles glued in the cluster at point $x = x^*(t)$, and

$$j(x,t) = \left\{\frac{\partial \bar{a}(x,t)}{\partial x}\right\}.$$

The braces are used to indicate that $j(x, t)$ is the derivative of function $\bar{a}(x, t)$ for all $x \neq x^*(t)$, where $\bar{a}(x, t)$ is a smooth function of variable x. At the point $x = x^*$ one can assign to $j(x^*, t)$ any finite value.

It is clear that if we insert the formula

$$\rho(x, t) = m(t)\delta(x - x^*(t)) + \rho_0(\bar{a}(x, t))j(x, t) \qquad (26)$$

in the functional

$$\int \rho(x, t)\phi(x)\, dx = \rho(x, t)[\phi(x)],$$

the relation (25) is recovered. It means that (26) gives the sought generalized density of the gas of sticky particles and takes into account the adhesion process. Recall that the first summand at the right-hand side of (26) is the singular density of the sticking particles' cluster, while the second summand is the smooth density of nonsticking particles outside the cluster (see Fig. 2.9.4).

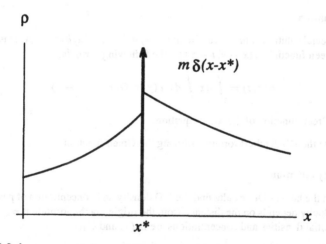

FIGURE 2.9.4
Plot of the generalized density (26). The vertical arrow indicates symbolically the singular density of the cluster of glued particles.

We stress that, as expected, the generalized density (24) satisfies the physical mass conservation law:

$$m = \int \rho(x, t)\, dx \equiv \int \rho_0(y)\, dy = const.$$

2.10 Exercises

Ordinary differential equations

1. Convert the homogeneous differential equation $\ddot{y} + \gamma \dot{y} + \omega^2 y = 0$ with the initial conditions $y(0) = a$, $\dot{y}(0) = b$ into a nonhomogeneous equation; the solution thereof satisfies the causality principle which, for $t > 0$, coincides with a solution of the original initial value problem.

2. What initial conditions should be imposed on the corresponding homogeneous equation in order that, for $t > 0$, its solution coincides with a solution (satisfying the causality principle) of equation $\ddot{y} + \gamma \dot{y} + \omega^2 y = \dot{\delta}(t)$?

3. Find an even solution of equation $\ddot{y} + y = \delta(t)$.

4. Find a solution of equation $\ddot{y} + y = a\chi(t) + b\dot{\delta}(t)$ satisfying the causality principle.

5. What equation can replace equation $\ddot{y} + y = (1/\varepsilon)f(x/\varepsilon)$ for "microscopically small" ε? ($\int f(x)dx = 1$.)

6. Find a general solution of equation $t\dot{y} = thy$.

Wave equation

7. A special solution of the wave equation $\partial^2 u/\partial t^2 = c^2 \partial^2 u/\partial x^2 + f(x, t)$ is expressed via the Green function $G(x - y, t - \tau)$ by the following formula:

$$u(x, t) = \int d\tau \int dy\, f(y, \tau) G(x - y, t - \tau).$$

Find the Green function of the wave equation.

8. Write the D'Alembert formula utilizing the Green function.

Continuity equation

9. With the help of Dirac delta find the 1-D density and concentration of passive tracer whose velocity depends on the distance from the origin via the formula $v = gx$. Assume that the initial densities and concentrations are $\rho_0(x)$ and $C_0(x)$.

10. Solve the previous problem in the case if the particles move by inertia, and for $t = 0$ their velocity depends linearly on their coordinate.

11. Find the dependence on altitude (measured as a distance from the cloud) of the density of rain drops assuming that their initial density is ρ_0, initial velocity is v, and that they fall freely under the gravity force acceleration g.

12. Obtain the same result as a solution of the continuity equation for densities $\rho(x)$ that stabilize in time, such that

$$\frac{\partial}{\partial x}\Big(v(x)\rho(x)\Big) = 0.$$

13. What happens if one assumes that $g < 0$?

14. Let the stream function Ψ of a 2-D incompressible fluid be independent of time, i.e., $\Psi = \Psi(x_1, x_2)$. What should be the passive tracer distribution so that its density would not depend on time?

15. One of possible distributions of the passive tracer described in the above problem has density $\rho(x) = a\delta(\Psi(x) - b)$. What is the dimension of factor a? Find the mass of passive tracer with that distribution.

16. Derive the continuity equation in the 6D phase space (x, v) for the density $f(x, v, t)$ of a collection of particles with positions and velocities satisfying equations

$$\frac{dX_i}{dt} = V_i, \quad \frac{dV_i}{dt} = g(X_i, V_i), \quad i = 1, 2, \ldots, n, \ldots$$

17. Solve the above equation by the Green function method in the case when particles move in a viscous medium ($g = -hv$) and their initial density is $f_0(x, v)$.

18. Consider again the above problem but under the assumption that all of the particles have mass m and at time $t = 0$ are concentrated at the origin with the initial density $f_0(x, v) = m\delta(x)w(v)$, where function $w(v)$ satisfies the norming condition $\int w(v)d^3v = 4\pi \int_0^\infty w(v)v^2dv = 1$. Find the density $\rho(x, t)$ of particles in the physical space $x \in \mathbf{R}^3$.

19. What is the weak limit of density $f(x, v, t)$ from the previous problem as $t \to \infty$.

20. In the simplest example of the inertial motion of particles the Liouville equation takes the form

$$\frac{\partial f}{\partial t} + (v \cdot \nabla_x)f = 0.$$

In the hydrodynamic approximation the initial condition for this equation is $f_0(x, v) = \rho_0(x)\delta(v - v_0(x))$. It expresses mathematically the fact that the particles with the same position x have to have the same velocity $v_0(x)$. We shall seek a solution of the Liouville equation in the similar form

$$f(x, v, t) = \rho(x, t)\delta(v - v(x, t)).$$

In this situation find equations for $\rho(x, t)$ and $v(x, t)$.

21. Assume that the velocities of particles in the hydrodynamic flow satisfy identical equations $DV/Dt = g(X, V, t)$. Write these equations for the velocity field $v(x, t)$ in Eulerian coordinates.

22. Density of particles $\rho(x, t) = \int f(x, v, t)d^3v$ always satisfies the continuity equation

$$\frac{\partial \rho}{\partial t} + \text{div } (v(x, t)\rho) = 0.$$

Find an expression for the velocity field $v(x, t)$ in terms of the density $f(x, v, t)$ in the 6-D phase space.

23. Let the initial 1-D density (see Fig. 2.3) be

$$\rho_0(x) = m \sum_{i=-\infty}^{\infty} \delta(x - \beta(i\Delta s)),$$

where $\beta(y) = 2y(|y| + s)/(2|y| + s)$. Find the averaged density $\bar{\rho}_0(x)$ ((see, (2.4.13)) in the limit case $\Delta \to 0$, $l \to 0$, $m = \Delta s\rho$, $\Delta = o(l)$.

Pragmatic approach

24. From the pragmatic viewpoint, what is the coefficient μ in the equality $\delta(f(x)) = \mu\delta(x)$, if $f(x) = \alpha x$ for $x \geq 0$ and $= \beta x$ for $x < 0$?

25. What pragmatic value would one assign to functional $f = \int f(x)\delta'(x)dx$, where $f(x)$ is the function from the previous exercise?

26. Find the matching (glueing) conditions for solution of the equation

$$\ddot{y} + \gamma\delta(t)\dot{y} + \omega^2 y = 0.$$

27. Find the matching (glueing) conditions for solution of the equation

$$y'' + \gamma(\delta(t)y)' + \omega^2 y = 0.$$

28. Find the distribution $R(y) = \delta(\alpha(x) - y)[\phi(x)]$ in the following cases:
(a) $\alpha(x) = x\chi(x)$;
(b) $\alpha(x) = (|x| + 1)\,\text{sign}\,(x)$;
(c) $\alpha(x) = x + (1/2)(|x - 1| - |x + 1|)$.

29. Find a generalized particle density if the particles stick at point $x = 0$, and up to that point move with constant velocity $v(y) = -w/4y$, $w > 0$, where y is the Lagrangian coordinate of the particle. Assume that the initial density of particles was the same everywhere and equal to ρ_0.

Part II

INTEGRAL TRANSFORMS AND DIVERGENT SERIES

Chapter 3

Fourier Transform

3.1 Definition and elementary properties

In this chapter we study the *Fourier transform* and investigate its properties for functions $f(t)$ depending on a single variable t which will be interpreted as time. The Fourier transform (or *Fourier image*) $\tilde{f}(\omega)$ of $f(t)$ is defined by

$$\tilde{f}(\omega) = \frac{1}{2\pi} \int f(t) e^{-i\omega t} dt, \tag{1}$$

whenever the integral on the right-hand side exists.

Notice that if f is absolutely integrable on the whole real line the above integral is well-defined. In particular, if $f(t)$ is bounded and decays at infinity faster than $1/|t|^{1+\epsilon}$ then the integrability condition is satisfied and the Fourier transform is well-defined as a function of ω. On the other hand, constant functions such as $f(t) = 1$ do not have well defined Fourier transforms in the above sense. We will return to this difficulty later on in the section on generalized Fourier transforms.

Formula (1) describes an operation $f \mapsto \tilde{f}$ called the *Fourier transformation* which transforms a function f of the time variable t into a function $\tilde{f}$ of another variable ω; in this context it will be called *angular frequency*. It is clear from the defining formula (1) that the Fourier transformation is a linear operation, that is, for any constants A, B,

$$(Af + Bg)^{\sim} = A\tilde{f} + B\tilde{g}. \tag{2}$$

Other connections between simple operations on the original function $f(t)$ and corresponding modifications of their Fourier images also follow immediately from the formula (1). So, if $f(t) \mapsto \tilde{f}(t)$, then

© Springer Nature Switzerland AG 2018
A. I. Saichev and W. Woyczynski, *Distributions in the Physical and Engineering Sciences, Volume 1*, Applied and Numerical Harmonic Analysis, https://doi.org/10.1007/978-3-319-97958-8_3

$$f(t + \tau) \longmapsto \tilde{f}(\omega)e^{i\omega\tau}, \tag{3a}$$

$$f(t)e^{i\Omega t} \longmapsto \tilde{f}(\omega - \Omega), \tag{3b}$$

$$f(-t) \longmapsto \tilde{f}(-\omega), \tag{3c}$$

$$f^*(t) \longmapsto \tilde{f}^*(-\omega). \tag{3d}$$

The first two equalities concern the shifts in variables t and ω and the last two describe what happens under the change of sign of the same variables. If function $f(t)$ is real-valued then its Fourier image satisfies the symmetry condition

$$\tilde{f}(-\omega) = \tilde{f}^*(\omega). \tag{4}$$

This formula permits us to deal with positive frequencies alone.

Numerous applications of the Fourier transform depend on its fundamental connection with the notions of *homogeneity of time and space*. Indeed, the complex Fourier integral kernel $f(t) = e^{i\omega t}$ satisfies the remarkable functional equation

$$f(t + \tau) = f(t)f(\tau). \tag{5}$$

Equation (5) reflects the invariance of the exponential function $\exp(t)$ under time shifts: when the time variable t is shifted by a fixed τ, function $\exp(t)$ does not change except for a constant multiplier $\exp(\tau)$. By contrast with a purely real solution e^{rt} of the same functional equation (5), increasing if $r > 0$ and decreasing if $r < 0$, for the complex exponential

$$|e^{i\omega t}| = 1,$$

that is, the modulus of function $e^{i\omega t}$ remains constant. Also, since for any integer k

$$e^{i2\pi k} = 1,$$

function $e^{i\omega t}$ is *invariant* (i.e. transformed into itself) under time-shifts $T = 2\pi k/\omega$, that is,

$$\exp(i\omega(t + 2\pi k/\omega)) = \exp(i\omega t).$$

In other words, function $e^{i\omega t}$ is periodic with period $T = 2\pi k/\omega$, and it can be viewed as a clock hand performing a full revolution in time T, or as a ruler of length T applied to the time axis.

The variable ω is usually interpreted as the *angular frequency* and in the physical sciences one often uses the quantity

$$\nu = \omega/2\pi$$

which is called *frequency*. In what follows, we will only utilize angular frequency ω which is more convenient from the view point of mathematical exposition, and—without danger of misunderstanding—we will refer to it as frequency ω.

Another remarkable property of the Fourier transform's kernel $e^{i\omega t}$ is particularly important in the analysis of linear systems: a linear combination of arbitrarily time shifted exponential functions is again, up to a constant coefficient, such a function. More precisely,

$$A \exp[i\omega(t + \tau_1)] + B \exp[i\omega(t + \tau_2)] = Ce^{i\omega t},$$

where

$$C = A \exp(i\omega\tau_1) + B \exp(i\omega\tau_2).$$

Consider a physical linear system. Recall that the system is said to be linear if its response to an outside input obeys the *superposition principle,* that is the system's response to the sum of input signals is equal to the sum of independent responses to the components of the sum. Mathematically, this means that the output signal, which will be denoted by $g(t)$ is a *linear operator* on the input signal $\phi(t)$. In the general case such an operator is of the form [1]:

$$g(t) = \int h(t, \tau)\phi(\tau)d\tau,$$

where $h(t, \tau)$ is a response to the δ-pulse input, which thus completely determines properties of the linear system. Suppose that properties of the system are invariant in time. Then, under an arbitrary time shift of the input signal, the output signal should not change except for the same time shift. Such a property of invariance with respect to time shifts will be satisfied if the response to the δ-pulse input depends only on one variable, i.e., $h(t, \tau) = h(t - \tau)$. In this case the relationship between the input and output signals is given by the familiar convolution integral

$$g(t) = \int h(\tau)\phi(t - \tau)d\tau.$$

If input signal is of the form $\phi(t) = e^{\gamma t}$, then the shape of the time-invariant system's response will not be distorted by the system as $g(t) = K\phi(t)$. The complex factor

$$K = \int h(\tau)\phi(-\tau)d\tau$$

---

[1]The area of mathematics studying such general operators is called functional analysis. In particular, it provides the above representation formula.

describes the attenuation of the output signal in comparison with the input signal. In the particular case of $\phi = e^{i\omega t}$, we see that $K = 2\pi \tilde{h}(\omega)$, so that—up to the factor 2π—we get the Fourier transform of $h(t)$. This property makes the Fourier transform technique effective in physical and engineering applications.

To get a better feel for the remarkable function $e^{i\omega t}$, we suggest contemplating the following erroneous line of reasoning which students are sometimes tempted to follow. Observe that $\exp(i2\pi \nu t) = [\exp(i2\pi)]^{\nu t}$, where $\nu = \omega/2\pi$ is the frequency (measured in Herzes). Since $\exp(i2\pi) = 1$, conclude that $\exp(i\omega t) = 1^{\nu t} = 1$, and replace the integral (1) by a simpler integral $(1/2\pi) \int f(t)\,dt$. What went wrong?

3.2 Smoothness, inverse transform and convolution

We begin with the relationship between the smoothness of function $f(t)$ and the asymptotic behavior of its Fourier transform $\tilde{f}(\omega)$ as $\omega \to \infty$. Assume that $f(t)$ is n-times continuously differentiable on **R** and absolutely integrable together with its first n derivatives. Multiply (3.1.1) by $(-i\omega)^n$ and note that $(-i\omega)^n e^{-i\omega t}$ is the nth derivative of function $e^{-i\omega t}$ with respect to t. Hence, integrating n times by parts, we arrive at the equality

$$(i\omega)^n \tilde{f}(\omega) = \frac{1}{2\pi} \int f^{(n)}(t)e^{-i\omega t}dt, \tag{1}$$

which expresses one of the most useful consequences of the invariance of the Fourier transform's kernel with respect to shifts of its variables: *differentiation of the original function corresponds to multiplication of its Fourier image by $i\omega$.* Absolute integrability of the integrand in (1) implies that the expression on the left-hand side is bounded so that the modulus $|\tilde{f}(\omega)|$ of the Fourier transform decays at infinity not slower than $|\omega|^{-n}$. This is the basic connection between the original function's degree of smoothness and its Fourier transform's speed of decay as $|\omega| \to \infty$. In view of the extraordinary importance of this property we will formulate it explicitly as the following principle:

Sufficiently smooth functions with absolutely integrable nth derivative have Fourier transforms that decay at infinity not slower than $|\omega|^{-n}$.

In view of the symmetry between the Fourier transformation and its inverse (to be established below), the inverse implication is also true:

If the Fourier transform of original function is a smooth function with an absolutely integrable nth derivative (with respect to ω), then the original function decays at infinity not slower than $|t|^{-n}$.

Now let us turn to the *inversion formula* which permits recovery of a function from its Fourier transform. We will assume that $f(t)$ is absolutely integrable and sufficiently smooth so that $\tilde{f}(\omega)$ is absolutely integrable as well. Multiplying (3.1.1) by $\tilde{\phi}(\omega)e^{i\omega\tau}$, where $\tilde{\phi}(\omega)$ is an absolutely integrable function, and integrating both sides of the equality with respect to ω, we obtain that

$$\int \tilde{\phi}(\omega)\tilde{f}(\omega)e^{i\omega\tau}\,d\omega = \frac{1}{2\pi}\int f(t)\left[\int \tilde{\phi}(\omega)\exp[i\omega(\tau - t)]d\omega\right]dt. \quad (2)$$

Note that we have changed the order of integration on the right-hand side. This is justified by the absolute integrability of integrands. Let us put

$$\tilde{\phi}(\omega) = \exp\left(-\frac{\epsilon^2\omega^2}{2}\right),$$

and evaluate the inner integral on the right-hand side using the well-known formula

$$\int \exp(-bx^2 + ikx)dx = \sqrt{\frac{\pi}{b}}\exp\left(-\frac{k^2}{4b}\right), \quad (3)$$

valid for any Re $b \geq 0$, $b \neq 0$. As a result, equality (2) is transformed into equality

$$\int \tilde{f}(\omega)\exp\left(-\frac{\epsilon^2\omega^2}{2} + i\omega\tau\right)d\omega = \int f(t)\frac{1}{\sqrt{2\pi}\epsilon}\exp\left(-\frac{(t - \tau)^2}{2\epsilon^2}\right)dt.$$

For $\epsilon \to 0$, the second function in the right-hand side integral weakly converges to a shifted Dirac delta $\delta((t - \tau))$, and its probing property recovers the value of $f(t)$ at $t = \tau$. On the left-hand side, in view of the absolute integrability of $\tilde{f}(\omega)$, we can just set $\epsilon = 0$. Finally, replacing τ by t, we arrive at the Fourier integral

$$f(t) = \int \tilde{f}(\omega)e^{i\omega t}\,d\omega, \quad (4)$$

which expresses function $f(t)$ through its Fourier transform.

We would like to stress that the direct Fourier transform (3.1.1) and the inverse Fourier transform (4) are symmetric in the sense that if $\tilde{f}(\omega)$ is the Fourier transform of $f(t)$ then $f(-\omega)/2\pi$ is the Fourier transform of function $\tilde{f}(t)$. Using the notation introduced in (3.1.2) we can write this statement as follows:

$$If \quad f(t) \mapsto \tilde{f}(\omega) \quad then \quad \tilde{f}(t) \mapsto f(-\omega)/2\pi. \quad (5)$$

The above relationships can generate a protest among engineers and physicists since frequency ω and time t have different dimensionalities. That is why, whenever necessary, we will assume that the frequency and the time are nondimensionalized.

In view of (4), the integral in the brackets in (2) can be replaced by $\phi(\tau - t)$, and we obtain that

$$\int \tilde{f}(\omega)\tilde{\phi}(\omega)e^{i\omega\tau}d\omega = \frac{1}{2\pi}\int f(t)\phi(\tau - t)dt. \tag{6}$$

The integral on the right-hand side is the *convolution of functions* f and ϕ:

$$\int f(t)\phi(\tau - t)dt = f(t) * \phi(t). \tag{7}$$

Comparing (6) with (4) we conclude that the Fourier transform of a convolution of functions is, up to the constant 2π, a product of their Fourier transforms, that is

$$f(t) * \phi(t) \longmapsto 2\pi \tilde{f}(\omega)\tilde{\phi}(\omega). \tag{8}$$

Multiplying (4) by $\phi(t)e^{i\omega t}$, integrating it with respect to t, and invoking the defining formula (3.1.1) to evaluate the integral on the right-hand side, we arrive at

$$\int f(t)\phi(t)e^{i\omega t}dt = 2\pi \int \tilde{f}(\Omega)\tilde{\phi}(\omega - \Omega)d\Omega, \tag{9}$$

dual to the equality (6). The formula implies, in particular, that the Fourier transform of a product of two functions is the convolution of their Fourier images:

$$f(t)\phi(t) \longmapsto \tilde{f}(\omega) * \tilde{\phi}(\omega). \tag{10}$$

We shall call formulas (6) and (9) *Parseval equalities,* although that name is often reserved for their special case which is obtained from (6) by setting $\tau = 0$ and substituting $f^*(-t)$ for $\phi(t)$. As a result, we get that

$$\int |f(t)|^2 dt = 2\pi \int |\tilde{f}(\omega)|^2 d\omega. \tag{11}$$

In engineering applications, $P(t) = |f(t)|^2$ often represents the signal's *power function,* so that the integral on the left-hand side of (11) is the signals' *total energy.* Thus, in view of the Parceval equality (11), the function $|\tilde{f}(\omega)|^2$ shows how energy is distributed over different frequencies.

3.3 Generalized Fourier transform

Let us try to extend the Fourier transform's domain beyond the class of absolutely integrable functions. Such a generalized Fourier transform will be introduced in the same way the distributions were in Chapter 1. Recall that the distribution was defined as a linear functional which assigned a number to each function ϕ from a certain set of test functions. In other words, you can tell a distribution by its action on test functions.

Equality (3.2.6) offers a similar opportunity to extend the Fourier transform's domain. We will call $\tilde{f}(\omega)$ the *generalized Fourier transform* of function $f(t)$ if the integral on the right-hand side is equal to the functional on the left-hand side for each test function $\phi(t)$ from a certain class S of test functions to be determined later. Since the convolution on the right-hand side is equally well defined for function f replaced by a distribution T, and it remains a smooth function in the latter case, we can define the generalized Fourier transform $\tilde{T}$ of a distribution T by the condition that

$$\int \tilde{T}(\omega)\tilde{\phi}(\omega)e^{i\omega\tau}\,d\omega = \frac{1}{2\pi}T*\phi(\tau). \tag{1}$$

Example 1. To find the Fourier transform of the Dirac delta, notice that the right-hand side of (1) becomes $\delta*\phi(\tau) = \phi(\tau)$. The next step is to look at the left-hand side and search for a $\tilde{T}$ that will make the left-hand side equal to $\phi(\tau)$. But the inversion formula (3.2.4) makes it clear that such a $\tilde{T}$ has to be the function identically equal to $1/2\pi$. Hence,

$$\tilde{\delta}(\omega) = \frac{1}{2\pi}. \tag{2}$$

This fact can be also symbolically written as:

$$\delta(t) = \frac{1}{2\pi}\int e^{i\omega t}\,d\omega. \tag{3}$$

Conversely, a function which gives value $\delta[\phi] = \tilde{\phi}(0)$ to the integral on the left-hand side of (3.2.6) is equal to the generalized Fourier transform of the constant function $f(t) = 1$. This is clear from comparison of the right-hand side of formula (3.2.6) with (3.1.1). In other words,

$$\tilde{1} = \delta(\omega). \qquad\blacksquare$$

A mathematical aside: tempered distributions[2]. Before we continue with examples of other generalized Fourier transforms, let us discuss some mathematical problems emerging here.

The original distribution space $\mathcal{D}'$ was introduced as a set of functionals T on the set $\mathcal{D}$ of infinitely differentiable test functions $\phi(t)$ with compact support. If the generalized Fourier transforms $\tilde{T}$ are to be defined by linear functionals on the left-hand side of equality (1) then Fourier transforms $\tilde{\phi}(\omega)$ will have to play the role of test functions. However, the Fourier transforms $\tilde{\phi}$, for $\phi \in \mathcal{D}$, are not a rich enough set to determine $\tilde{T}$ in the above mentioned sense. In particular, they do not have compact supports. As a result we loose the symmetry of the generalized Fourier transform. Indeed, it is impossible to assign to each distribution $T \in \mathcal{D}'$ its generalized Fourier transform which would be a continuous linear functional on the set $\{\tilde{\phi}(\omega) : \phi \in \mathcal{D}\}$.

The solution is to expand the space of test functions $\{\phi(t)\}$, by requiring a symmetry in equality (1), or more precisely, by demanding that the sets of $\phi(t)$'s and of $\tilde{\phi}(\omega)$'s be equal. It turns out that the right space of test functions happens to be the set $\mathcal{S}$ of all infinitely differentiable functions $\phi(x)$, which decrease at infinity, together with all their derivatives, faster than arbitrary power function $|x|^{-n}$. Such functions are often called *rapidly decreasing*. In other words, $\phi \in \mathcal{S}$ if, for any positive integers $n, m \geq 0$, we can find constants K_{mn} such that, for arbitrary x,

$$|x^n \phi^{(m)}(x)| < K_{nm}. \tag{4}$$

It is easy to show that if function $\phi(t) \in \mathcal{S}$ then its Fourier transform $\tilde{\phi}(\omega)$ is also infinitely differentiable and rapidly decreasing. Indeed, according to (3.2.1), the infinite differentiability of $\phi(t)$ implies that its Fourier transform decays faster than an arbitrary power $|\omega|^{-n}$, and vice versa, the speed of decay of $\phi(t)$ as $|t| \to \infty$ implies that its Fourier transform is infinitely differentiable.

Expansion of the test function space from $\mathcal{D}$ to $\mathcal{S}$ obviously narrows the set of corresponding distributions from $\mathcal{D}'$ to $\mathcal{S}' \subset \mathcal{D}'$. Traditionally, the set $\mathcal{S}'$ of continuous linear functionals on $\mathcal{S}$ with convergence related to the conditions (4) is called the space of *tempered distributions*.

Notice that the selection of this new and smaller distribution space does not affect us too seriously. All the examples of distributions from $\mathcal{D}'$ discussed so far are also continuous functionals on set $\mathcal{S}$. All the distributions with compact support, including the Dirac delta and all its derivatives, are tempered distributions. In other words,

$$\mathcal{S}' \subset \mathcal{D}'.$$

Also, any function $f(t)$ which grows slower than a certain power of t, defines a continuous functional on $\mathcal{S}$ (that is a distribution $T_f \in \mathcal{S}'$) by the formula $\int f(t)\phi(t)dt$, $\phi \in \mathcal{S}$. Hence, the name tempered distributions.

To get a better feel for the narrowness of the set of distributions introduced by the above expansion of the set of test functions, notice that both function $\exp(t^2)$ and e^t represent distributions in the space $\mathcal{D}'$ but not in $\mathcal{S}'$. On the other hand, for any distribution in $\mathcal{S}'$ one can define its Fourier transform which is also a distribution in $\mathcal{S}'$, and this is the real reason for the usefulness of tempered distributions.

All the operations applicable to ordinary Fourier transforms remain valid for their generalized cousins. So, in view of (3.1.3a), the shifted Dirac delta $\delta(t - \tau)$ has Fourier transform $e^{-i\omega\tau}/2\pi$, and (3.2.1) implies that the Fourier transform

---

[2]This material may be skipped by the first time reader

of $\delta^{(n)}(t)$ is equal to $(i\omega)^n/2\pi$. Two additional formulas involving generalized Fourier transforms will be useful:

$$t^n e^{ist} \longmapsto (i)^n \frac{\partial^n}{\partial \omega^n} \delta(\omega - s), \qquad (5a)$$

$$\frac{\partial^n}{\partial t^n} \delta(t - \tau) \longmapsto \frac{1}{2\pi}(i\omega)^n e^{-i\omega\tau}. \qquad (5b)$$

As was the case for ordinary distributions, the generalized Fourier transforms could be defined as weak limits of regular Fourier transforms. For example, let us show that the functions

$$\tilde{f}(\omega, \lambda) = \frac{1}{\pi} \frac{\sin(\omega\lambda)}{\omega}, \qquad (6)$$

dependent on parameter λ, converge to $\delta(\omega)$ as $\lambda \to \infty$. For that purpose consider the functional

$$F(\lambda) = \int \tilde{\phi}(\omega) \frac{\sin(\omega\lambda)}{\omega} d\omega,$$

which is called the *Dirichlet integral*. Differentiating the above equality with respect to λ, we get that

$$F'(\lambda) = \int \tilde{\phi}(\omega) \cos(\omega\lambda) d\omega = \frac{1}{2}\big[\phi(+\lambda) + \phi(-\lambda)\big].$$

Now, taking definite integrals of both sides over the interval $(0, \lambda)$, and noticing that $F(0) = 0$, we get that

$$F(\lambda) = \frac{1}{2} \int_{-\lambda}^{\lambda} \phi(t) dt.$$

If the Fourier transform $\tilde{\phi}(\omega)$ is sufficiently smooth, then the original function $\phi(t)$ is absolutely integrable and the above integral converges, as $\lambda \to \infty$, to

$$F(\infty) = \frac{1}{2} \int \phi(t) dt = \pi \tilde{\phi}(0),$$

which is the value of the test function at 0 multiplied by π. This proves that, weakly,

$$\frac{1}{\pi} \frac{\sin(\omega\lambda)}{\omega} \to \delta(\omega)\text{sign}\,(\lambda) \quad (\lambda \to \infty), \qquad (7)$$

although the same function does not converge to zero for any $\omega \neq 0$, as it "fills out" the area between the four branches of the hyperbola $\pm 1/\pi |\omega|$ (see Fig. 3.3.1).

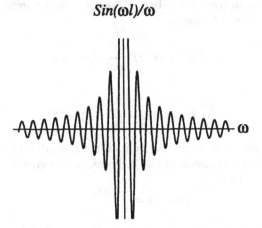

FIGURE 3.3.1
The graph of kernels (6) in the Dirichlet integral. The kernels weakly converge to $\pi \delta(\omega)$ as $\lambda \to \infty$, filling up the area between hyperbolas $\pm 1/|\omega|$.

3.4　Transport equation

Many applications of the Fourier transform are based on the fact that it often simplifies solving functional equations. In particular, it reduces some differential and integral equations to algebraic equations. Numerous examples of this method will be discussed in the following chapters. For now, to give a simple illustration of what we have in mind, we will just consider a 1-D *transport equation*

$$\frac{\partial f}{\partial t} + v \frac{\partial f}{\partial x} = \int g(u) f(x, v - u, t) du, \tag{1}$$

with the initial condition

$$f(x, v, t = 0) = f_0(x, v).$$

Similar equations arise in scattering theory. *The collision (scattering) integral* on the right-hand side takes into account the process of interaction of particles with the medium, and function $f(x, v, t)$ is the particle density in the *phase space* (x, v). The density of particles in the actual physical space, in our case—the x axis, can then be found by integrating f over the velocities

$$\rho(x, t) = \int f(x, v, t) dv. \tag{2}$$

To begin with, we will show an important consequence of equation (1). Integrating it over all v's gives

$$\frac{\partial \rho}{\partial t} + \frac{\partial}{\partial x} \int v f(x, v, t) dv = \rho(x, t) \int g(v) dv.$$

Integrating it next over all x's, we obtain an equation for the total mass of particles

$$M = \int \rho(x, t) dx,$$

and for its rate of change

$$\frac{\partial M}{\partial t} = M \int g(v) dv.$$

Thus, to fulfill the mass conservation law, it is necessary to impose the condition

$$\int g(v) dv = 0. \tag{3}$$

Let us now return to the transport equation (1). Its right-hand side contains a convolution integral, whose Fourier transform is equal to the product of Fourier images of the factors under the integral sign. That means that by passing to the Fourier image

$$F(x, \mu, t) = \frac{1}{2\pi} \int f(x, v, t) e^{-i\mu v} dv \tag{4}$$

of f with respect to v, we transform the integro-differential equation (1) into a purely differential second-order equation

$$\frac{\partial F}{\partial t} + i \frac{\partial^2 F}{\partial x \partial \mu} = 2\pi \tilde{g}(\mu) F(x, \mu, t).$$

Here

$$\tilde{g}(\mu) = \frac{1}{2\pi} \int g(v) e^{-i\mu v} dv. \tag{5}$$

An application of one more Fourier transformation

$$\Phi(\kappa, \mu, t) = \frac{1}{2\pi} \int F(x, \mu, t) e^{-i\kappa x} dx, \qquad (6)$$

this time with respect to x, gives the first-order partial differential equation

$$\frac{\partial \Phi}{\partial t} - \kappa \frac{\partial \Phi}{\partial \mu} = 2\pi \tilde{g}(\mu) \Phi(\kappa, \mu, t), \qquad (7)$$

with the initial condition

$$\Phi(\kappa, \mu, t = 0) = \Phi_0(\kappa, \mu).$$

This equation can be easily solved by the *method of characteristics* described in Section 2.6. Recall that the method assumes that the independent variable μ is a function of t, and that it satisfies equation

$$\frac{d\mu}{dt} = -\kappa, \quad \mu(t = 0) = \nu. \qquad (8)$$

In this case, the left-hand side of equation (7) turns out to be a total derivative of $\Phi(t) = \Phi(\kappa, \mu(t), t)$ with respect to time t:

$$\frac{d\Phi}{dt} = -2\pi \tilde{g}(\mu) \Phi(t), \qquad (9)$$

$$\Phi(t = 0) = \Phi_0(\kappa, \nu).$$

The solution of the *characteristic equations* (8) and (9) is of the form

$$\mu = \nu - \kappa t, \qquad (10)$$

$$\Phi(\kappa, \nu, t) = \Phi_0(\kappa, \nu) \exp\left(2\pi \int_0^t \tilde{g}(\nu - \kappa\tau) d\tau \right). \qquad (11)$$

Notice that (11) is the solution of equation (7) along the line $\mu = \nu - \kappa t$ in the (μ, t) plane (Fig. 3.4.1).

As the parameter ν varies, we obtain a family of foregoing lines which cover the entire plane. Point (μ, t) corresponds to a line with parameter $\nu = \mu + \kappa t$. Substituting in (11), we find the solution of equation (7) with given μ and t:

$$\Phi(\kappa, \mu, t) = \Phi_0(\kappa, \mu + \kappa t) \exp\left(2\pi \int_0^t \tilde{g}(\mu + \kappa\tau) d\tau \right). \qquad (12)$$

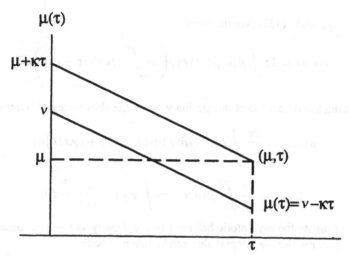

FIGURE 3.4.1
Characteristic lines for equation (7).

To get the desired solution of the original equation (1), we need to calculate the double inverse Fourier transform

$$f(x, v, t) = \int\int \Phi(\kappa, \mu, t) \exp(i\mu v + i\kappa x) d\kappa \, d\mu.$$

However, if we are interested only in finding the density of particles on the x axis, there is no need to evaluate this double integral. Indeed, it suffices to put $\mu = 0$ in (12), which, as is obvious from (4), corresponds to integration of f over all v's, and then just to find the inverse transform with respect to κ:

$$\rho(x, t) = 2\pi \int \Phi_0(\kappa, \kappa t) \exp\left(2\pi \int_0^t \tilde{g}(\kappa\tau)d\tau + i\kappa x\right) d\kappa. \tag{13}$$

Let us take a look at properties of the density function in the case of a particle stream with the initial density $\rho(x)$. In addition, we shall assume that at each stream point the particle velocities are distributed with a normalized density $p(v)$. Thus

$$f_0(x, v) = \rho(x)p(v), \quad \int p(v)dv = 1.$$

Respectively,

$$\Phi_0(\kappa, \mu) = \tilde{\rho}(\kappa)\tilde{p}(\mu), \quad \tilde{p}(0) = \frac{1}{2\pi},$$

and the expression (13) takes the form

$$\rho(x,t) = 2\pi \int \tilde{\rho}(\kappa)\tilde{p}(\kappa t)\exp\left(2\pi \int_0^t \tilde{g}(\kappa\tau)d\tau + i\kappa x\right)d\kappa. \tag{14}$$

Introducing a new variable of integration $\gamma = \kappa t$, the above integral is transformed into

$$\rho(x,t) = \frac{2\pi}{t}\int \tilde{\rho}(\gamma/t)\tilde{p}(\gamma)\exp\left(tG(\gamma) + i\gamma x/t\right)d\gamma \tag{15}$$

where

$$G(\gamma) = \frac{2\pi}{\gamma}\int_0^\gamma \tilde{g}(\mu)d\mu = \int g(v)\frac{1 - e^{-i\gamma v}}{i\gamma v}dv. \tag{16}$$

To investigate the asymptotic behavior of the density as $t \to \infty$, observe that the first factor under the integral sign can be taken outside

$$\rho(x,t) = \frac{1}{t}M\int \tilde{p}(\gamma)\exp\left(tG(\gamma) + i\gamma x/t\right)d\gamma,$$

and, as far as the function $G(\gamma)$ is concerned, we can restrict ourselves to the first nonvanishing term of the power series expansion

$$G(\gamma) = \langle 1 \rangle - \langle v \rangle\frac{i}{2}\gamma - \frac{1}{6}\langle v^2 \rangle\gamma^2 + \dots \tag{17}$$

In the above formula

$$\langle v^n \rangle = \int v^n g(v)dv.$$

In view of (3), the first summand on the right-hand side of (17) is equal to zero. Assuming that the collision integral kernel $g(v)$ is symmetric, the second term in (17) vanishes as well. Let us keep in the expansion (17) only the quadratic term:

$$G(\gamma) = -\frac{1}{6}\langle v^2 \rangle\gamma^2.$$

It corresponds to the so-called *diffusion approximation* which replaces kernel $g(v)$ in the original equation (1) by the distribution

$$g(v) = \langle v^2 \rangle\delta''(v).$$

Then the transport equation becomes the differential equation

$$\frac{\partial f}{\partial t} + v\frac{\partial f}{\partial x} = \langle v^2 \rangle\frac{\partial^2 f}{\partial v^2},$$

and the expression for density (15) takes the form

$$\rho(x, t) = \frac{1}{t} M \int \tilde{p}(\gamma) \exp\left(-t\frac{1}{6}\langle v^2 \rangle \gamma^2 + i\gamma \frac{x}{t}\right) d\gamma.$$

For very large times, it asymptotically converges to

$$\rho(x, t) = M \sqrt{\frac{3}{2\pi \langle v^2 \rangle t^3}} \exp\left(-\frac{3x^2}{2\langle v^2 \rangle t^3}\right). \tag{18}$$

To complete the picture, we will provide another useful form of the density which follows from equality (13). First, notice that setting $\tilde{g} = 0$ in (13) gives the particle density in the absence of scattering. Denote it by $\rho^0(x, t)$. The corresponding density $f^0(x, v, t)$ in the phase space satisfies equation

$$\frac{\partial f^0}{\partial t} + v\frac{\partial f^0}{\partial x} = 0, \quad f^0(x, v, t = 0) = f_0(x, v).$$

Here, the use of the Fourier transform is not necessary since, obviously,

$$f^0(x, v, t) = f_0(x - vt, v).$$

Hence, the density of nonscattered particles

$$\rho^0(x, t) = \int f_0(x - vt, v) dv.$$

In particular, if all the particles were motionless at $t = 0$, that is, if

$$\rho^0(x, t) = \rho_0(x)\delta(v),$$

then

$$\rho^0(x, t) = \rho_0(x).$$

On the other hand, according to (13), for $\tilde{g} \equiv 0$,

$$\rho^0(x, t) = 2\pi \int \Phi_0(\kappa, \kappa t) e^{i\kappa x} dx.$$

Taking the inverse Fourier transform we get that

$$2\pi \Phi_0(\kappa, \kappa t) = \frac{1}{2\pi} \int \rho^0(y, t) e^{-i\kappa y} dy,$$

and substituting this expression for $\Phi_0(\kappa, \kappa t)$ in (13), we get the convolution integral

$$\rho(x, t) = \int \rho^0(y, t)G(x - y, t)dy, \tag{19}$$

where function

$$G(x, t) = \frac{1}{2\pi} \int \exp\left(2\pi \int_0^t \tilde{g}(\kappa \tau)d\tau + i\kappa x\right)d\kappa \tag{20}$$

describes influence of the scattering processes on the density. The behavior of $G(x, t)$ depends on the form of function $\tilde{g}(\mu)$ (5), which expresses the nature of scattering. In the diffusion approximation, $G(x, t)$ is given by the right-hand side of the equality (18), with M deleted.

3.5 Exercises

1. Find the Fourier transform of function $f(t) = \chi(t)e^{-\gamma t}$, $\gamma > 0$.

2. Find the Fourier transform of function $f(t) = e^{-\gamma |t|}$, $\gamma > 0$.

3. Taking into account symmetry between the Fourier transform and its inverse, and the answer to the preceding problem, find the Fourier transform of $f(t) = 1/(\gamma^2 + t^2)$.

4. Find the Fourier transform of function $f(t) = t^2/(\gamma^2 + t^2)$.

5. Find the Fourier transform of function

$$f(t) = \sum_{n=0}^{\infty} \delta(t + n)e^{-\gamma n}, \ \gamma > 0.$$

6. Find the Fourier transforms of the odd component $f_o(t) = (1/2)[f(t) - f(-t)]$ and of the even component $f_e(t) = (1/2)[f(t) + f(-t)]$ of function f from the previous exercise.

7. Find $r(\gamma) = \max \tilde{f}_e / \min \tilde{f}_e$ and study its behavior as $\gamma \to 0+$.

8. Find a function $f(t)$ such that $\tilde{f}(\omega) = \sin \omega / \omega$. (Hint. First find $f'(t)$.)

9. Find the Fourier transform of function $f(t) = 1 - t^2$ for $|t| \le 1$, and $= 0$ for $|t| > 1$. (Hint: Find first $f''(t)$.)

10. Find the Fourier image of the function

$$f(t) = \begin{cases} 2 - t^2, & \text{for } |t| < 1; \\ (2 - t)^2, & \text{for } 1 \le |t| < 2; \\ 0, & \text{for } 2 \le |t|. \end{cases}$$

Hint: Begin with calculation of the Fourier image of the third derivative of $f(t)$.

11. The values of function $f(t)$ are known at points $t_n = \Delta n$, $-\infty < n < \infty$. Find the Fourier image of the linear interpolation function

$$f_l(t) = f(n\Delta) + [f((n+1)\Delta) - f(n\Delta)]\frac{t - n\Delta}{\Delta}, \qquad n\Delta < t < (n+1)\Delta.$$

Hint. Begin with calculation of the Fourier image of the second derivative of the interpolated function.

12. Find the Fourier image of the function $f(t) = \int h(\tau)h(\tau + t)\,d\tau$, where

$$h(t) = \begin{cases} t, & \text{for } 0 < t \le \theta; \\ 0, & \text{for } \theta \le t \text{ and } t \le 0. \end{cases}$$

13. Taking into account the form of the solution to the preceding exercise find the 4th order derivative of $f(t)$.

14. Using the fact that the function

$$f(t; p) = \begin{cases} \cos^p(t), & \text{for } |t| < \pi/2; \\ 0, & \text{for } |t| \ge \pi/2; \end{cases}$$

satisfies, for $p \ge 2$, the recurrence relation

$$\frac{d^2}{dt^2} f(t; p) + p^2 f(t; p) = p(p-1)f(t; p-2),$$

find the Fourier image $\tilde{f}(\omega; p)$ for any integer $p \ge 1$.

Chapter 4

Asymptotics of Fourier Transforms

In Chapter 3 we demonstrated that the Fourier transform $\tilde{f}(\omega)$ of a smooth function $f(t)$ rapidly decays to zero as $\omega \to \infty$. However, smoothness is rare in natural phenomena and one often encounters processes that are either discontinuous or violate the smoothness assumption in other ways. Such phenomena include, for example, shock fronts generated by large amplitude acoustic waves, ocean waves, or desert dunes with their characteristic sharp crests. These and many other examples explain the importance of the Fourier analysis of nonsmooth processes.

Roughly speaking, values of the Fourier transform $\tilde{f}$ at angular frequency ω are determined by the behavior of the function $f(t)$ at time scales of the order $2\pi/\omega$. The latter quantity decreases as ω increases. Hence, violations of smoothness which, by their very nature have a local character, are related mostly to the behavior of the Fourier transform at large values of ω. The larger the ω is, the more the impact of nonsmoothness is felt. Mathematically speaking, the nonsmoothness of the original function dictates the asymptotic behavior of its Fourier transform as $\omega \to \infty$. In the present chapter we will study these asymptotics.

4.1 Asymptotic notation, or how to get a camel to pass through a needle's eye

We begin by recalling the standard, and widely used asymptotic notation.

- If the fraction $\phi(x)/\psi(x)$ converges to 1 as $x \to \infty$, then this fact is denoted in the form of an *equivalence relation*

$$\phi(x) \sim \psi(x), \quad (x \to \infty),$$

and we say that $\phi(x)$ is *asymptotically equivalent* to function $\psi(x)$ for $x \to \infty$.

© Springer Nature Switzerland AG 2018
A. I. Saichev and W. Woyczynski, *Distributions in the Physical and Engineering Sciences, Volume 1*, Applied and Numerical Harmonic Analysis, https://doi.org/10.1007/978-3-319-97958-8_4

If, for a positive constant a,

$$\phi(x) \sim a\psi(x), \quad (x \to \infty),$$

then we shall say that $\phi(x)$ and $\psi(x)$ are *of the same order* at infinity.

• If the fraction $\phi(x)/\psi(x)$ converges to 0 as $x \to \infty$, then we write

$$\phi(x) = o\{\psi(x)\} \quad (x \to \infty),$$

and say that function $\phi(x)$ is at infinity *of the order smaller than* $\psi(x)$. In particular, the notation $\phi(x) = o\{1\}$ $(x \to \infty)$, means that function $\phi(x)$ converges to 0 at infinity.

• Finally, if there exist positive constants a and M such that

$$|\phi(x)/\psi(x)| \le M, \quad \text{for} \quad a \le x < \infty,$$

then we write that

$$\phi(x) = O\{\psi(x)\} \quad (x \to \infty),$$

and say that $\phi(x)$ is at infinity *of the order not greater than* $\psi(x)$.

In a similar fashion, analogous asymptotic notation can be introduced for $x \to 0$, or for any other limit point.

We shall illustrate how the asymptotic notation works by solving two simple, and somewhat light-hearted problems. The intuitively surprising solutions are obtained via standard asymptotic analysis.

Example 1. Camel passing through a needle's eye. Let us start with a familiar elementary school mind teaser. Assume that Earth is an ideal ball of radius $R = 6400$ km and that it is wrapped tightly at the Equator with a length of rope. Cut the rope in one place, splice into it another piece of rope of length $L = 1$ m, and stretch it, keeping it at a uniform height h above the Earth's surface. The question usually asked is: What is that height? The obvious answer is that $h = L/2\pi = 16$ cm. In particular, h is independent of the radius of the Earth (it would be the same on the Moon, Jupiter, tennis ball, etc.). So, with just an extra 1m piece of rope, cats all over the Equator would have the freedom to walk underneath the rope.

A young friend of ours spent a sleepless night puzzling over the philosophical ramifications of the above answer. But he was even more perplexed by the solution to the following related problem. So, perhaps, the insomniacs should skip this page and the rest should travel to the Sahara Desert and hoist the spliced rope above the Earth surface as high as possible. The question is: How high would that be? Denote this height by H. The situation is schematically pictured in Fig. 4.1.1.

To better expose the mathematical contents of the problem, let us assume initially that the Earth' radius is equal to 1. Then the height η depends on the angle θ by

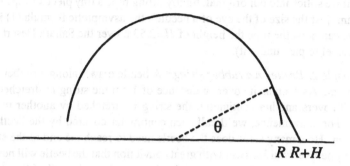

FIGURE 4.1.1
A schematic picture of the spliced rope hoisted above the Earth's surface at one point of the Equator.

the formula

$$\eta = \frac{1}{\cos\theta} - 1,$$

and the length l of the additional piece of rope, in terms of angle θ, is

$$l = 2(\tan\theta - \theta).$$

Our case corresponds to the situation where the length of the additional piece of rope is very small in comparison to the Earth's radius ($l \ll 1$). So we can replace the above exact expressions by an asymptotically equivalent expressions valid for $\theta \to 0$ to get

$$2\eta \sim \theta^2, \quad \frac{3}{2}l \sim \theta^3.$$

Eliminating θ from these equivalence relations, we get a relationship connecting η and l:

$$\eta \sim \mu l^\alpha, \quad l \to 0, \tag{1}$$

where

$$\alpha = \frac{2}{3}, \quad \mu = \frac{1}{2}\left(\frac{3}{2}\right)^\alpha \approx 0.655.$$

It is clear from (1) that $l = o(\eta)$, $(\eta \to 0)$, that is, l is of an order smaller than η as $\eta \to 0$. Actually, this means that the height η of the rope hoisted over the Sahara Desert is much larger than the length l of the inserted piece of rope, and the ratio of these two quantities gets larger as l gets smaller.

Finally, going back to our original "real" Earth example and taking $l = L/R = 1.5625 \cdot 10^{-7}$, we get from the asymptotic relation (1) that the sought height

$$H = \eta \cdot R = 121.6\,\text{m}.$$

Now, if we splice into our original, tightly fitting rope a tiny piece of rope of length $L=3$mm (just the size of the eye of a needle), the asymptotic formula (1) indicates that we can hoist the rope the height of $H=2.53$m over the Sahara Desert (enough for a camel to pass under it).   ∎

Example 2. Beetle on a rubber string. A beetle crawls along a rubber string of length 1m. As soon as it covers a distance of 1 cm the string is stretched by 1m. After it covers another centimeter, the string is stretched by another meter, and so on. For convenience, we'll call each centimeter covered by the beetle a step. Question: How many steps does the beetle need to reach the end of the string?

Many people would answer with great conviction that the beetle will never reach the end of the string and that, in fact, its distance from the end of the string will increase indefinitely. Asymptotic analysis shows, however, that the number of needed steps is finite and helps to evaluate it with high precision. The legend has it that Andrei Sakharov, father of Russian H-bomb and later a famous dissident, solved the problem in one minute during a solemn anniversary celebration of the Soviet Academy, working on the back of the invitation card. This record is not included in the Guiness Book.

Let δ be the fraction of the string length covered by the beetle by the time the string is first stretched. In the above setting $\delta=0.01$. Our goal is to find an asymptotic formula, as $\delta \to 0$, for the required number N of steps as a function of δ. In the nth step the beetle covers a fraction of the string length equal to δ/n. Hence the total fraction of the string length covered by the beetle in N steps is

$$L(N) = \delta \left(1 + \frac{1}{2} + \frac{1}{3} + \frac{1}{4} + \frac{1}{5} + \cdots + \frac{1}{N} \right). \tag{2}$$

The *harmonic series* on the right-hand sides diverges and $L(N) \to \infty$ as $N \to \infty$, so the beetle will reach the string's end in a finite number of steps. Its arrival at the string's end in N steps means that $L(N) \sim 1$ for a certain large N. To find that N approximately, we will need an asymptotic formula for the partial sum

$$H(N) = 1 + \frac{1}{2} + \frac{1}{3} + \frac{1}{4} + \frac{1}{5} + \cdots + \frac{1}{N}$$

of the harmonic series. For that purpose consider an auxiliary numerical sequence

$$u(n) = 1/n - \ln[(n+1)/n].$$

Summing up the first N terms of this sequence gives

$$\sum_{n=1}^{N} u(n) = H(N) - \ln(N+1). \tag{3}$$

Let us show that, as $N \to \infty$, the corresponding series converges absolutely. Indeed, observe that, for any x such that $-1 < x < \infty$,

$$\ln(1 + x) \leq x.$$

Hence

$$\frac{1}{n} \geq \ln\left(\frac{1 + n}{n}\right) \geq \frac{1}{n + 1}.$$

Therefore,

$$0 \leq u(n) \leq \frac{1}{n} - \frac{1}{n + 1} < \frac{1}{n^2}.$$

The majorizing series $\sum 1/n^2$ converges absolutely, and so does $\sum u(n)$. As a result, we obtain the following asymptotic relation:

$$\sum_{n=1}^{N} u(n) = \gamma + o\{1\} \quad (N \to \infty),$$

where γ is the limit value of the partial sums. Hence, the right-hand side of equality (3) has to satisfy the same asymptotic relation, and we get that

$$H(N) \sim \ln(N + 1) + \gamma. \tag{4}$$

It is known that

$$\gamma = 0.57721566490\ldots, \tag{5}$$

and, traditionally, it is called the *Euler constant*. It is a transcendental number, like the more familiar constants π and e.

Substituting (4) into (2), we arrive at the asymptotic formula

$$L(N) \sim \delta[\ln(N) + \gamma],$$

since replacement of $\ln(N + 1)$ by $\ln(N)$ does not affect the asymptotics. Finally, the right-hand side is approximately 1 when

$$N \sim \exp\left(\delta^{-1} - \gamma\right). \tag{6}$$

For $1/\delta = 100$ we get that the beetle needs $N \approx 1.52 \cdot 10^{43}$ to reach the end of the string.

At this point, people who guessed that the beetle would never reach the end of the string could argue that, after all, they were "almost" right and that the above

N is physically as good as "never," given that the age of the universe is some 20 billion years $= 6.31 \cdot 10^{17}$ s, and that the common educated guess is that it will be in existence only for another 20 billion odd years. If the beetle moved with the speed of 1cm/s, and started out at the Beginning of Time, by now it would have covered only 0.415 of the string's length. Even a bionic Superbeetle, traveling at jet fighter speed, would not be able to complete the journey before the End of the World.

Let us conclude the analysis of the example with a remark that the approximate formula (6) is amazingly accurate. Derived as an asymptotic formula for $\delta \to 0$, it also works very well for arguments of order 1. For example, for $\delta = 1/2$, the exact formula (2) gives $N = 4$, whereas the asymptotic formula (6) gives $n \sim 4.15$. Similarly, for $\delta = 1/3$, formula (2) gives $N = 11$ and (6) gives $N \sim 11.27$. For $\delta = 1/4$, we get respectively $N = 31$ and $N \sim 30.65$.   ∎

4.2   Riemann-Lebesgue Lemma

In this section we turn to the principal topic of the present chapter: a study of the asymptotic behavior of Fourier transforms of nonsmooth functions. The following *Riemann-Lebesgue Lemma*, formulated below in a simplified—sufficient for physical applications—version, is the key.

Assume that function $f(t)$ is continuous on a finite interval $[0, T]$. Then,

$$\int_0^T f(t)e^{-i\omega t}\,dt = o\{1\} \quad (\omega \to \infty), \tag{1}$$

To prove the above, recall that a continuous function $f(t)$ on a closed interval $[0, T]$ is uniformly continuous. This means that, for any $\varepsilon > 0$, one can find a $\delta = T/n > 0$, $n < \infty$, such that, for all $m = 0, 1, \ldots, n - 1$,

$$|f(t) - f(m\delta)| < \varepsilon/2T \quad \text{whenever} \quad m\delta < t < (m+1)\delta. \tag{2}$$

Splitting the interval of integration in (1) into n subintervals $(m\delta, (m + 1)\delta)$, we obtain

$$\int_0^T f(t)e^{-i\omega t}\,dt = \sum_{m=0}^{n-1} f(m\delta) \int_{m\delta}^{(m+1)\delta} e^{-i\omega t}\,dt$$

$$+ \sum_{m=0}^{n-1} \int_{m\delta}^{(m+1)\delta} e^{-i\omega t}\,[f(t) - f(m\delta)]\,dt. \tag{3}$$

Conditions (2) imply that the integrals on the right-hand side of (3) satisfy inequalities

$$\left| \int_{m\delta}^{(m+1)\delta} e^{-i\omega t} \left[f(t) - f(m\delta) \right] dt \right| < \frac{\varepsilon\delta}{2T}.$$

Adding up these n terms, and taking into account the fact that $n\delta = T$, we obtain that the second sum on the right-hand side of (3) is less than $\varepsilon/2$. The integrals in the first sum are easily evaluated to give

$$\int_{m\delta}^{(m+1)\delta} e^{-i\omega t} dt = i \frac{e^{-i\omega(m+1)\delta} - e^{-i\omega m\delta}}{\omega}.$$

The moduli of these integrals are bounded from above by $2/|\omega|$. By the *First Weierstrass Theorem*, a continuous function on a closed interval is bounded so that

$$|f(t)| < F < \infty, \quad t \in [0, T],$$

and we obtain that the first sum in (3) is less than $2Fn/|\omega|$. Therefore, the total integral in (1) satisfies the inequality

$$\left| \int_0^T f(t) e^{-i\omega t} dt \right| < \frac{2Fn}{|\omega|} + \frac{\varepsilon}{2}.$$

Now, it is clear that, for an arbitrary $\varepsilon > 0$, we can select a finite $\Omega = 4Fn/\varepsilon$ such that the investigated integral is less than ε for any $\omega > \Omega$. This proves the Riemann-Lebesgue Lemma. ∎

Additional remarks are in order here:

Remark 1. In physical applications, the upper limit in the integral (1) is often infinite, and it is useful to generalize the Riemann-Lebesgue Lemma to that case. This can be easily done if function $f(t)$ is assumed not only continuous on $(0, \infty)$, but also absolutely integrable on the half-line. Then, for any $\varepsilon > 0$, we can find a T such that

$$\int_T^\infty |f(t)| dt < \frac{\varepsilon}{3}.$$

Replacing $\varepsilon/2$ in (2) by $\varepsilon/3$, and putting $\Omega = 6Fn/\varepsilon$, we can immediately check the validity of Riemann-Lebesgue Lemma in the case of the infinite upper integration limit.

Remark 2. The above argument indicates that the absolute convergence of the above improper integral is not necessary for the validity of the Riemann-Lebesgue Lemma. A sufficient condition is that the integral

$$\int_T^\infty f(t)e^{-i\omega t}\,dt$$

converges uniformly for large ω's. Recall, that the integral converges uniformly for $|\omega| > \Omega$, if, for any $\varepsilon > 0$, one can find a number $L(\varepsilon) > T$, independent of ω and such that, for all $|\omega| > \Omega$ and $l > L$,

$$\left| \int_l^\infty f(t)e^{-i\omega t}\,dt \right| < \varepsilon.$$

We will omit the proof of this statement but the following example illuminates the situation.

Example 1. Consider the integral

$$\int_0^\infty \frac{\cos(\omega t)}{\sqrt{1+t^2}}\,dt = K_0(\omega),$$

which is called the *modified Bessel function of the third kind*. The integral does not converge absolutely since the integrand decays at infinity too slowly (only as $1/t$). However, it is not difficult to show that the integral converges uniformly for $\omega > 0$, so that the Riemann-Lebesgue Lemma is applicable. Indeed, in the theory of Bessel functions one demonstrates that the above integral rapidly converges to 0 as $\omega \to \infty$. More precisely,

$$K_0(\omega) \sim \sqrt{\frac{\pi}{2\omega}}e^{-\omega} \quad (\omega \to +\infty).$$

The absence of absolute convergence of the integral is reflected by its behavior for $\omega \to 0$, where $K_0(\omega) \sim -\ln(|\omega|)$.

4.3   Functions with jumps

4.3.1. Discontinuities of the first kind. Having armed ourselves with the
Riemann-Lebesgue Lemma, we can now proceed with an analysis of the asymptotic
behavior of Fourier transforms of functions with *discontinuities of the first kind*
(that is, jumps). We begin with a look at the simplest situation where the original
function $f(t)$ has a jump at point $t = \tau$ and is continuous, together with its first
derivative, for $t \neq \tau$. The additional assumption is that both function $f(t)$ and its
derivative $f'(t)$ are absolutely integrable over the entire real axis.

Let us split the Fourier integral at the jump point to get

$$\tilde{f}(\omega) = \frac{1}{2\pi} \int f(t)e^{-i\omega t}\,dt = \frac{1}{2\pi} \int_{\tau}^{\infty} f(t)e^{-i\omega t}\,dt + \frac{1}{2\pi} \int_{-\infty}^{\tau} f(t)e^{-i\omega t}\,dt, \quad (1)$$

and then integrate by parts each of the two terms on the right-hand side. The first
term

$$\frac{1}{2\pi} \int_{\tau}^{\infty} f(t)e^{-i\omega t}\,dt = f(\tau + 0)\frac{e^{-i\omega \tau}}{2\pi i \omega} + \frac{1}{2\pi i \omega} \int_{\tau}^{\infty} f'(t)e^{-i\omega t}\,dt. \quad (2)$$

In view of the assumptions, function $f'(t)$ is continuous and absolutely integrable.
Hence, the Riemann-Lebesgue Lemma is applicable to the second term on the
right-hand side of (2), and implies that, as $\omega \to \infty$,

it is of the order smaller than function $1/\omega$. Thus, we have the asymptotic
relation

$$\frac{1}{2\pi} \int_{\tau}^{\infty} f(t)e^{-i\omega t}\,dt = f(\tau + 0)\frac{e^{-i\omega \tau}}{2\pi i \omega} + o\left\{\frac{1}{\omega}\right\}.$$

A similar relation

$$\frac{1}{2\pi} \int_{-\infty}^{\tau} f(t)e^{-i\omega t}\,dt = -f(\tau - 0)\frac{e^{-i\omega \tau}}{2\pi i \omega} + o\left\{\frac{1}{\omega}\right\}$$

holds true for the second integral in (1). Putting these two results together we get
the desired asymptotic formula for the Fourier transform of a function with a jump
at point τ:

$$\tilde{f}(\omega) = \lfloor f \rfloor \frac{e^{-i\omega \tau}}{2\pi i \omega} + o\left\{\frac{1}{\omega}\right\} \qquad (\omega \to \infty), \quad (3)$$

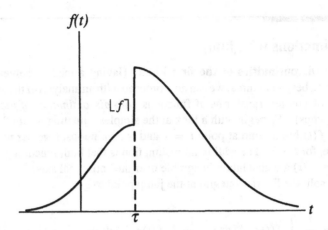

FIGURE 4.3.1
Function $f(t)$ is smooth outside $t = \tau$ and has a discontinuity of the first kind
at $t = \tau$.

where we used the notation

$$\lfloor f(\tau)\rfloor = f(\tau + 0) - f(\tau - 0)$$

to denote the jump size of f at the point τ.

Remark 1. There is a clear-cut connection between the asymptotic formula
(3) and the generalized Fourier transform. Formula (3) describes the asymptotic
behavior of the classical Fourier transform of a discontinuous function. However,
if we multiply both sides in (3) by $i\omega$ we obtain that

$$i\omega \tilde{f}(\omega) = \lfloor f\rfloor \frac{e^{-i\omega\tau}}{2\pi} + o\{1\},$$

which, according to (3.3.2), describes the asymptotic behavior of the generalized
derivative $f'(t)$. Indeed, its first term corresponds to the Fourier transform of the
shifted Dirac delta $\lfloor f(\tau)\rfloor \delta(t - \tau)$.

4.3.2. Remainder terms of the asymptotics. Similar asymptotic relations are
often encountered in physical and engineering problems. However, applying them
in practice, we immediately face the following fundamental dilemma: How large
need ω be so that we can ignore the summand of order $o\{1/\omega\}$, and retain on the
right-hand side of (3) only the "principal" first term? A common sense physical
answer is: The values of ω have to be much larger than 1. This obviously is not a
rigorous answer, and on one particular occasion one of the authors observed two

distinguished physicists seriously arguing whether number 4 is much larger than 1.

The rigorous approach to the problem reformulates the question and asks for estimates of the *remainder term*'s magnitude, which are specific to each asymptotic formula. To explain the notion of remainder term, let us replace formula (3) by an exact equality

$$\tilde{f}(\omega) = \lfloor f \rfloor \frac{e^{-i\omega\tau}}{2\pi i\omega} + R_1(\omega). \tag{4}$$

Here, R_1 is called the remainder term and it is equal to the difference between the accurate value of $\tilde{f}$ and the value of its principal asymptotic term. In our case,

$$R_1(\omega) = \frac{1}{2\pi\omega} \int \{f'(t)\} e^{-i\omega t} dt. \tag{5}$$

By $\{f'(t)\}$ we mean a function which is equal to the derivative of $f(t)$ for all values $t \neq \tau$, where it exists in the classical sense, and which is arbitrarily defined at $t = \tau$ (this does not influence the integral's value). Then the question of estimation of the remainder term's magnitude is reduced to searching for inequalities of the type

$$|R_1(\omega)| < M(\omega),$$

where, hopefully, the function $M(\omega)$ has a simpler structure than $R_1(\omega)$. If that is the case, the *majorant* $M(\omega)$ permits a quantitative evaluation of the error committed by using just the principal term in the asymptotic formula. Sometimes the search for an accurate majorant, which does not overexaggerate the true error of an asymptotic formula, requires a lot of mathematical virtuosity. However, it is often possible to use the standard tool of *asymptotic expansions* which will be described below.

4.3.3. Asymptotic expansions. Consider the Fourier transform of a function $f(t)$ with a single jump at $t = \tau$. The additional assumption is that, for $t \neq \tau$, $f(t)$ has n continuous derivatives which are absolutely integrable on $(-\infty, \tau)$ and (τ, ∞). In this case, by repeated integration by parts, we obtain that

$$\tilde{f}(\omega) = \frac{e^{-i\omega\tau}}{2\pi} \sum_{m=0}^{n-1} \left(\frac{1}{i\omega}\right)^{m+1} \lfloor f^{(m)} \rfloor + R_n(\omega), \tag{6}$$

with the remainder term

$$R_n(\omega) = \frac{1}{2\pi} \left(\frac{1}{i\omega}\right)^n \int \{f^{(n)}(t)\} e^{-i\omega t} dt. \tag{7}$$

It is clear that the above formula generalizes formula (4). The familiar symbol $\lfloor f^{(m)}(\tau) \rceil$ denotes the jump size of $f^{(m)}(t)$ at $t = \tau$.

By the Riemann-Lebesgue Lemma, the remainder term in (6) is of order smaller than the last term in the sum, which is of order $O\{1/\omega^n\}$. However, the Riemann-Lebesgue Lemma does not provide a recipe for the quantitative estimate of the remainder term. The situation becomes simpler if, as in practical computations, one retains only terms up to $(n-1)$th power of the quantity $1/\omega$. Then one obtains a rough estimate of the remainder term based on the inequality

$$\left| \int \{f^{(n)}(t)\} e^{-i\omega t} dt \right| \leq \int |\{f^{(n)}(t)\}| dt = I_n.$$

The integral on the right-hand side is completely independent of ω and, as a rule, it is easily evaluated either analytically or numerically to a desired degree of accuracy. If number I_n is known then the remainder term has an explicit estimate

$$|R_n(\omega)| \leq M(\omega) = I_n / 2\pi \omega^n.$$

Roughness of the estimate is caused by the fact that the majorizing function $M(\omega)$ is of the same order of magnitude as the last term $1/\omega^n$ in (6). However, putting $M(\omega)$ together with the last term of the sum, we can find a majorant for the remainder term $R_{n-1}(\omega)$ which is of order smaller than the order of the last term of the truncated asymptotic formula

$$\tilde{f}(\omega) \sim \frac{e^{-i\omega\tau}}{2\pi} \sum_{m=0}^{n-2} \left(\frac{1}{i\omega} \right)^{m+1} \lfloor f^{(m)} \rceil. \tag{8}$$

In particular, a majorant of the remainder term in (4) found in this fashion, which contains only the *principal term of asymptotic expansion*, gives

$$|R_1(\omega)| \leq \frac{1}{2\pi\omega^2} \{ |\lfloor f' \rceil| + I_1 \} = O\left\{ \frac{1}{\omega^2} \right\}.$$

The inequality demonstrates that, as ω increases, the first term in (4), which is $O\{1/\omega\}$, gives a better and better description of the asymptotic behavior of the Fourier transform.

Example 1. Consider two real-valued integrals which can be evaluated in closed form:

$$S = \frac{1}{\pi} \int_0^\infty e^{-ht} \sin(\omega t) dt = \frac{1}{\pi} \frac{\omega}{\omega^2 + h^2} \tag{9}$$

and

$$C = \frac{1}{\pi} \int\limits_{0}^{\infty} e^{-ht} \cos(\omega t) dt = \frac{1}{\pi} \frac{h}{\omega^2 + h^2}. \tag{10}$$

We shall find their asymptotics for large ω by expanding the algebraic expressions on the right-hand sides into power series in $1/\omega$. Note, that their principal asymptotic terms are, respectively, $1/\pi\omega$ and $h/\pi\omega^2$. Thus, as $\omega \to \infty$, the second integral turns out to be of order smaller than that of the first integral.

Let us take a closer look at that disparity of asymptotics of the two seemingly similar integrals. Like an old war veteran who, on a rainy day, feels that his amputated leg is still there, our integrals "feel" the influence of the full Fourier integrals

$$S = \frac{1}{2\pi} \int ie^{-h|t|} \operatorname{sign}(t) e^{-i\omega t} dt, \quad C = \frac{1}{2\pi} \int e^{-h|t|} e^{-i\omega t} dt. \tag{11}$$

Functions which are being Fourier-transformed in (11) are, respectively, odd and even extensions to the half-line $(-\infty, 0)$ of the function e^{-ht} from (9-10). Note that, in the first case, the integrand has a jump discontinuity of the size $2i$ at $t = 0$. Therefore, according to the asymptotic formula (4), the principal asymptotic term of the integral S is $1/\pi\omega$. On the other hand, the even extension of e^{-ht} in the integral C is everywhere continuous, but has a discontinuous derivative with a jump of size $\lfloor f'(0) \rfloor = -2h$. This means that, in our case, the first nonvanishing (principal) term in the asymptotic expansion (6) of C in a power series in $1/\omega$ is the second term, which is equal to $1/\pi\omega^2$.   ∎

Example 2. A contrast is even stronger in the behavior of integrals

$$C = \int\limits_{0}^{\infty} \exp(-t^2) \cos(\omega t) dt = \tfrac{1}{2}\sqrt{\pi} \exp\left(-\omega^2/4\right),$$
$$S = \int\limits_{0}^{\infty} \exp(-t^2) \sin(\omega t) dt. \tag{12}$$

The first integral, which is proportional to the Fourier transform of an infinitely differentiable function, decays to 0 (as ω increases) faster than the exponential function. The second integral, in view of the above asymptotic formulas, satisfies the asymptotic relation

$$S \sim \frac{1}{\omega}. \tag{13}$$

It has a milder power-type decay at infinity (see Fig. 4.3.2).   ∎

4.3.4. Log-log scales and power-type behavior. The above differences in asymptotic behavior are well illustrated on Fig. 4.3.2 by the graphs of C and S as functions

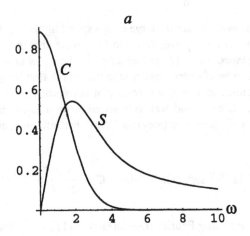

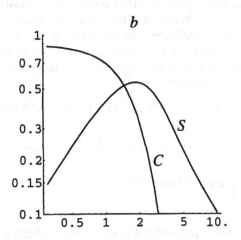

FIGURE 4.3.2
Dependence of integrals C and S from formula (12) on ω. (a) Linear scales,
(b) Logarithmic scales.

of ω. The graph (a) is in the usual, linear scales. However, in the areas of physics where power functional relations are common, it is more convenient to employ the logarithmic scales which are used on the graph (b).

The logarithmic scales are of extraordinary importance in the physical sciences and engineering, and deserve a few detailed comments. For the sake of concreteness, we will illustrate the situation in the case of acoustic pressure P which is often measured in units called *decibels* (or, in short, dB), and found from the formula

$$D = 10 \log(P^2/P_0^2).$$

A threshold value P_0 is selected on the basis of physical considerations. For example, in acoustics it is selected to be equal to $2 \cdot 10^{-5}$ Pa. The squares of the quantities under the logarithm are introduced by the physicists so that the energetic characteristics of the processes, which are proportional to the square of the pressure, can be measured in decibels. Also, the squares ensure that the decibels are well defined for arbitrary, even negative values of P. Taking advantage of the logarithm's properties, the above formula for the number of decibels can be rewritten in the form

$$D = 20 \log |P| - 20 \log P_0.$$

Assume that an experiment found two values of acoustic pressure, P_1 and P_2, generated by a car noise and a jet aircraft noise. The difference between the corresponding numbers of decibels does not depend on the selection of units of pressure for the initial measurements, be they Pascals or millimeters of mercury. The difference is also independent of the threshold value P_0. This is the reason why the decibels are so useful in comparing measurement results. If only the relative, and not the absolute number of decibels is important, then one can utilize a truncated formula

$$D = 20 \log(|P|).$$

As an example, consider a power-type Fourier transform

$$\tilde{f}(\omega) = A/\omega^r,$$

of a time-dependent function $f(t)$. In decibels (that is in the logarithmic scale) its values are expressed by

$$D = -20r \log(\omega) + 20 \log(|A|).$$

If the frequency ω itself is also measured in the logarithmic scale, that is, if we introduce $s = \log(\omega)$, then in the (D, s)-plane the power law is represented by a straight line with the slope equal to $-20r$ determined by the power exponent r. Hence, if the graphs are drawn in the logarithmic coordinate scales, then the presence, and even the magnitude of the exponent r can be readily detected. Fig. 4.3.2 (b) clearly demonstrates, for large ω, the absence of the power law for C, and its presence for S. The graph of S looks a little bit like a boomerang with two linear pieces symmetrically angled towards the S axis. The decreasing piece corresponds to the power law asymptotics (13), and the linear growth for small ω is connected to the principal asymptotics $\tilde{S}(\omega) \sim \omega/2$ of the integral S in (12) for $\omega \to 0$.

4.3.5. Fourier transforms of pulse functions and optimization of directional antennas. A study of the properties of Fourier transforms of discontinuous functions would be incomplete without mentioning rectangular functions describing pulse signals or indicator functions of intervals. They are the simplest discontinuous functions with an obvious symmetry, useful in applications. Since we are not going to discuss direct engineering problems here, assume that the argument t is dimensionless, and concentrate on the specific example.

Example 3. Define the function

$$\Pi(t) = [\chi(t+1) - \chi(t-1)] = \begin{cases} 1, & |t| \leq 1; \\ 0, & |t| > 1. \end{cases}$$

Its Fourier transform

$$\tilde{\Pi}(\omega) = \frac{1}{2\pi} \int_{-1}^{1} e^{-i\omega t}\, dt = \frac{1}{\pi}\, \text{sinc}\left(\frac{\omega}{\pi}\right), \tag{14}$$

where, in a commonly accepted notation,

$$\text{sinc}(\omega) = \frac{\sin(\pi\omega)}{\pi\omega}.$$

The graph of the absolute value of function sinc ω is plotted in Fig. 4.3.2. A slow decay of the local maxima of $|\text{sinc}\,\omega|$, as ω increases, reflects the slow decay asymptotics ($\sim 1/\omega$) of the Fourier transform of $\Pi(t)$ related to the jumps of $\Pi(t)$ at $t = \pm 1$.

In the theory of antennas, the graph on Fig. 4.3.3 represents the antenna's directional pattern as a function of an azimuth-dependent coordinate ω. Existence of the far away "lobes" with large amplitudes is not desirable as it lessens the angular resolution of a radar system. For that reason, in real systems, one tries to dampen the lobes, while preserving some of the character of the original function $f(t)$. ∎

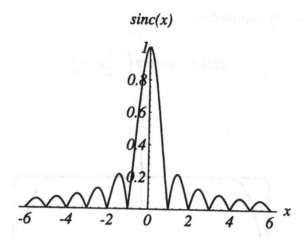

FIGURE 4.3.3
Graph of the absolute value of function sinc x.

4.3.6. Smoothing and filtering. We shall mention two methods which help to accelerate the decay of slowly decreasing tails of Fourier transforms of discontinuous functions. They both rely on the idea of *discontinuity smoothing*.

The first smoothing method, which is commonly used in linear filtration of pulse signals, relies on the convolution

$$g(t) = \int h(\tau) f(t - \tau) d\tau.$$

of original function $f(t)$ with a filtering function $h(t)$. Its Fourier transform, according to (3.2.8), is equal to

$$\tilde{g}(\omega) = 2\pi \tilde{h}(\omega) \tilde{f}(\omega). \tag{15}$$

So if the normalized Gaussian function

$$h(t) = \frac{1}{\sqrt{2\pi}\varepsilon} \exp\left(-\frac{t^2}{2\varepsilon^2}\right), \tag{16}$$

is taken as the filtering function, its Fourier transform is

$$\tilde{h}(\omega) = \frac{1}{2\pi} \exp\left(-\frac{1}{2}\varepsilon^2\omega^2\right),$$

and formula (15) takes the form

$$\tilde{g}(\omega) = \tilde{f}(\omega) \exp\left(-\frac{1}{2}\varepsilon^2\omega^2\right).$$

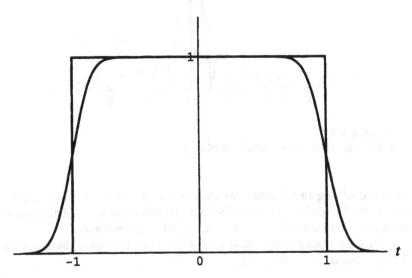

FIGURE 4.3.4
Graph of the pulse function $\Pi(t)$ before and after smoothing.

This particular filtering procedure does not significantly change the form of the Fourier transform $\tilde{f}(\omega)$ for not too large frequencies (say, for $|\omega| < 1/\varepsilon$), but it dramatically dampens the slowly decaying maxima of the lobes for large frequencies (say, $|\omega| \gg 1/\varepsilon$).

The filtration method, so effective in signal processing, is not optimal for antenna problems. Filtration washes out the function, and makes $g(t)$ "last longer" than $f(t)$. An antenna has a finite spatial extent, and the function $f(t)$—which describes the distribution of sources on the antenna as a function of the spatial parameter t—must be zero outside it. In mathematical terms, the engineering problem can be formulated as follows: Find a smooth function, with support $[-1, 1]$, which optimizes—according to a chosen criterion—the Fourier transform $\tilde{f}(\omega)$.

In the engineering practice, the experience would dictate a selection appropriate for a given concrete situation. Here, without getting too deeply involved in mathematical intricacies, we will provide a few examples of functions which have a fast decaying Fourier transforms while preserving some of the characteristics of the impulse function.

Example 4. Consider a triangular function

$$f(t) = \begin{cases} 1 - |t|, & |t| < 1, \\ 0, & |t| \geq 1. \end{cases} \tag{17}$$

It is a continuous function, but its derivative has jumps of size 1 at $t = \pm 1$, and of size -2 at $t = 0$. Therefore, by analogy with formula (8),

$$\tilde{f}(\omega) \sim -\frac{1}{2\pi\omega^2} \left(e^{-i\omega} - 2 + e^{i\omega} \right).$$

Since the second derivative of the triangular function is zero outside the points $t = -1, 0, 1$, the remainder term $R_2(\omega)$ (7) is equal to zero as well. This means that the asymptotic formula

$$\tilde{f}(\omega) = \frac{1}{2\pi} \operatorname{sinc}^2 \left(\frac{\omega}{2\pi} \right) \tag{18}$$

is exact. This fact could have been guessed if we had noticed that the triangular function (17) is the convolution of the rectangular function $\Pi(2t)$ with itself. Hence, the Fourier transform of the triangular function is equal to the square of the Fourier transform of $\Pi(2t)$ (multiplied by 2π), which turns out to be equal to $\tilde{\Pi}(\omega/2)/2$. ∎

The absence of discontinuities in the triangular function guarantees a relatively strong, by comparison with the rectangular function, damping of the Fourier transform's lobes. Nevertheless, in applications one often selects even smoother substitutes of the pulse function.

Example 5. Consider function

$$f(t) = \begin{cases} \cos^2(\pi t/2) = [1 + \cos(\pi t)]/2, & \text{for } |t| < 1; \\ 0 & \text{for } |t| \geq 1. \end{cases} \tag{19}$$

As an exercise, we shall go through a detailed computation of the Fourier transform of $f(t)$. First, let us get rid of the constant term in (19) by taking the derivative

$$g(t) = f'(t) = -\frac{\pi}{2} \sin(\pi t) = i\frac{\pi}{4} [e^{i\pi t} - e^{-i\pi t}],$$

for $|t| < 1$. It follows from formula (3.1.2) that

$$\tilde{g}(\omega) = i\frac{\pi}{4} \left[\tilde{\Pi}(\omega - \pi) - \tilde{\Pi}(\omega + \pi) \right] = \frac{i}{4} \left[\frac{\sin(\omega - \pi)}{(\omega - \pi)} - \frac{\sin(\omega + \pi)}{(\omega + \pi)} \right]$$

is proportional to the difference of shifted Fourier transforms of rectangular functions. Taking $\sin(\omega \pm \pi) = -\sin(\omega)$ outside the brackets, and writing the remaining fractions over the common denominator, we get that

$$\tilde{g}(\omega) = i\frac{\pi}{2}\frac{\sin(\omega)}{\pi^2 - \omega^2}.$$

Returning to the Fourier transform of the original function (19), we find that

$$\tilde{f}(\omega) = \frac{\tilde{g}(\omega)}{i\omega} = \frac{\pi}{2}\frac{\sin(\omega)}{\omega(\pi^2 - \omega^2)}. \tag{20}$$

For $\omega \to \infty$, this Fourier transform decays as $1/\omega^3$, the better behavior resulting from the function itself having discontinuities only in the second derivative.

Similarly, the function

$$f(t) = \begin{cases} \cos^4\left(\frac{\pi t}{2}\right) = \frac{3}{8} + \frac{1}{2}\cos(\pi t) + \frac{1}{8}\cos(2\pi t), & \text{for } |t| < 1; \\ 0, & \text{for } |t| \geq 1. \end{cases} \tag{21}$$

which has discontinuities only in the fourth derivative, has the Fourier transform

$$\tilde{f}(\omega) = \frac{3}{2}\pi^3\frac{\sin\omega}{\omega(\omega^2 - \pi^2)(\omega^2 - 4\pi^2)}, \tag{22}$$

which decays at infinity as $1/\omega^5$. Graphs of functions (19) and (21) are plotted on Fig. 4.3.5. It is almost impossible to differentiate them by naked-eye inspection and tell which one is smoother in the neighborhood of $t = \pm 1$. On the other hand, their Fourier transforms, shown in Fig 4.3.6 with the logarithmic scales on the ordinate axes, are very sensitive to the existence of "hidden" discontinuities of the original functions. ∎

4.4 Gamma function and Fourier transforms of power functions

The previous section discussed Fourier transforms of functions with isolated discontinuities of the first kind (jumps); finite, albeit different at some points, one-sided limits were assumed to exist. If a function has a discontinuity at an isolated point t which is not of the first kind, then we say that it is a *discontinuity of the second kind*. At such a point, either the one-sided limits are infinite (as in $\lim_{t\to 0+} t^{-1}$ or they do not exist (as in $\lim_{t\to 0+} \sin(t^{-1})$).

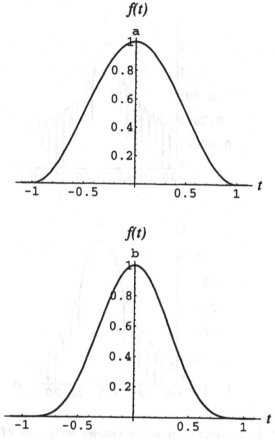

FIGURE 4.3.5
(a) Graph of function (19). (b) Graph of function (21).

In this section we will study the Fourier transforms of functions $f(t) = 0$ for $t \le 0$, smooth for $t > 0$ and sufficiently rapidly decaying for $t \to \infty$, with the asymptotics at the origin

$$f(t) \sim t^{\alpha-1} \quad (t \to 0+), \tag{1}$$

where

$$\alpha > 0. \tag{2}$$

For $0 < \alpha < 1$, the functions f themselves have a discontinuity of the second kind, and for fractional $\alpha > 1$, it is their derivatives of order $n = \lfloor \alpha \rfloor$ (the greatest integer less than or equal to α) and greater that are discontinuous. Our analysis,

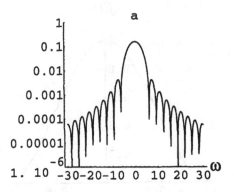

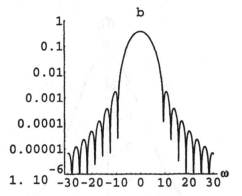

FIGURE 4.3.6
(a) Graph of the absolute value of the Fourier image (20) of function (19).
(b) Graph of the Fourier image (22) of function (21).

with obvious adjustments, applies equally to functions with shifted singularities of the type

$$f(t) \sim (t - \tau)^{\alpha-1}, \quad (t \to \tau + 0).$$

Condition (2) guarantees that $f(t)$ is locally integrable in any neighborhood of point $t = 0$, and that the Fourier transform of $f(t)$ exists in the classical sense.

Integration by parts, which was so effective in finding asymptotics for Fourier transforms of step functions, is not helpful in this case. More efficient is a comparison of the asymptotic laws of the Fourier transforms of (1) with asymptotics of certain "gauge" functions. To construct these gauge functions we need to recall properties of the *Gamma function*

$$\Gamma(s) = \int_0^\infty e^{-t} t^{s-1} dt. \tag{3}$$

The integral on the right-hand side is also called the *Euler integral*.

The Gamma function provides an interpolation of the factorial function $(n-1)!$ to noninteger arguments. Indeed, for positive integer arguments, one can check by induction that

$$\Gamma(n+1) = 1 \cdot 2 \cdot 3 \cdot \ldots \cdot n = n!$$

For general (even complex) values of its argument, the Gamma function satisfies the recurrence relation

$$\Gamma(z+1) = z\Gamma(z).$$

Also, we have a *symmetrization formula*

$$\Gamma(z)\Gamma(1-z) = \pi / \sin(\pi z),$$

from which it follows immediately that $\Gamma(1/2) = \sqrt{\pi}$. So, in this sense,

$$(-1/2)! = \sqrt{\pi}.$$

To study the Fourier transform of (1), we will begin with a more general line integral

$$\int_C z^{\alpha-1} e^{-pz} dz.$$

over a contour C in the complex z-plane which is closely related to the Gamma function. Here, α and p are positive numbers. The integrand is analytic in an arbitrary bounded domain of the complex plane, which does not contain $z = 0$.

As contour C, see Fig. 4.4.1, we select the contour formed by a segment of the real axis from ε to R, followed by a segment of the circular arc from $z = R\exp(i0) = R$ to $z = R\exp(i\beta)$, then by a radial segment from $z = R\exp(i\beta)$ to $z = \varepsilon\exp(i\beta)$ at an angle β to the real axis $(0 < \beta < \pi/2)$, and completed by a small arc from $z = \varepsilon\exp(i\beta)$ to $z = \varepsilon\exp(i0) = \varepsilon$. We travel along the contour in the positive direction, leaving the enclosed domain (which does not contain 0!) on the left-hand side.

By the *Cauchy Theorem* (see Section 6.4), the integral of $z^{\alpha-1} e^{-pz}$ over C vanishes. Splitting it into four obvious pieces corresponding to the above segments

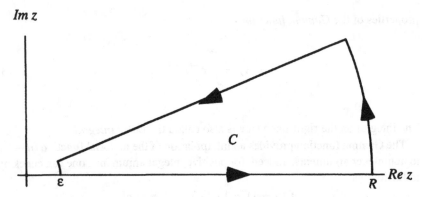

FIGURE 4.4.1
Contour of integration leading to formula (4).

of contour C we get that

$$0 = \int_\varepsilon^R t^{\alpha-1} e^{-pt} dt + i R^\alpha \int_o^\beta \exp\left(i\alpha\varphi - pRe^{i\varphi}\right) d\varphi$$

$$-e^{i\beta\alpha} \int_\varepsilon^R t^{\alpha-1} \exp\left(-pe^{i\beta}t\right) dt - i\varepsilon^\alpha \int_o^\beta \exp\left(i\alpha\varphi - p\varepsilon e^{i\varphi}\right) d\varphi. \qquad (4)$$

On the first segment, which is a subset of the real axis, we replaced the variable of integration z by the real variable of integration t, on the first arc we made a substitution $z = R\exp(i\varphi)$, $0 \le \varphi \le \beta$, on the radial segment—$z = t\exp(i\beta)$, and on the small arc— $z = \varepsilon\exp(i\varphi)$.

Let $\varepsilon \to 0$ and $R \to \infty$. In the limit, the integral along the small arc vanishes because in the factor $\varepsilon^\alpha \to 0$, parameter α is positive.

As $R \to \infty$, the integral over the large arc also converges to 0, and a proof of this fact is equivalent to a proof of the well known *Jordan Lemma*. We will sketch it beginning with an obvious inequality

$$R^\alpha \left| \int_o^\beta \exp(i\alpha\varphi - pRe^{i\varphi}) d\varphi \right| \le R^\alpha \int_o^\beta \exp(-pR\cos\varphi) d\varphi. \qquad (5)$$

The convexity of the graph of $\cos\varphi$ for $0 \le \varphi \le \pi/2$ implies a geometrically obvious (see Fig. 4.4.2) inequality

$$\cos\varphi \ge 1 - \frac{1 - \cos\beta}{\beta}\varphi, \quad 0 \le \varphi \le \beta \le \pi/2.$$

which, together with inequality (5), gives

$$R^\alpha \left| \int_0^\beta \exp(i\alpha\varphi - pRe^{i\varphi})d\varphi \right| \le R^\alpha \exp(-pR) \int_0^\beta \exp\left(pR\frac{1-\cos\beta}{\beta}\varphi \right) d\varphi$$

$$= \frac{R^{\alpha-1}\beta}{p(1-\cos\beta)}\left[\exp(-pR\cos\beta) - \exp(-pR)\right].$$

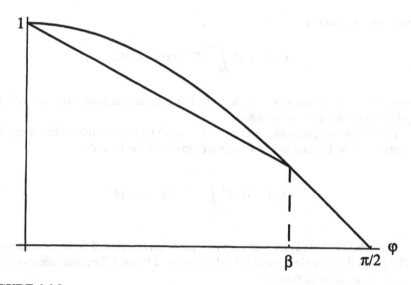

FIGURE 4.4.2
Convexity of the function $\cos\varphi$.

For $|\beta| < \pi/2$, we have $\cos\beta > 0$, and the function in brackets decays exponentially to zero as R increases, thus offsetting the polynomial increase of the factor $R^{\alpha-1}$. This guarantees that, for any α, the integral along the large arc converges to 0 as $R \to \infty$. A useful comment is in order here: For $\beta = \pi/2$, when the first summand in brackets is equal to 1, and for $\alpha < 1$, the integral over the large arc also converges to 0 because of the factor $R^{\alpha-1}$.

So, in the limits $\varepsilon \to 0$ and $R \to \infty$, equality (4) takes the form

$$\int_0^\infty t^{\alpha-1}e^{-pt}dt = e^{i\beta\alpha}\int_0^\infty t^{\alpha-1}\exp(-pe^{i\beta}t)dt.$$

Multiplying both sides by p^α, introducing on the left-hand side a new variable of integration $\tau = pt$, and noticing that the transformed integral on the left-hand side coincides with the Gamma function (3), we get an equality

$$\Gamma(\alpha) = p^\alpha e^{i\beta\alpha} \int_0^\infty t^{\alpha-1} \exp(-pe^{i\beta}t)dt,$$

which, after a change of variables,

$$u = pe^{i\beta} = h + i\omega, \quad h = p\cos\beta, \quad \omega = p\sin\beta, \tag{6}$$

becomes an equality

$$\Gamma(\alpha) = u^\alpha \int_0^\infty t^{\alpha-1} \exp(-ut)dt, \tag{7}$$

valid for $\alpha > 0$, and $h = \operatorname{Re} u > 0$. For a fractional power function u^α, its principal branch must be selected.

For $0 < \alpha < 1$, we can put $h = 0$ ($\beta = \pi/2$) (see the comment following the analysis of the integral over the large arc) to obtain the formula

$$\Gamma(\alpha) = (i\omega)^\alpha \int_0^\infty t^{\alpha-1} \exp(-i\omega t)dt. \tag{8}$$

Formulas (7) and (8) have an easy interpretation in terms of Fourier transforms. The integral on the right-hand side of (8), up to a factor $1/2\pi$, coincides with the Fourier transform of function

$$g(t; \alpha) = \chi(t)t^{\alpha-1}, \quad 0 < \alpha < 1, \tag{9}$$

so, its Fourier transform

$$\tilde{g}(\omega, \alpha) = \frac{\Gamma(\alpha)}{2\pi(i\omega)^\alpha}. \tag{10}$$

In particular, for $\alpha = 1/2$, we have a remarkable formula

$$\frac{\chi(t)}{\sqrt{t}} \longmapsto \frac{1}{(1+i)\sqrt{2\pi\omega}}. \tag{11}$$

Remark 1. Note, that it is easy to obtain the power law $\tilde{g}(\omega) \sim 1/\omega^\alpha$ in (10) from dimensional analysis, although this type of argument will not yield the precise value $\Gamma(\alpha)/2\pi$ of the numerical coefficient.

Remark 2. The validity of the same power law across the whole frequency range of the Fourier transform reflects two, fundamentally different, behaviors of the original time function. Its validity for $\omega \to \infty$ indicates the presence of a discontinuity of the second kind of the original function at time $t = 0$, while its validity for $\omega \to 0$, is a consequence of moderate ($\sim t^{\alpha-1}$) decay of the tail of the original function as $t \to \infty$.

Remark 3. Equalities (7-8) were derived with the help of an integration contour in the upper half-plane (see Fig. 4.4.1). Hence (see (6)), frequency $\omega = p \cos \beta$ appearing in these formulas is positive. However, it is easy to show that the above proof remains in force if the contour is reflected into the lower half-plane. Thus, the formulas (7-11) remain valid for all frequencies $-\infty < \omega < \infty$. In particular,

$$\frac{\chi(t)}{\sqrt{t}} \longmapsto \begin{cases} (1-i)/\sqrt{8\pi|\omega|}, & \text{for } \omega > 0; \\ (1+i)/\sqrt{8\pi|\omega|}, & \text{for } \omega < 0. \end{cases} \tag{12}$$

Remark 4. Time reversal in the original function results in the frequency changing sign in the Fourier transform (see (3.1.3c)). Hence, it follows from (12) that

$$\frac{\chi(-t)}{\sqrt{|t|}} \longmapsto \begin{cases} (1+i)/\sqrt{8\pi|\omega|}, & \text{for } \omega > 0; \\ (1-i)/\sqrt{8\pi|\omega|}, & \text{for } \omega < 0. \end{cases}$$

Combining this relationship with (12), we find the Fourier transform of a symmetric in time function $1/\sqrt{|t|}$, displaying a discontinuity of the second kind:

$$\frac{1}{\sqrt{|t|}} \longmapsto \frac{1}{\sqrt{2\pi|\omega|}}. \tag{13}$$

Now, let us return to the discussion of the formula (7). With the help of notation introduced in (6), we can rewrite (7) in the form

$$\Gamma(\alpha) = (h + i\omega)^\alpha \int_0^\infty t^{\alpha-1} \exp(-ht - i\omega t)\, dt. \tag{14}$$

In other words, function

$$g(t; \alpha, h) = \chi(t)t^{\alpha-1}e^{-ht} \tag{15}$$

has the Fourier transform

$$\tilde{g}(\omega; \alpha, h) = \frac{\Gamma(\alpha)}{2\pi(h + i\omega)^\alpha}. \tag{16}$$

Its principal asymptotics at infinity is described by the relation

$$\tilde{g}(\omega; \alpha, h) \sim \frac{\Gamma(\alpha)}{2\pi(i\omega)^\alpha}, \quad (|\omega| \to \infty), \tag{17}$$

which has a form identical to (10), but is valid for any $\alpha > 0$. For fractional α this asymptotic formula is a consequence of the original function's (see (15)) discontinuities of the second kind, or of similar discontinuities of its derivatives. For the integer values of α the formula agrees with the asymptotic behavior of Fourier transforms of functions with explicit (in the function itself), or hidden (in the derivatives) discontinuities of the first kind which were discussed in Section 4.3.

Despite asymmetry of the function (15) (it is identically equal to 0 for $t < 0$) one can still consider its even and odd components. For an arbitrary function $g(t)$, these two components are given by

$$g_{even}(t) = \frac{1}{2}[g(t) + g(-t)], \quad g_{odd}(t) = \frac{1}{2}[g(t) - g(-t)]. \tag{18}$$

Clearly,

$$g_{even}(t) + g_{odd}(t) = g(t).$$

It follows from properties (3.1.3c) and (3.1.4) of the Fourier transform of real functions that the Fourier transforms of even and odd components of g correspond, respectively, to the real and the imaginary parts of the Fourier transform of the original function g. More formally,

$$\tilde{g}_{even}(\omega) = \text{Re } \tilde{g}(\omega), \quad \tilde{g}_{odd}(\omega) = i \text{ Im } \tilde{g}(\omega). \tag{19}$$

Separating the real and imaginary parts of the Fourier transform (16), we obtain that

$$\tilde{g}_{even}(\omega) = \frac{\Gamma(\alpha)\cos(\alpha\kappa(\gamma))}{2\pi(h^2 + \omega^2)^{\alpha/2}}, \quad \tilde{g}_{odd}(\omega) = i\frac{\Gamma(\alpha)\sin(\alpha\kappa(\gamma))}{2\pi(h^2 + \omega^2)^{\alpha/2}}. \tag{20}$$

The argument κ of a complex number $h + i\omega$ introduced above depends on the dimensionless frequency γ via the formula

$$\kappa = \arctan\gamma, \quad \gamma = \omega/h. \tag{21}$$

For $\gamma \to \pm\infty$ and $\kappa \to \pm\pi/2$, the Fourier transforms of even and odd parts of the original function (15) have the following principal asymptotics:

$$\tilde{g}_{even}(\omega) \sim \frac{\Gamma(\alpha)\cos(\alpha\pi/2)}{2\pi\omega^\alpha}, \tag{22a}$$

$$\tilde{g}_{odd}(\omega) \sim i \frac{\Gamma(\alpha)\sin(\alpha\pi/2)}{2\pi\omega^\alpha} \text{ sign }(\omega), \qquad (|\omega| \to \infty). \qquad (22b)$$

Remark 5. In engineering and physical applications, these asymptotic formulas are much more useful and important than the exact formulas (20). Formulas (20) give Fourier transforms of a narrow class of gauge functions, whereas formulas (22) describe asymptotics of Fourier transforms of a much broader class of functions which, at some arbitrary instants of time t_k, have local singularities $\sim (t - t_k)^{\alpha-1}$.

Remark 6. For odd values of α, the cosine in asymptotic formula (22a) becomes 0, and for even α the sine in (22b) vanishes. This means that the asymptotics of the corresponding functions is of order smaller than $1/\omega^\alpha$. This phenomenon is similar to the one already encountered for functions C and S in (4.3.4-5). Its essence can be explained with the help of two functions: $\chi(t)t^2$ and $\chi(t)t^3$. The former, extended to an odd function becomes sign $(t)\, t^2$, which has discontinuities of the second derivative, whereas its even extension is an infinitely differentiable function t^2. The latter becomes infinitely differentiable under the odd extension but has a discontinuity in the third derivative under the even extension.

The above functions have Fourier transforms only in the generalized, distributional sense. However, the general principle stating that Fourier transforms of infinitely differentiable functions decay, for $\omega \to \infty$, faster than any power of ω, extends to them as well. Consequently, the generalized Fourier transforms of an even function t^2 and the odd function t^3 do not have power asymptotics. In other words, all the coefficients in their asymptotic expansions in powers of $1/\omega$ are equal to 0. This will become clear when we recall the generalized Fourier transforms (3.3.5) of the above two functions.

In the case of fractional α, the principal asymptotics, as $|\omega| \to \infty$, of Fourier transforms of both even and odd parts of function (9) are the same and of order $1/\omega^\alpha$. Either extension of function (9) to the negative half-line does not remove its characteristic absence of smoothness at the origin. Here, the reader may feel that the asymptotics of the Fourier transform of function (9) established rigorously above for any α, contradicts geometric common sense which seems to be telling us that the graphs of even and odd components of (9) are qualitatively different (see Fig. 4.4.3 (a) and (b)). The graph of the even component of

$$g(t) = \chi(t)\sqrt{t}e^{-ht}, \qquad (23)$$

with a characteristic cusp at $t = 0$, gives us an impression of a function that is much less smooth than that of the odd component, although the latter also has a vertical slope at $t = 0$. Nevertheless, in spite of our geometric intuition, both of them have the same Fourier transform asymptotics $1/\omega^{3/2}$. The Fourier transform of function (23) can be calculated explicitly with the help of formula (20), but an alternative derivation provided below is more direct.

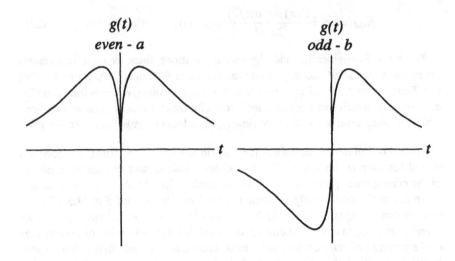

FIGURE 4.4.3
Graphs of even (a) and odd (b) components of function $g(t) = \chi(t)\sqrt{t}e^{-ht}$,
corresponding to $\alpha = 3/2$.

Since $\Gamma(3/2) = (1/2)! = \sqrt{\pi}/2$, it follows from (16) that

$$\tilde{g}(\omega) = \sqrt{\frac{1}{\pi h^3}} \frac{1}{4(1+i\gamma)\sqrt{1+i\gamma}}, \quad \gamma = \omega/h.$$

Separating the real and imaginary parts is easy if we use algebraic identities instead of the general relation (22) which contains trigonometric functions. Squaring both sides of equality $\sqrt{1+i\gamma} = x + iy$, and solving the resulting algebraic equations with respect to x and y, we obtain the main branch of the radical

$$\sqrt{1+i\gamma} = \sqrt{\frac{1+\sqrt{1+\gamma^2}}{2}}\left[1 + \frac{i\gamma}{1+\sqrt{1+\gamma^2}}\right].$$

Further, standard but tedious calculations give that $\tilde{g}(\omega)$ is equal to

$$\sqrt{\frac{1}{32\pi h^3}}\sqrt{\frac{1+\sqrt{1+\gamma^2}}{(1+\gamma^2)^3}} \times$$

$$\times\left[\left(1 - \frac{\gamma^2}{1+\sqrt{1+\gamma^2}}\right) - i\gamma\left(1 + \frac{1}{1+\sqrt{1+\gamma^2}}\right)\right].$$

4.5 Generalized Fourier transforms of power functions

In the previous section we demonstrated (4.4.15-16) that

$$\chi(t)t^{\alpha-1}e^{-ht} \longmapsto \frac{\Gamma(\alpha)}{2\pi(h+i\omega)^\alpha}.$$

In the limit $h \to 0$ the above formula generates the whole family of generalized Fourier transforms, many of them not encountered thus far. We shall study them in this section.

Recall (Section 3.2), that $\tilde{g}(\omega)$ is said to be a generalized Fourier transform of function $g(t)$ if, for any test function $\phi(t) \in S$, and its Fourier transform $\tilde{\phi}(\omega)$ (also in S),

$$\int \tilde{g}(\omega)\tilde{\phi}(\omega)d\omega = \frac{1}{2\pi}\int g(t)\varphi(-t)dt.$$

Note that the above formula corresponds to formula (3.2.6) with $\tau = 0$, but is sufficient to uniquely determine distribution $\tilde{g}(\omega)$.

By definition, the Fourier transform of function g from (4.4.15) is equal to

$$\tilde{g}(\omega; a, h) = \int_0^\infty t^{\alpha-1}e^{-ht-i\omega t}\, dt. \qquad (1)$$

Multiplication of both sides of (1) by a function $\tilde{\phi} \in S$, and integration over the entire ω-axis, gives that

$$2\pi \int \tilde{g}(\omega; a, h)\tilde{\phi}(\omega)d\omega = \int_0^\infty t^{\alpha-1}e^{-ht}\left[\int \tilde{\phi}(\omega)e^{-i\omega t}d\omega\right]dt.$$

The integral in the brackets is equal to function $\phi(-t) \in S$ which is absolutely integrable and rapidly (that is, faster than any power) decreases to 0 at infinity. For that reason, the integral on the right-hand side of the equality

$$2\pi \int \tilde{g}(\omega; \alpha, h)\tilde{\phi}(\omega)d\omega = \int_0^\infty t^{\alpha-1}e^{-ht}\phi(-t)dt$$

exists in the classical sense, for any $h \geq 0$ and $\alpha > 0$. In particular, for $h = 0$ we arrive at a symbolic equality

$$2\pi \int \tilde{g}(\omega; \alpha)\tilde{\phi}(\omega)d\omega = \int\limits_{0}^{\infty} t^{\alpha-1}\phi(-t)dt. \tag{2}$$

The left-hand side is a functional of a product of the generalized Fourier transform $\tilde{g}(\omega; \alpha)$ and a test function $\tilde{\phi}(\omega)$. The right-hand side is a regular linear continuous functional of function $\phi \in S$. Hence, for $h \to 0+$, the distribution $\tilde{g}(\omega; \alpha, h)$ weakly converges to the distribution $\tilde{g}(\omega; \alpha) \in S'$, which is determined as a functional on S by the right-hand side of equality (2). For this distribution, in analogy with (4.4.16), we employ a symbolic notation

$$\tilde{g}(\omega; \alpha) = \frac{\Gamma(\alpha)}{2\pi(i\omega + 0)^{\alpha}}, \qquad (\alpha > 0).$$

The above distribution is a generalized Fourier transform of function $\chi(t)t^{\alpha-1}$ and we record this fact in the form of relation

$$\chi(t)t^{\alpha-1} \longmapsto \frac{\Gamma(\alpha)}{2\pi(i\omega + 0)^{\alpha}}, \qquad (\alpha > 0). \tag{3}$$

Let us establish some properties of the distribution $\tilde{g}(\omega; \alpha)$. Differentiating both sides of (1) m times with respect to ω we arrive at

$$(i)^{m}\frac{d^{m}}{d\omega^{m}}\tilde{g}(\omega; \alpha, h) = \tilde{g}(\omega; \alpha + m, h),$$

which, as $h \to 0$, becomes (in the sense of weak convergence) a distributional equality

$$(i)^{m}\frac{d^{m}}{d\omega^{m}}\tilde{g}(\omega; \alpha) = \tilde{g}(\omega; \alpha + m). \tag{4}$$

It shows that, for values $\alpha > 1$, one can express distributions $\tilde{g}(\omega; \alpha)$ by distributions $\tilde{g}(\omega; \alpha)$ with parameter α values in the interval $0 < \alpha < 1$.

The case $\alpha = 1$; Fourier transform of the Heaviside function. The limit case of $\alpha = 1$, when

$$\tilde{g}(\omega; 1) = \frac{1}{2\pi(i\omega + 0)}. \tag{5}$$

has to be considered separately. It is convenient to establish the functional action of this distribution on test functions by a detailed analysis of the corresponding regular Fourier transforms

$$\tilde{g}(\omega; 1, h) = \frac{1}{2\pi(i\omega + h)}$$

for $h > 0$. Separating their real and imaginary parts we get that

$$\tilde{g}(\omega; 1, h) = \frac{1}{2\pi} \frac{h}{h^2 + \omega^2} - \frac{1}{2\pi} \frac{i\omega}{h^2 + \omega^2}.$$

The regular limit, as $h \to 0$, of functions on the right-hand side does not exist. However, this is not a serious obstacle as we are interested in the weak convergence, that is, in the result of integration of these functions against an arbitrary test function $\tilde{\phi}(\omega)$. In that sense, the real part, which is the familiar Lorentz function, converges to $\delta(\omega)/2$. The imaginary part, when integrated against an arbitrary smooth and absolutely integrable function $\tilde{\phi}(\omega)$, gives that

$$\lim_{h \to o} \int \frac{\omega \tilde{\phi}(\omega)}{h^2 + \omega^2} d\omega = \mathcal{PV} \int \frac{\varphi(\omega)}{\omega} d\omega,$$

where $\mathcal{PV} \int$ stands for the *principal value of the integral*. At this point, we will not dwell on the notion of the principal value of the integral as it is going to be discussed in depth (together with its physical applications) in Chapter 6. We shall only show that the above principal value integral induces a new distribution which, symbolically, will be denoted

$$\mathcal{PV}\frac{1}{\omega}.$$

In this notation, distribution $\tilde{g}(\omega; 1)$ is given by equality

$$\tilde{g}(\omega; 1) = \frac{1}{2}\delta(\omega) + \frac{1}{2\pi}\mathcal{PV}\frac{1}{i\omega}, \tag{6}$$

which is often written in the form

$$\frac{1}{i\omega + 0} = \mathcal{PV}\frac{1}{i\omega} + \pi\delta(\omega). \tag{7}$$

For $\alpha = 1$, our general original function $\chi(t)t^{\alpha-1}$ degenerates to the Heaviside function $\chi(t)$, so that (6) becomes the generalized Fourier transform of Heaviside function. In this case, its real part is equal to the Fourier transform of the even component of the Heaviside function, which is simply equal to 1/2, and the imaginary part is the Fourier transform of the odd component, which is sign$(t)/2$.

Utilizing the recurrence formulas (4), for any integer m, we can express the generalized Fourier transforms of functions $\chi(t)t^{m-1}$ via the Fourier transform of Heaviside function

$$\tilde{g}(\omega; m+1) = (i)^m \frac{d^m}{d\omega^m} \tilde{g}(\omega; 1).$$

For $m = 2$, this formula gives the Fourier transform

$$\tilde{g}(\omega; 3) = -\frac{1}{2}\delta''(\omega) + i\frac{1}{2\pi}\frac{d^2}{d\omega^2}\mathcal{PV}\frac{1}{\omega}.$$

of function $\chi(t)t^2$ discussed in Remark 4.4.6. Its real part is the generalized Fourier transform of infinitely differentiable function t^2. It is identically equal to 0 for arbitrary $\omega \neq 0$, and , as was discussed before, does not have a power asymptotics for $\omega \to \infty$.

The case $\alpha = 0$; Fourier transform of function $\chi(t)/t$—a physical approach. The case of the Fourier transform of function $\chi(t)/t$, corresponding to the value $\alpha = 0$, has to be considered separately from other generalized Fourier transforms of power functions. Copying formally the approach that was so successful in analyzing of generalized Fourier transforms for $\alpha > 0$, we will try to determine the Fourier transform of $\chi(t)/t$ as the weak limit (for $h \to 0$) of the integral

$$\tilde{g}(\omega; 0, h) = \frac{1}{2\pi}\int_0^\infty \frac{1}{t}\exp(-ht - i\omega t)dt. \tag{8}$$

Just a passing glance at (8) permits an observation that, for any ω and any h, the integral on the right-hand side is infinite, in view of the nonintegrable singularity at the lower limit. Nevertheless, neither mathematicians nor physicists throw up their hands in despair in such a situation. Mathematicians introduce new distributions that assign well defined values to integrals (8). Physicists quote additional physical arguments, which also give a finite answer. The ideas of mathematicians and physicists, although different in details and method of argumentation, are similar in essence. Various ways of computing integrals of type (8) can be grouped under a unifying umbrella of *renormalization techniques*. Most often, renormalization techniques are applied in quantum electrodynamics where the computation of physical quantities by the perturbation method leads to divergent integrals.

Without reference to the physical processes that are described by function $\chi(t)/t$, we will use only very general renormalization ideas. A reasonably accurate measurement of the Fourier transform of a real physical process at frequency ω requires much longer time than the period $T = 2\pi/\omega$ of the corresponding oscillation. For that reason, the Fourier transform at frequency $\omega = 0$ is not an observable quantity as it would require for its measurement an infinitely long observation interval. Hence, it is natural to exclude it from considerations, reading its value off other values of the Fourier transform. For $\omega = 0$, the Fourier transform (8)

$$\tilde{g}(0; 0, h) = \frac{1}{2\pi} \int_0^\infty \frac{1}{t} \exp(-ht)dt$$

is infinite. Nevertheless, subtracting it from (8), we arrive at a renormalized Fourier transform

$$\tilde{f}(\omega; h) = \frac{1}{2\pi} \int_0^\infty \frac{1}{t} e^{-ht} \left[e^{-i\omega t} - 1 \right] dt, \tag{9}$$

which assumes finite values and correctly reflects the dependence of the Fourier transform $\tilde{g}$ on frequency.

Passing in (9) to a new variable of integration $\tau = \omega t$,

$$\tilde{f}(\omega; h) = \frac{1}{2\pi} \int_0^\infty \frac{1}{\tau} e^{-\mu\tau} \left[e^{-i\tau} - 1 \right] d\tau,$$

where $\mu = h/\omega$ is a dimensionless parameter. Differentiating both sides of the above equality with respect to that parameter, we obtain that

$$\frac{d\tilde{f}}{d\mu} = -\frac{1}{2\pi} \int_0^\infty \left[e^{-\tau(\mu+i)} - e^{-\tau\mu} \right] d\tau = \frac{1}{2\pi} \left[\frac{1}{\mu} - \frac{1}{\mu+i} \right].$$

Utilizing the fact that if $\mu \to \infty$ then $\tilde{f} \to 0$, we can compute $\tilde{f}$ from

$$\tilde{f} = -\frac{1}{2\pi} \int_\mu^\infty \left[\frac{1}{s} - \frac{1}{s+i} \right] ds = \text{Re } \tilde{f} + i\text{Im } \tilde{f}, \tag{10}$$

where

$$\text{Re } \tilde{f} = \frac{1}{2\pi} \ln \left(\frac{|\mu|}{\sqrt{1+\mu^2}} \right), \tag{11a}$$

$$\text{Im } \tilde{f} = -\frac{1}{4} \left[1 - \frac{1}{2\pi} \arctan(\mu) \right] \text{sign}(\mu). \tag{11b}$$

The absolute value under the logarithm and function sign(μ) in the imaginary part make these expression valid for negative μ as well.

Now, to find a generalized renormalized Fourier transform of function $\chi(t)/t$, it suffices to let $h \to 0$ in expressions (10) and (11). At the beginning, we will do that for the imaginary part. Observing that in this case $\mu \to 0+$ for $\omega > 0$, and $\mu \to 0-$ for $\omega < 0$, we get that

$$\mathrm{Im}\ \tilde{f} = -\frac{1}{4}\ \mathrm{sign}\ (\omega).$$

The real part of the renormalized Fourier transform $\chi(t)/t$ requires a more thoughtful treatment. Recall that $\mu = h/\omega$, and observe that h in the first equality in (11) can not be taken to converge to 0 as its right hand side then becomes infinite. For that reason, we will impose a restriction $\omega \gg h$. Then $\mu \ll 1$, and we can utilize a simpler approximate expression

$$\mathrm{Re}\ \tilde{f} = \frac{1}{2\pi}\ \ln(|\mu|) = -\frac{1}{2\pi}\ \ln(|\omega|) + C, \quad C = \frac{1}{2\pi}\ \ln(h).$$

The constant C above will be called the *calibrating* constant, as its value should be selected on the basis of comparison with results of the measurements and the choice of a frequency scale.

Combining the last two formulas we arrive at a remarkable relation

$$\frac{\chi(t)}{t} \longmapsto -\frac{1}{2\pi}\ \ln(|\omega|) + C - i\frac{1}{4}\ \mathrm{sign}\ (\omega). \tag{12}$$

The above "physical" approach to calculating the Fourier integral proved successful even in the case which, taken at its face value, diverged for any frequency.

The case $\alpha = 0$; Fourier transform of $1/t$—a mathematical approach. We shall now show how one can deal with this situation from the mathematical viewpoint. For simplicity, we shall restrict ourselves to a calculation of the Fourier transform of an even function $1/|t|$. In this situation, one defines a new distribution

$$T = \mathcal{PV}\frac{1}{|t|},$$

directly as a functional on test functions. Its continuity will be guaranteed by an exclusion of singularities in the corresponding integral. In the case under consideration, this functional is defined by the equality

$$T[\phi] = \int_{-1}^{1} \frac{\phi(t) - \phi(0)}{|t|}dt + \int_{|t|>1} \frac{\phi(t)}{|t|}dt. \tag{13}$$

Such regularization of divergent integrals is justified, from the view-point of math-ematicians, by the fact that for any test function $\phi(t) \in S$ with $\phi(0) = 0$, the value of the function $T[\phi]$ (13) coincides with the original integral $\int \phi(t)/|t|\, dt$.

Let $\tilde{T}(\omega)$ be the Fourier transform of our distribution. By Parseval formula (3.2.6) (for $\tau = 0$), it has to satisfy equality

$$\tilde{T}[\tilde{\phi}] = \frac{1}{2\pi} \int PV \frac{1}{|t|}\phi(-t)dt = \frac{1}{2\pi}\left[\int\limits_{-1}^{1} \frac{\phi(-t) - \phi(0)}{|t|}dt + \int\limits_{|t|>1} \frac{\phi(-t)}{|t|}dt\right].$$

If we transform the integrals in the brackets by expressing test function

$$\phi(-t) = \int \tilde{\phi}(\omega)e^{-i\omega t}dt$$

in terms of its own Fourier transform, and change the order of integration, we get that

$$\tilde{T}[\tilde{\phi}] = \frac{1}{\pi}\int \tilde{\phi}(\omega)\left[\int\limits_{0}^{1} \frac{\cos(\omega t) - 1}{t}dt + \int\limits_{1}^{\infty} \frac{\cos(\omega t)}{t}dt\right]d\omega.$$

Passing to the new variable of integration $\tau = \omega t$ in the inner integrals gives

$$\tilde{T}[\tilde{\phi}] = -\frac{1}{\pi}\int \tilde{\phi}(\omega)\left[\text{Cin}(\omega) + \text{Ci}(\omega)\right]d\omega,$$

where

$$\text{Ci}(z) = -\int\limits_{z}^{\infty} \frac{\cos(s)}{s}ds$$

is a special function called the *integral cosine*, and

$$\text{Cin}(z) = \int\limits_{0}^{z} \frac{1 - \cos(s)}{s}ds$$

is another related special function. Neither of them, separately, can be expressed in terms of elementary functions, but their sum, up to a constant, is equal to the logarithmic function. Indeed, if we write function Cin of a real argument ω in the form

$$\text{Cin}(\omega) = \int\limits_{0}^{\omega} \frac{1 - \cos s}{s}ds = \int\limits_{0}^{1} \frac{1 - \cos s}{s}ds + \int\limits_{1}^{\omega} \frac{1 - \cos s}{s}ds,$$

and the last integral above in the form

$$\int\limits_{1}^{\omega} \frac{1-\cos s}{s}ds = \int\limits_{1}^{\omega} \frac{1}{s}ds - \int\limits_{1}^{\omega} \frac{\cos s}{s}ds = \ln(|\omega|) - \int\limits_{1}^{\infty} \frac{\cos s}{s}ds - \mathrm{Ci}(\omega),$$

then we see that

$$\mathrm{Cin}(\omega) = \ln(|\omega|) + \gamma - \mathrm{Ci}(\omega),$$

where constant

$$\gamma = \int\limits_{0}^{1} \frac{1-\cos s}{s}ds - \int\limits_{1}^{\infty} \frac{\cos s}{s}ds.$$

One can prove that constant γ coincides with the Euler constant (4.1.5). Hence

$$\mathrm{Cin}(\omega) + \mathrm{Ci}(\omega) = \ln(|\omega|) + \gamma,$$

which gives the following addition to our tables of generalized Fourier transforms:

$$\mathcal{PV}\frac{1}{|t|} \longmapsto -\frac{1}{\pi}\Big[\ln(|\omega|) + \gamma\Big].$$

4.6 Discontinuities of the second kind

Let us return to one of the main topics of this chapter: analysis of the asymptotics of Fourier transforms of functions with discontinuities of the second kind through a study of the gauge functions (4.4.9) and (4.4.15). Consider the integral

$$\int\limits_{0}^{\infty} f(t)e^{-i\omega t}dt,$$

where, for $t > 0$, $f(t)$ is a sufficiently smooth function, and for $t \to 0$, it has a singularity of the type

$$f(t) \sim At^{\alpha-1}, \quad \alpha > 0.$$

To make the situation more concrete assume that $f(t) \equiv 0$ for $t < 0$, so that the above integral is equal, up to a factor of $1/2\pi$, to the Fourier transform of function $f(t)$. In the previous two sections, a detailed analysis of gauge functions (4.4.9) and (4.4.15) gave us some hints that the Fourier transform $\tilde{f}(\omega)$ of function $f(t)$

has, for $\omega \to \infty$, principal asymptotics of the order $1/\omega^\alpha$. This fact has yet to be proved rigorously, and explicit necessary conditions on function $f(t)$ have to be spelled out.

We will adopt a "patching-up" method which relies on removing singularities from the original function by superposing on it a "patching-up" function with the same singularity. More exactly, we will consider an auxiliary function

$$v(t) = f(t) - Ag(t; \alpha, h). \tag{1}$$

Since $g(t; \alpha, h) \sim t^{\alpha-1}$ ($t \to 0$), function $v(t)$ is of order smaller than $f(t)$, that is, $v(t) = o\{t^{\alpha-1}\}$ ($t \to 0$). Hence, it is natural to expect asymptotics of its Fourier transform to be of order smaller than the expected principal asymptotics of the Fourier transform $\tilde{f}(\omega)$, that is, $\tilde{v}(\omega) = o\{1/\omega^\alpha\}$ ($\omega \to \infty$). If this is indeed the case, then the principal asymptotics of function

$$f(t) = Ag(t; \alpha, h) + v(t)$$

coincides with the main asymptotics (4.4.17) of the gauge (or patching-up) function $g(t; \alpha, h)$ multiplied by A, and the Fourier transform of function $f(t)$ has asymptotics

$$\tilde{f}(\omega) = A\frac{\Gamma(\alpha)}{2\pi(i\omega)^\alpha} + o\left\{\frac{1}{\omega^\alpha}\right\} \qquad (\omega \to \infty). \tag{2}$$

The rigorous proof is based on the following modification of the more general *Riemann-Lebesgue Lemma*:

Let function $v(t)$ be continuous for $t > 0$ and absolutely integrable on the infinite interval $(0, \infty)$. Furthermore, assume the same is true for all the derivatives of $v(t)$ of order up to $n = \lfloor \alpha + 1 \rfloor \geq \alpha > 0$. Additionally, assume that, for $t \to 0$, function $v(t)$, and all its derivatives up to order n, satisfy the asymptotic relations

$$v^{(m)} = o\{t^{\alpha-m-1}\} \quad (t \to 0). \tag{3}$$

Then the integral

$$F(\omega) = \int\limits_0^\infty v(t)e^{-i\omega t} dt \tag{4}$$

has the asymptotics

$$F(\omega) = o\{1/\omega^\alpha\} \quad (|\omega| \to \infty). \tag{5}$$

First, let us sketch a proof of this modification. Asymptotic relations (3) are equivalent to the following statement: For any $\varepsilon > 0$ we can find a $\kappa > 0$ such that, for $t < \kappa$,

$$|v^{(m)}(t)| \leq \varepsilon t^{\alpha - m - 1}, \quad m = 1, 2, \ldots, n. \tag{6}$$

We shall apply these inequalities to the integral (4). Select $\omega > 1/\kappa$, and split the interval of integration into subintervals $(0, 1/\omega)$ and $(1/\omega, \infty)$. For the integral over the first subinterval,

$$\left| \int_0^{1/\omega} v(t)e^{-i\omega t} dt \right| \leq \int_0^{1/\omega} |v(t)| dt \leq \varepsilon \int_0^{1/\omega} t^{\alpha - 1} dt = \frac{\varepsilon}{\alpha} \frac{1}{\omega^\alpha}, \tag{7}$$

which means that this piece of integral (4) is, as $|\omega| \to \infty$, of order smaller than $1/\omega^\alpha$.

The second piece of the integral (4) over interval $(1/\omega, \infty)$ will be transformed by repeated integration by parts. The first integration by parts gives that

$$\int_{1/\omega}^\infty v(t)e^{-i\omega t} dt = \frac{e^{-i}}{i\omega} v\left(\frac{1}{\omega}\right) + \frac{1}{i\omega} \int_{1/\omega}^\infty v'(t)e^{-i\omega t} dt.$$

In view of (6), the first summand on the right-hand side is $\leq \varepsilon/\omega^\alpha$ and, as a result, it is of order smaller than $1/\omega^\alpha$ for $|\omega| \to \infty$. Repeating the integration by parts another $n - 1$ times, and making a similar observation to the effect that the nonintegral terms are $o\{1/\omega^\alpha\}$, we are lead in the end to an asymptotic equality

$$\int_{1/\omega}^\infty v(t)e^{-i\omega t} dt = \left(\frac{1}{i\omega}\right)^n \int_{1/\omega}^\infty v^{(n)}(t)e^{-i\omega t} dt + o\left\{\frac{1}{\omega^\alpha}\right\}.$$

Let us split the integral on the right-hand side

$$\int_{1/\omega}^\infty v^{(n)}(t)e^{-i\omega t} dt = \int_{1/\omega}^{1/\kappa} v^{(n)}(t)e^{-i\omega t} dt + \int_{1/\kappa}^\infty v^{(n)}(t)e^{-i\omega t} dt.$$

In view of (6), the contribution of the first summand

$$\left| \left(\frac{1}{i\omega}\right)^n \int_{1/\omega}^{1/\kappa} v^{(n)}(t)e^{-i\omega t} dt \right| \leq \frac{\epsilon}{\omega^n} \int_{1/\omega}^{1/\kappa} t^{\alpha - n - 1} dt < \frac{\epsilon}{n - \alpha} \cdot \frac{1}{\omega^\alpha}.$$

So

$$\int_{1/\omega}^\infty v(t)e^{-i\omega t} dt = \left(\frac{1}{i\omega}\right)^n \int_{1/\kappa}^\infty v^{(n)}(t)e^{-i\omega t} dt + o\left\{\frac{1}{\omega^\alpha}\right\}.$$

It follows from the Riemann-Lebesgue Lemma of Section 4.2 that the remaining integral, which contains a continuous and absolutely integrable function $v^{(n)}(t)$, converges to 0 as $|\omega| \to \infty$. This, together with (7), proves the validity of asymptotic relation (5). ∎

Now, the proof of (2) immediately follows from the above modification of the Riemann-Lebesgue Lemma, since the auxiliary function $v(t)$ in (1) satisfies conditions of the lemma.

Example 1. Consider the asymptotics of Fourier transform of function

$$f(t) = \exp(-|t|^{\alpha}), \quad \alpha > 0. \tag{8}$$

In contrast to functions considered above, for $t = 0$, it assumes a non-zero finite value $f(0) = 1$. As a result, the auxiliary integral

$$\int_0^\infty f(t)e^{-i\omega t} dt$$

has the principal asymptotics of the order $O(1/\omega)$ which is absent in the actual asymptotics of the full Fourier transform of function (8). To exclude this asymptotics and to find the asymptotic behavior of the full Fourier transform, first consider the derivative of function (8)

$$f'(t) = -\alpha|t|^{\alpha-1} \exp(-|t|^{\alpha}) \operatorname{sign}(t). \tag{9}$$

Let us form an auxiliary function equal to 0 for $t < 0$ and, for $t > 0$, given by equality

$$v(t) = f'(t) + \alpha g(t; \alpha, h).$$

It is easy to see that it satisfies all the requirements of the lemma proved above. Hence, the principal asymptotics, for $|\omega| \to \infty$, of the Fourier transform of the "one-sided" derivative $f'(t)\chi(t)$ is described by formula (2) with $A = -\alpha$. The actual derivative (9) of the original function (8) is an odd function of t. Consequently, the asymptotics of its Fourier transform is twice the imaginary part of the asymptotics (2), that is

$$2i\alpha \frac{\Gamma(\alpha)\sin(\pi\alpha/2)}{2\pi|\omega|^{\alpha}} \operatorname{sign}(\omega).$$

Finally, the asymptotics of Fourier transform of the original function (8) can be found by dividing the above expression by $i\omega$, thus obtaining

$$\tilde{f}(\omega) \sim \frac{\Gamma(\alpha+1)\sin(\pi\alpha/2)}{\pi|\omega|^{\alpha+1}}. \tag{10}$$

∎

4.7 Exercises

1. Find the main power asymptotics (as $x \to 0$) of function $f(x) = (\sin 2x - 2\sin x)/x$.

2. Investigate asymptotic behavior, as $x \to 0$, and as $x \to \infty$, of function $f(x) = (x - \tanh x)/x^2$.

3. Investigate asymptotic behavior, as $x \to 0$, of function $f(x) = (1 - x\cot x)/x$.

4. Utilizing answer to the above exercise find asymptotic behavior, as $a \to \infty$, of the root of the transcendental equation $a - x = \cot x$, $0 < x < \pi/2$.

5. Investigate asymptotic behavior, as $N \to \infty$, of the expression $f(N) = \prod_{n=1}^{N}(1 + \alpha_n/N)$, where $\{\alpha_n\}$ is a bounded sequence.

6. Find the asymptotics (as $N \to \infty$) of expression $\prod_{n=1}^{N}(1 + \Delta\varphi(n\Delta))$, where $\Delta = t/N$, and $\varphi(\tau)$ is a function integrable in the interval $\tau \in (0, t)$. Provide an upper estimate of the remainder term in the obtained asymptotic formula.

7. Determine the character of convergence to 0, as $\omega \to \infty$, of the following two integrals:

$$S(\omega) = \int_{0}^{\infty} \sin\omega t \sin\left(\frac{\alpha t}{1 + \beta t^4}\right) dt, \; C(\omega) = \int_{0}^{\infty} \cos\omega t \sin\left(\frac{\alpha t}{1 + \beta t^4}\right) dt.$$

8. Find the principal asymptotics, for $\omega \to \infty$, of the Fourier transform of function $f(t) = \exp(-\alpha|t|^3)$, and evaluate the infinitesimal order of the remainder term.

9. Assume that $f(t)$ is an even, infinitely differentiable function on the interval $t \in (-\tau, \tau)$, which is identically equal to zero outside this interval. Furthermore, suppose that there exists limit $\lim_{t \to \tau - 0} f(t)/(\tau - t)^n = A$, where $A > 0$ and n is positive integer. Find the principal asymptotics of the Fourier transform of this function for $\omega \to \infty$.

10. Find the principal asymptotics, for $\omega \to \infty$, of the integral

$$J(\omega) = \int_{-1}^{1} \ln\left(\frac{2+t^2}{1+2t^2}\right) \cos\omega t \, dt.$$

11. Find the principal asymptotics, for $\omega \to \infty$, of the Fourier transform of function $f(t)$, with the graph shown in Fig. 4.7.1.

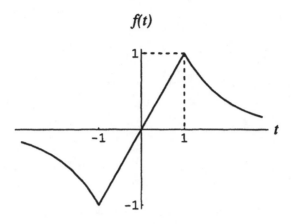

f(t)

FIGURE 4.7.1

12. What is the principal asymptotics (as $\omega \to \infty$) of the Fourier transform of function

$$f(t) = e^{-t^2} \begin{cases} 1, & t < 0; \\ 1 - (t/(1+t))^\beta, & t > 0, \end{cases} \qquad (0 < \beta < 1).$$

Its graph, reminiscent of the profile of an ocean wave (or a sand dune), appears in Fig. 4.7.2.

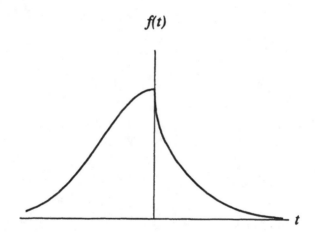

f(t)

FIGURE 4.7.2

13. Find the principal asymptotics, for $\omega \to \infty$, of the Fourier transform of the semi-circle function $f(t)$, equal to zero outside the interval $t \in (-1, 1)$, and equal to $\sqrt{1 - t^2}$ inside that interval.

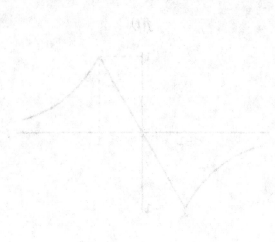

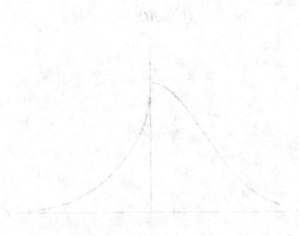

Chapter 5

Stationary Phase and Related Methods

In this chapter we will use methods developed in Chapter 4 to provide a general scheme for finding asymptotics. The remarkable Kelvin's method of *stationary phase* will be employed as well.

5.1   Finding asymptotics: a general scheme

Consider the integral

$$I = I(x) = \int_0^B f(t) \exp[-ixp(t)]\, dt, \tag{1}$$

where $f(t)$ is a continuously differentiable function on the interval $[0, B]$ and such that $f(0) = f \neq 0$. The function $p(t)$ appearing in the exponent will be assumed twice differentiable on $(0, B)$ and monotonically increasing with $p'(t) > 0$.

To apply to (1) the standard methods of asymptotic analysis described in Chapter 4, we will change the variable of integration to

$$s = p(t) - p(0).$$

Denote the monotonically increasing inverse of the above function by

$$t = q(s) \quad (q(Q) = B).$$

After this change of variables, the integral (1) assumes familiar form of the Fourier integral

$$I = \exp[-ixp(0)] \int_0^Q F(s)q'(s)e^{-ixs}\, ds,$$

© Springer Nature Switzerland AG 2018
A. I. Saichev and W. Woyczynski, *Distributions in the Physical and Engineering Sciences, Volume 1*, Applied and Numerical Harmonic Analysis, https://doi.org/10.1007/978-3-319-97958-8_5

where $F(s) = f(q(s))$ is continuously differentiable on the interval $[0, Q]$ function such that $F(0) = f \neq 0$. Let us investigate the asymptotic behavior of I as $x \to \infty$.

First, observe that the factor in front of the integral has modulus 1 and has no effect on asymptotics. So, from now on, we will omit it (putting $p(0) = 0$) and consider

$$I = \int_0^Q F(s)q'(s)e^{-ixs} \, ds. \tag{2}$$

The function $q'(s)$ is also continuously differentiable on the interval $(0, Q)$. If, in addition, it had a finite limit for $s \to 0$, then a further study of the integral I would repeat the previously developed asymptotic analysis of Fourier images of functions with discontinuities of the first kind.

More interesting, and physically more important, is the case of functions $p(t)$ which have the asymptotics

$$p(t) \sim Pt^\alpha \qquad (t \to 0) \tag{3}$$

for a certain $\alpha > 0$, which we will consider in some detail. In this situation

$$q(s) \sim \left(\frac{s}{P}\right)^\beta, \qquad q'(s) \sim Gs^{\beta-1} \qquad (s \to 0), \tag{4}$$

where

$$G = \beta \left(\frac{1}{P}\right)^\beta, \qquad \beta = \frac{1}{\alpha}.$$

The above asymptotics ($s \to 0$) of the integrand and experience gained in the previous chapter suggest the following asymptotics for the integral (2):

$$I \sim \Gamma(\beta)fG \left(\frac{1}{ix}\right)^\beta = f\frac{1}{\alpha}\Gamma\left(\frac{1}{\alpha}\right)\left(\frac{1}{Pix}\right)^{1/\alpha} \qquad (|x| \to \infty). \tag{5}$$

On the other hand, if $F(Q) = f(B) \neq 0$, then the behavior of the integrand close to the upper limit gives the asymptotics

$$I \sim f(B)q'(Q)/ix \qquad (|x| \to \infty),$$

which, for $\alpha > 1$, is of order smaller than (5). Thus, the main term of the asymptotic expansion of integral (2) is given by formula (5). Reinserting the factor omitted earlier, and returning to the notation of the original integral (1), we can finally write that, for any $\alpha > 1$, as $|x| \to \infty$,

$$\int_A^\infty f(t) \exp[-ixp(t)] \, dt \sim \Gamma\left(\frac{1}{\alpha} + 1\right)\left(\frac{1}{Pix}\right)^{1/\alpha} f(A) \exp[-ixp(A)]. \tag{6}$$

The upper limit in (6) was deliberately set to be infinite to avoid distraction caused by smaller order asymptotics generated by a finite upper limit. The lower limit was kept arbitrary for the sake of generality.

Observe that the assumption that function $p(t)$ be strictly increasing is not necessary for the above result. A strict monotonicity suffices as the two cases can be transformed into each other by a nonessential replacement of i into $-i$.

Let us indicate a number of consequences of formula (6) that are important in applications:

(1) If function $p(t)$ is symmetric in the neighborhood of A, then the asymptotics

$$p(t) \sim P|t - A|^{\alpha}, \tag{7}$$

for an $\alpha > 0$, implies the doubled asymptotics (6) for the integral with infinite limits:

$$\int f(t) \exp[-ixp(t)] \, dt \sim 2\Gamma\left(\frac{1}{\alpha}+1\right)\left(\frac{1}{Pix}\right)^{1/\alpha} f(A) \exp[-ixp(A)]. \tag{8}$$

(2) If $p(t)$ is antisymmetric in the neighborhood of A, then only the real part is preserved in the asymptotics and

$$\int f(t) \exp[-ixp(t)] \, dt \sim \Gamma\left(\frac{1}{\alpha}+1\right) Re\left(\frac{1}{Pix}\right)^{1/\alpha} f(A) \exp[-ixp(A)]. \tag{9}$$

(3) In terms of distribution theory the equality (8) means that, for any $\alpha > 1$, the family of functions of variable t

$$\exp\left[-ixp(t)|t - A|^{\alpha}\right] \bigg/ 2\Gamma\left(\frac{1}{\alpha}+1\right)\left(\frac{1}{Pix}\right)^{1/\alpha}, \qquad x \in \mathbf{R},$$

weakly converges to the Dirac delta $\delta(t - A)$ as $x \to \infty$. The particular case $\alpha = 2$ corresponds to the familiar function (1.3.5).

(4) If $f(t)$ is constant and $p(t) = P(t - A)^{\alpha}$, then the asymptotic relation (6) becomes an exact equality. Specifying $f = 1, A = 0$, and $x = 1$, and separating the real and imaginary parts we arrive at the following standard integral formulas valid for $\alpha > 1$ and $P > 0$:

$$\int_0^\infty \cos(Pt^{\alpha}) \, dt = \Gamma\left(\frac{1}{\alpha}+1\right)\left(\frac{1}{P}\right)^{1/\alpha} \cos\left(\frac{\pi}{2\alpha}\right), \tag{10a}$$

$$\int_0^\infty \sin(Pt^{\alpha}) \, dt = \Gamma\left(\frac{1}{\alpha}+1\right)\left(\frac{1}{P}\right)^{1/\alpha} \sin\left(\frac{\pi}{2\alpha}\right). \tag{10b}$$

5.2   Stationary phase method

Assumptions (5.1.3) and (5.1.7) which secured the above asymptotics may seem artificially chosen, just to make mathematics rigorous. Actually, many of them, and in particular the case $\alpha = 2$, emerge perfectly naturally in the physical phenomena. Consider the integral

$$\int_A^B f(t) \exp[-ixp(t)] \, dt, \tag{1}$$

where $p(t)$ is an arbitrary function twice differentiable on the interval of integration. Function $f(t)$ will be assumed continuously differentiable. It turns out that in this fairly general situation the asymptotics of (1) corresponds to the special case $\alpha = 2$.

We shall begin the asymptotic ($x \to \infty$) analysis of (1) by finding *stationary points* of $p(t)$ where

$$p'(t) = 0.$$

Denote the roots of this equation by τ_m, $m = 1, \ldots, N < \infty$. Assume that all of them correspond to simple extrema of function $p(t)$ with $p''(\tau_m) \neq 0$. In their neighborhood, $p(t)$ has a parabolic behavior

$$p(t) - p(\tau_m) \sim p''(\tau_m)(t - \tau_m)^2/2, \qquad (t \to \tau_m). \tag{2}$$

Consequently, the integral (1) has automatically the symmetric asymptotics (5.1.7) with

$$\alpha = 2, \quad P = p''(\tau_m)/2.$$

Let us partition the interval of integration into disjoint intervals, each containing just one of the stationary points. In our case $\alpha = 2 > 1$, and the asymptotics contributed by the boundary points of the intervals are of order smaller than the asymptotics generated by the stationary points (5.1.8). For this reason, in final formulas only the latter appear. Summing contributions of all the stationary points we arrive at the asymptotic ($x \to \infty$) formula

$$\int_A^B f(t) \exp[-ixp(t)] \, dt \sim \sum_{m=1}^N f(\tau_m) \sqrt{\frac{2\pi}{ixp''(\tau_m)}} \exp[-ixp(\tau_m)]. \tag{3}$$

If some stationary points coincide with the endpoints of the interval of integration then the corresponding summands will appear with coefficient 1/2.

5.3 Fresnel approximation

Let us take a look at the asymptotics (5.2.3) from a slightly different viewpoint and focus our attention on the case where there is only one stationary point τ and the limits of integration are infinite. Then (5.2.3) reduces to the asymptotic equality

$$\int f(t) \exp[-ixp(t)]\,dt \sim f(\tau)\sqrt{\frac{2\pi}{ixp''(\tau)}} \exp[-ixp(\tau)], \quad (x \to \infty). \quad (1)$$

We shall attempt to find the "hidden springs" of the stationary phase method by analyzing this example in some depth. The totally rigorous mathematical derivation of (1) seems bland and incomplete to physicists and engineers if it is given without that extra insight that comes from a perhaps imprecise but revealing heuristic arguments. Actually, mathematicians also often gain a deeper understanding of their subject by accumulating a store of sometimes imprecise analogies acquired in "real-life" experiences and physical "thought" experiments.

The stationary phase method can also be elucidated by such "real-life" arguments: the fast oscillation

$$\exp[-ixp(t)] \quad (2)$$

has current (time-dependent) frequency $\omega = xp'(t)$ and period $T = 2\pi/|\omega|$ which decays like $1/x$. If $f(t)$ has a characteristic scale a then, in the domain of integration where $T \ll a$, the adjacent crests and troughs of the integrated process compensate, the better the bigger x, and only close to the stationary point $t = \tau$, where $\omega(\tau) = 0$, does that compensation becomes less effective. As a result, a small and shrinking with the growth of x neighborhood of point τ gives the main contribution to the integral. In the neighborhood of the simple stationary point τ, function $p(t)$ is well approximated by the parabola

$$p(t) = p(\tau) + r(t - \tau)^2/2, \qquad r = p''(\tau).$$

Outside that small neighborhood, function (2) can be replaced by

$$\exp\left[-ip(\tau) - ixr\frac{(t - \tau)^2}{2}\right],$$

without changing the value of the integral significantly, since both functions oscillate quickly and give a small contribution to the integral. For this reason, the original integral (1) can be replaced by an asymptotically equivalent expression

$$\exp[-ip(\tau)]\int f(t) \exp\left[-ixr\frac{(t - \tau)^2}{2}\right]dt. \quad (3)$$

Furthermore, observe that for $x \to \infty$, in the neighborhood of the stationary point essential for the integral, function $f(t)$ practically coincides with constant $f(\tau)$. The latter can be taken outside the integral sign and the remaining integral can be calculated with the help of the standard formula

$$\int \exp[ix^2] \, dx = \sqrt{i\pi}.$$

Remark 1. Physicists often stop short of the asymptotic relation (1) and operate with the integral (3). Analogously with optics, where such integrals appear in the so-called "Fresnel approximation", the approximation of the integral on the left-hand side of (1) by the integral (3) will be also called the *Fresnel approximation*. Later on, discussing optics applications, we shall show that the asymptotic formulas (5.2.3) and (1) correspond to the crude *geometric optics* approximation.

Remark 2. The Fresnel approximations of integrals of type (3) are closely related to the *Fresnel sine and cosine integral* special functions

$$C(z) = \int_0^z \cos(\pi t^2/2) \, dt, \tag{4a}$$

$$S(z) = \int_0^z \sin(\pi t^2/2) \, dt, \tag{4b}$$

often encountered in wave problems. They are both odd functions of z, with limits

$$C(\infty) = S(\infty) = 1/2. \tag{5}$$

The graphs of Fresnel integral functions are shown in Fig. 5.3.1.

5.4 Accuracy of the stationary phase approximation

To enhance the practical value of the main-asymptotics formulas (5.2.3) and (5.3.1) for integrals of rapidly oscillating functions we need estimates of the remainder terms. Their magnitude strongly depends on functions $f(t)$ and $p(t)$. Moreover, no universal method of finding precise estimates is known. For that reason we will analyze the accuracy of asymptotics (5.3.1) in just two generic cases, hoping that detailed analysis of a few concrete examples will better illuminate the essence of the problem than plowing through a laborious general argument.

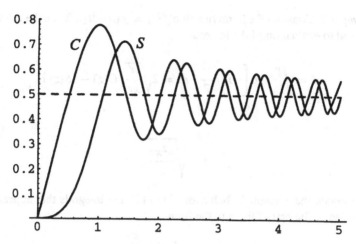

FIGURE 5.3.1
Graphs of the Fresnel integral functions $C(z)$ and $S(z)$.

Example 1. First, let us find the magnitude of error created by replacing the integral (5.3.3) by the right-hand side of (5.3.1) in the case when $f(t)$ is a continuously differentiable (smooth) Gaussian function. Then (5.3.3) becomes the standard integral

$$\int \exp\left[-\frac{t^2}{2a^2} - ixr\frac{t^2}{2}\right] dt = \sqrt{\frac{2\pi}{ixr}} \frac{1}{\sqrt{1 - (i/a^2 xr)}}. \tag{1}$$

For simplicity's sake put $\tau = 0$; consideration of the more complex general case contributes little to the understanding of the essence of the situation. First factor on the right-hand side of (1) corresponds to the main asymptotics (5.3.1), and the second describes the deviation from it. Thus, the relative error is of the order (obtained via formula (5.3.1))

$$\left|1 - \frac{1}{\sqrt{1 - (i/a^2 xr)}}\right| \sim \frac{1}{2a^2 xr}, \qquad (x \to \infty). \tag{2}$$

For a general smooth function $f(t)$ the accuracy of formulas like (5.3.1) can be estimated by replacing a with the characteristic scale of function $f(t)$—an admittedly nonrigorous but heuristically useful approach. ∎

Example 2. Consider the jump function $f(t) = \chi(a - |t|)$. Then the calculation is reduced to evaluation of the integral

$$I = \int_{-a}^{a} \exp\left[-ixr\frac{t^2}{2}\right] dt = 2\sqrt{\frac{\pi}{xr}}\left(C(z) - iS(z)\right), \qquad (3)$$

where

$$z = \sqrt{\frac{a^2 xr}{\pi}}.$$

To investigate the asymptotic behavior of the Fresnel integrals that appear in (3) we shall write the first of them in the form

$$C(z) = \frac{1}{2} - c(z),$$

where

$$c(z) = C(\infty) - C(z) = \int_{z}^{\infty} \cos(\pi t^2/2)\, dt.$$

Changing to a new variable of integration $y = t^2$, we get that

$$c(z) = \frac{1}{2}\int_{z^2}^{\infty} \frac{1}{\sqrt{y}} \cos(\pi y/2)\, dy,$$

and integration by parts gives

$$c(z) = -\frac{1}{\pi z}\sin(\pi z^2/2) + \frac{1}{2\pi}\int_{z^2}^{\infty} \frac{1}{y\sqrt{y}} \sin(\pi y/2)\, dy.$$

Another integration by parts shows that the remaining integral is $o(1/z)$ as $z \to \infty$. Therefore, $C(z)$ satisfies the asymptotic relation

$$C(z) = \frac{1}{2} + \frac{1}{\pi z}\sin(\pi z^2/2) + o(1/z), \quad (z \to \infty).$$

An analogous relation is valid for $S(z)$:

$$S(z) = \frac{1}{2} - \frac{1}{\pi z}\cos(\pi z^2/2) + o(1/z), \quad (z \to \infty).$$

Substituting these asymptotics into (3) we get that

$$I = \sqrt{\frac{2\pi}{ixr}} \left(1 + \frac{\sqrt{2i}}{\pi z} \exp[i\pi x z^2/2] + o(1/z)\right), \quad (z \to \infty).$$

Now it is clear that the relative error of replacing I by the right-hand side of (5.3.1) is

$$\sim \frac{\sqrt{2}}{\pi z} = \frac{1}{a}\sqrt{\frac{2}{\pi xr}}, \quad (z \to \infty).$$

Note that it is of the order $\sim 1/\sqrt{x}$ and not $\sim 1/x$, the latter being the case for smooth functions $f(t)$ (see (2)).   ∎

5.5 Method of steepest descent

Formula (5.2.3) contains the main ingredient of the stationary phase method. It is related to the *steepest descent method* (or *Laplace's method*) which is applicable to purely real integrals

$$\int_A^B f(t) \exp[-xp(t)] \, dt. \tag{1}$$

Without repeating considerations that led us to (5.2.3) we shall give the final formula for the main asymptotics of integral (1).

Let $p(t)$ be a sufficiently smooth function which has only simple minima at points τ_m located inside the interval (A, B). It turns out that just rewriting (5.2.3) without the imaginary unit i gives the correct asymptotics

$$\int_A^B f(t) \exp[-xp(t)] \, dt \sim \sum_{m=1}^N f(\tau_m) \sqrt{\frac{2\pi}{xp''(\tau_m)}} \exp[-xp(\tau_m)], \quad (z \to \infty).$$

$$\tag{2}$$

Despite its superficial similarity, the above formula differs from formula (5.2.3) in an essential way. First of all, the summation in it is not over all extrema but just over all minima of the exponent. Secondly, and this is the main point here, for different values of the minima, the exponential factors in the sum have different magnitudes, and these differences increase with the growth of x. For that reason the main asymptotics of integral (2) contains only one term corresponding to the *absolute minimum* of function $p(t)$ in the interval of integration

$$\int_A^B f(t) \exp[-xp(t)] \, dt \sim f(\tau) \sqrt{\frac{2\pi}{xp''(\tau)}} \exp[-xp(\tau)], \qquad (x \to \infty), \quad (2)$$

where $p(t) \le p(\tau)$, for all $t \in [A, B]$.

5.6 Exercises

1. The stationary phase method is often useful in problems of wave propagation. In particular, in Chapter 9 we will encounter the integral

$$G(\rho) = -\frac{1}{2\pi} \int_0^\infty \exp[-ik\rho \cosh(t)] \, dt,$$

which describes complex amplitude of a cylindrical wave. Analyze this integral using the stationary phase method.

2. The real *Anger function* $J_\omega(z)$ and *Weber function* $E_\omega(z)$ (here, we assume that ω and z are real) are uniquely determined by the equation

$$D_\omega(z) = J_\omega(z) - iE_\omega(z) = \frac{1}{\pi} \int_0^\pi \exp[-i(\omega\phi - z\sin\phi)] \, d\phi \tag{1}$$

and are often encountered in problems of mathematical physics. Find the main asymptotics of Anger and Weber functions for a fixed z and $\omega \to \infty$.

3. Find the main asymptotics of Anger and Weber functions (see Exercise 2) for fixed ω and $z \to \infty$.

4. Find the Fresnel approximation of function $D_\omega(z)$ for $z \gg 1$.

5. In Exercises 3 and 4 we have explored the asymptotic behavior of Anger and Weber functions along the ω and z axes of the (ω, z)-plane. What happens in the rest of the plane? More precisely, study the asymptotic behavior of Anger and Weber functions along the rays $z = \rho\omega$ for $0 < \rho < 1$.

6. Study the asymptotics of (1) for $\rho = 1$, $\omega \to \infty$. *Hint:* As $\rho \to 1$ the expression on the right-hand side of the relation (5.5) in the Answers and Solutions chapter diverges to infinity. This fact indicates that for $\rho = 1$ the asymptotics is of a different order.

7. Complete investigation of the integral $D_\omega(\rho\omega) = (1/\pi) \int_0^\pi e^{-i\omega p(\phi)} d\phi$, where $p(\phi) = \phi - \rho\sin\phi$, by checking its asymptotic behavior for $\omega \to \infty$ if $\rho > 1$.

Remarks on Exercises 2-7: The integral in Exercise 7 clearly has two different types of asymptotic behavior in the (z, ω)-plane: one in the octant $0 < z < \omega$ ($\rho < 1$) and another in the octant $0 < \omega < z$ ($\rho > 1$). Moreover, its asymptotic behavior on the boundary $z = \omega$ ($\rho = 1$) of these octants is qualitatively different from its asymptotic behavior in either of them. The total picture can be summarized as follows: the integral in Exercise 7 obeys the asymptotic power law of order $\alpha(\rho)$ where

$$\alpha(\rho) = \begin{cases} 1, & \text{if } 0 \le \rho < 1; \\ 1/3, & \text{if } \rho = 1; \\ 1/2, & \text{if } 1 < \rho. \end{cases} \tag{2}$$

The function $\alpha(\rho)$ which determines asymptotics of the integral in Exercise 7 has jumps as we move from one (z, ω)-region to another. An infinitesimal change of parameter ρ can cause a major change in the asymptotic behavior of that integral. Such phenomena are called *phase transitions* and they correspond to the physical phase transitions like melting, evaporation or crystallization, where small changes in temperature (or other physical parameters) can cause large and sudden changes in the physical properties of matter.

At first sight, the phase transition for the integral in Exercise 7 in the vicinity of the critical point $\rho = 1$ could be puzzling. The next exercise provides and additional insight into why it occurs.

8. Consider the integral

$$I(\omega) = \frac{1}{2\pi} \int_0^\infty \frac{e^{-i\omega t} dt}{\sqrt{t+\tau}}, \qquad \tau \ge 0. \tag{3}$$

It follows from (4.3.3) that, for any $\tau > 0$,

$$I(\omega) \sim \frac{1}{2\pi i \omega \sqrt{\tau}}, \qquad (\omega \to \infty) \tag{4}$$

i.e., (3) obeys the asymptotic power decay law of the order $\alpha = 1$. On the other hand, it follows from (4.6.2) that, for $\tau = 0$,

$$T(\omega) = \frac{1}{2\sqrt{\pi i \omega}}, \tag{5}$$

i.e., (3) obeys the asymptotic power decay law of the order $\alpha = 1/2$. Study the "phase transition" in the asymptotic behavior of (3) as $\tau \to 0$.

9. Utilize the method of steepest descent sketched in Section 5.5 to derive the *Stirling's approximate formula* for the factorial:

$$n! = n \cdot (n-1) \cdot \ldots \cdot 2 \cdot 1 \sim \sqrt{2\pi n}\, n^n e^{-n} \qquad (n \to \infty).$$

10. Consider the *Riemann equation*

$$\frac{\partial v}{\partial t} + v \frac{\partial v}{\partial x} = 0, \qquad v(x, t = 0) = v_0(x), \tag{6}$$

where $v_0(x)$ is an infinitely differentiable and absolutely integrable function whose derivative attains its minimum at a certain point z, i.e.,

$$\inf_{-\infty < x < \infty} v_0'(x) = v_0'(z) = -u, \qquad u > 0. \tag{7}$$

The equation arises in various physical problems (see Volume 2, Chapter 12 on nonlinear partial differential equations). The solution of this equation is known only in the *implicit* form

$$v(x, t) = v_0(x - tv(x, t)). \tag{8}$$

Find the asymptotics ($\kappa \to \infty$) at the time $t = -1/u$ of the spatial Fourier transform

$$\tilde{v}(\kappa, t) = \frac{1}{2\pi} \int v(x, t) e^{-i\kappa x} dx \tag{9}$$

of the solution v. As an example consider the initial condition

$$v_0(x) = -x \exp(-x^2). \tag{10}$$

Chapter 6

Singular Integrals and Fractal Calculus

This chapter is devoted to integrals similar to the familiar divergent Cauchy integral

$$\int \frac{\varphi(s)}{s-x} ds. \tag{1}$$

Such integrals are often encountered in physical applications. If the function $\varphi(s)$ does not vanish at $s = x$ then the integrand in (1) has a nonintegrable singularity at that point. In practice physicists, using their intuition as a guide, often assign certain finite values to these integrals anyway. Then, it is a mathematician's job to justify rigorously these "renormalizations of infinities", translate additional physical requirements into mathematical terms and point out how different assumptions lead to different values of integral (1). The situation is fairly typical in collaboration of physicists and mathematicians.

The first two sections of this chapter will survey various "recipes" for evaluation of singular integrals and later a typical physical example will be provided. At that time, natural physical considerations will help us select the unique solution.

6.1 Principal value distribution

Integral (6.0.1) will be studied via the notion of its *principal value* which, for $x = 0$, is defined as a symmetric limit

$$\mathcal{PV} \int \frac{\varphi(s)}{s} ds = \lim_{\varepsilon \to 0} \left[\int_{-\infty}^{-\varepsilon} + \int_{+\varepsilon}^{\infty} \right] \frac{\varphi(s)}{s} ds, \tag{1}$$

with an analogous definition for other values of x. The letters $\mathcal{PV}$ in front of the integral indicate that the integral is taken in the sense of its principal value. In

© Springer Nature Switzerland AG 2018
A. I. Saichev and W. Woyczynski, *Distributions in the Physical and Engineering Sciences, Volume 1*, Applied and Numerical Harmonic Analysis, https://doi.org/10.1007/978-3-319-97958-8_6

terms of the distribution theory, equality (1) defines distribution $T_{1/s}$, which acts on a test function φ using the expression on the right-hand side of (1).

Another way to calculate the principal value of integral (6.0.1) can be proposed if we restrict our attention to a test function $\varphi(x) \in \mathcal{D}$ which has compact support. Then one can find a number R such that $\varphi(x) \equiv 0$ for $|x| > R$ and

$$\mathcal{PV} \int \frac{\varphi(s)}{s} ds = \mathcal{PV} \int_{-R}^{R} \frac{\varphi(s)}{s} ds.$$

Representing function $\varphi(x)$ in the form

$$\varphi(x) = [\varphi(x) - \varphi(0)] + \varphi(0),$$

and noticing that the principal value of the integral containing the constant term $\varphi(0)$ is zero, we get that

$$\mathcal{PV} \int \frac{\varphi(s)}{s} ds = \int_{-R}^{R} \frac{\varphi(s) - \varphi(0)}{s} ds. \tag{2}$$

Now the integral on the right-hand side can be understood in the ordinary sense since its singularity at 0 has been removed. In other words, the integrand has been *regularized*. This immediately follows from the fact that any test function $\varphi(s) \in \mathcal{D}$ satisfies the *Lipschitz condition*

$$|\varphi(a) - \varphi(b)| < K|a - b|. \tag{3}$$

If function $\varphi(s)$ does not have compact support, then one can still use equality

$$\mathcal{PV} \int \frac{\varphi(s)}{s} ds = \lim_{R \to \infty} \int_{-R}^{R} \frac{\varphi(s) - \varphi(0)}{s} ds,$$

instead of (2) to define the principal value of integrals with infinite limits.

Finally, let us point out another obvious but useful representation of the principal-value integral (1):

$$\mathcal{PV} \int \frac{\varphi(s)}{s} ds = \int_{0}^{\infty} \frac{\varphi(s) - \varphi(-s)}{s} ds.$$

Standard operations on "well-behaved" convergent integrals, such as the change of variables, differentiation with respect to a parameter, etc., can also be used in analysis of principal-value integrals. The following example illustrates the situation.

Example 1. Consider the singular integral

$$I(x,b) = \mathcal{PV} \int \frac{\exp(-b^2 s^2)}{s-x} ds, \tag{4}$$

and observe that a change of variables $\tau = bs$ transforms it into the integral

$$I(x,b) = I(bx),$$

where

$$I(x) = \mathcal{PV} \int \frac{\exp(-\tau^2)}{\tau - x} d\tau.$$

Introducing variable of integration $y = \tau - x$, we obtain that

$$I(x) = \exp(-x^2) J(x),$$

where

$$J(x) = \mathcal{PV} \int \frac{\exp(-y^2 - 2xy)}{y} dy.$$

Differentiation of the above expression with respect to parameter x gives

$$J'(x) = -2 \int \exp(-y^2 - 2yx) dy = -2\sqrt{\pi} \exp(x^2).$$

Taking into account the fact that $J(0) = 0$, one can get that

$$I(x) = -2\sqrt{\pi} D(x),$$

where

$$D(x) = \exp(-x^2) \int_0^x \exp(y^2) dy$$

is the so-called *Dawson integral*. Thus, finally, we arrive at the formula

$$\mathcal{PV} \int \frac{\exp(-b^2 s^2)}{s-x} ds = -2\sqrt{\pi} D(bx).$$

Let us remark that for $|x| \to 0$, we have that $D(x) \sim x$, and that for $|x| \to \infty$ the Dawson integral has the asymptotics $D(x) \sim 1/2x$.

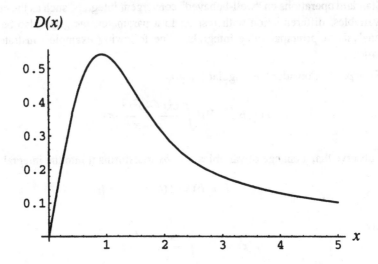

FIGURE 6.1.1
Graph of the Dawson integral.

6.2   Principal value of Cauchy integral

Another approach to evaluation of integral (6.0.1) depends on its interpretation as the limit

$$\int \frac{\varphi(s)}{s-x}ds = \lim_{y\to o} \int \frac{\varphi(s)}{s-z}ds, \tag{1}$$

where

$$z = x + iy$$

is a complex parameter. Moving into the complex plane vicinity of the real axis removes the singularity.

Let us find the above limit by separating explicitly the real and the imaginary parts of the expression

$$\frac{1}{s-z} = \frac{s-x}{(s-x)^2+y^2} + \frac{iy}{(s-x)^2+y^2}.$$

Substituting this sum into (1) and noticing that the integral of the real part converges to the principal value as $y \to 0$, we obtain that

$$\lim_{y\to 0} \int \frac{\varphi(s)}{s-z}ds = \mathcal{PV} \int \frac{\varphi(s)}{s-z}ds + i \lim_{y\to 0} \int \frac{y}{(s-x)^2+y^2}\varphi(s)ds. \tag{2}$$

Notice that the factor in front of function $\varphi(s)$ in the last integral coincides, up to π, with the familiar Lorentz curve (1.3.3)

$$\frac{1}{\pi} \frac{y}{(x-s)^2 + y^2},$$

which weakly converges to $\delta(x-s)$ as $y \to 0+$. Hence, the evaluation of the integral (6.0.1) using the limit procedure (1) leads to an identity

$$\int \frac{\varphi(s)}{s-x} ds = \mathcal{PV} \int \frac{\varphi(s)}{s-x} ds \pm i\pi\varphi(x).$$

The plus sign corresponds to the limit $y \to 0+$ with y's restricted to the upper half-plane and the minus—to the limit $y \to 0-$ with y's restricted to the lower half-plane. Thus equality (2) determines two distributions:

$$\frac{1}{s-x-i0} = \mathcal{PV}\frac{1}{s-x} + i\pi\delta(x-s) \tag{3}$$

and

$$\frac{1}{s-x+i0} = \mathcal{PV}\frac{1}{s-x} - i\pi\delta(x-s). \tag{4}$$

Although assigning complex values to real-valued integrals may seem strange at the first glance, these formulas often give the correct physical answer. The point is that their imaginary parts reflect the *causality principle* which was not spelled out explicitly when the original physical problem was posed but which, as we will see later on, plays an important role. Obvious physical arguments then permit us to indicate which of the formulas (3-4) exactly corresponds to the physical problem under consideration.

6.3   A study of monochromatic wave

Formulas of Section 2 give, for example, the correct physical answer in the problem of radiation by a monochromatic wave source with complex amplitude. For simplicity, we will discuss only the 1-D case. Then, wave radiation is described by the nonhomogeneous wave equation

$$\frac{1}{c^2}\frac{\partial^2 E}{\partial t^2} = \frac{\partial^2 E}{\partial x^2} - D(x,t).$$

If the wave source is monochromatic, that is

$$D(x, t) = w(x) \cos \omega t = \operatorname{Re} w(x) e^{i\omega t},$$

then the radiated wave is also monochromatic and can be written in the form

$$E(x, t) = \operatorname{Re} u(x) e^{i\omega t},$$

where the complex amplitude $u(x)$ of the propagating wave satisfies the Helmholtz equation

$$u'' + k^2 u = w(x). \tag{1}$$

Here, $k = \omega/c$ is the so-called *wavenumber*. The equation can be solved with the help of Green's function—an approach that will be discussed later on. Here, we will solve it by passing to the frequency domain and considering the Fourier transform

$$\tilde{u}(\kappa) = \frac{1}{2\pi} \int u(x) e^{-i\kappa x} dx.$$

The inverse Fourier transform is then given by

$$u(x) = \int \tilde{u}(x) e^{i\kappa x} dx.$$

Taking the Fourier transforms of both sides of equation (1) we get

$$\tilde{u}(\kappa) = \frac{\tilde{w}(\kappa)}{k^2 - \kappa^2} = \frac{\tilde{w}(\kappa)}{2k} \left[\frac{1}{\kappa + k} - \frac{1}{\kappa - k} \right],$$

where $\tilde{w}(\kappa)$ is the Fourier transform of the source function $w(x)$. We will assume that function $\tilde{w}(\kappa)$ is sufficiently smooth and rapidly decaying to 0 for $|\kappa| \to \infty$. The sought complex amplitude of the propagating wave can now be found by the inverse Fourier transform:

$$u(x) = \frac{1}{2k} \left[\int \frac{\tilde{w}(\kappa)}{\kappa + k} e^{i\kappa x} d\kappa - \int \frac{\tilde{w}(\kappa)}{\kappa - k} e^{i\kappa x} d\kappa \right].$$

Changing the variable of integration κ in the first integral to $-\kappa$ and observing that $\tilde{w}(-\kappa) = \tilde{w}^*(\kappa)$, where the asterisk denotes the complex conjugate, we obtain that

$$u(x) = -\frac{1}{k} \int \frac{\operatorname{Re}[\tilde{w}(\kappa) e^{i\kappa x}]}{\kappa - k} d\kappa. \tag{2}$$

First of all, let us calculate the principal value

$$PV[u(x)] = -\frac{1}{k} Re\left[PV \int \frac{\tilde{w}(\kappa)}{\kappa - k} e^{i\kappa x} d\kappa \right]$$

of that integral by splitting it into the sum of two components

$$PV \int \frac{\tilde{w}(\kappa)}{\kappa - k} e^{i\kappa x} d\kappa$$

$$= e^{ikx}\left[PV \int \tilde{w}(\kappa) \frac{\cos[(\kappa - k)x]}{\kappa - k} d\kappa + iPV \int \tilde{w}(\kappa) \frac{\sin[(\kappa - k)x]}{\kappa - k} d\kappa \right], \quad (3)$$

and studying the asymptotic behavior of each of them separately as $|x| \to \infty$. Notice that, according to (3.3.7),

$$\frac{\sin[(\kappa - k)x]}{\kappa - k} \to \pi \delta(\kappa - k) \text{ sign } (x),$$

weakly as $|x| \to \infty$. Consequently,

$$\lim_{x \to \infty} PV \int \tilde{w}(\kappa) \frac{\sin[(\kappa - k)x]}{\kappa - k} d\kappa = \pi \tilde{w}(k) \text{ sign } (x).$$

Now consider the first integral on the right-hand side of (3). Adding and subtracting number $\tilde{w}(k)$ from function $\tilde{w}(\kappa)$ we arrive at the equality

$$PV \int \tilde{w}(\kappa) \frac{\cos[(\kappa - k)x]}{\kappa - k} d\kappa$$

$$= \int \psi(\kappa, k) \cos[(\kappa - k)x] d\kappa + \tilde{w}(k) PV \int \frac{\cos[(\kappa - k)x]}{\kappa - k} d\kappa,$$

where

$$\psi(\kappa, k) = \frac{\tilde{w}(\kappa) - \tilde{w}(k)}{\kappa - k}.$$

By previous assumptions, $\psi(\kappa, k)$ is a continuous function of κ. The second integral on the right-hand side of the above equality vanishes because the integrand is odd. The first integral converges uniformly for $|x| > 0$, and by the Riemann-Lebesgue Lemma, its value converges to 0 as $x \to \infty$. Thus we arrive at the asymptotic formula

$$PV \int \tilde{w}(\kappa) \frac{\cos[(\kappa - k)x]}{\kappa - k} d\kappa \sim o(1), \qquad (|x| \to \infty),$$

which permits us to drop the corresponding, converging to zero, term to get that

$$\mathcal{PV}[u(x)] \sim \frac{\pi}{k}\operatorname{Im}[\tilde{w}(k)e^{ikx}\operatorname{sign}(x)], \qquad (|x| \to \infty). \qquad (4)$$

The above expression contradicts the radiation condition and is physically not acceptable.[1] To save the situation we will turn to formulas of Section 2 which give that

$$u(x) = \mathcal{PV}[u(x)] \mp i\frac{\pi}{k}\operatorname{Re}[\tilde{w}(k)e^{ikx}]. \qquad (5)$$

Formulas (4) and (5) imply that if we select the plus sign in the formulas of Section 2 then we shall arrive at the asymptotic formula

$$u(x) \approx \begin{cases} i\frac{\pi}{k}\tilde{w}^{*}(k)e^{-ikx}, & x > 0, \\ i\frac{\pi}{k}\tilde{w}(k)e^{ikx}, & x < 0, \end{cases}$$

which does satisfy the radiation condition. Here, the physically acceptable answer corresponds to the distribution

$$\frac{1}{\kappa - k + i0} = \mathcal{PV}\frac{1}{\kappa - k} - i\pi\delta(\kappa - k). \qquad (6)$$

Let us consider the physical arguments in favor of such a choice. There are no purely monochromatic wave sources in nature emitting radiation for infinitely long times. All real-world phenomena begin and end in finite time. The fact that the source was turned on sometime in the past can be taken into account by assuming that its time dependence reflects asymptotically negligible intensity of the source at time $-\infty$. That is exactly what the replacement of $k = \omega/c$ in the preceding formulas by

$$k = \frac{\omega}{c} - i\frac{\gamma}{c} = k - i0, \qquad (7)$$

accomplishes and what justifies utilization of the distribution (6).

Our choice was based on the *causality principle* which asserts that it is impossible to receive the wave before the wave source is turned on. The same result can be obtained if a *principle of infinitesimal relaxation* is applied. According to this principle, any medium (even the vacuum) damps waves. This means that, for example, the propagating to the right monochromatic wave $\exp(i\omega t - ikx)$ is attenuated with the growth of x. Again we are led to the conclusion that the real k in (2) should be replaced by the formula (7).

---

[1]Notice that the radiation condition demands that far from the source (i.e., for $|x| \to \infty$) there only exist waves running away from the source, like $\exp(i(\omega t - k|x|))$.

6.4   The Cauchy formula

The results of Section 2 are closely related to the *Cauchy formula* from the theory
of functions of a complex variable. The formula asserts that for any function $f(z)$,
analytic in a simply connected domain D in the complex plane $\mathbf{C}$ and continuous
on its closure $\bar{D}$ including the boundary contour C,

$$f(z) = \frac{1}{2\pi i} \int_C \frac{f(\zeta)d\zeta}{\zeta - z},$$

where z is a point in the interior of D and contour C is oriented counterclockwise.
If z is an arbitrary point of the complex plane then the above integral defines a new
function

$$F(z) = \frac{1}{2\pi i} \int_C \frac{f(\zeta)d\zeta}{\zeta - z}. \tag{1}$$

For z inside the contour of integration

$$F(z) = f(z), \qquad z \in D.$$

If $z \notin \bar{D}$ then the integrand is analytic everywhere in D and, by the *Cauchy's
Theorem*, $F(z) \equiv 0$. For a boundary point $z = \zeta_0 \in C$ the integral (1) is singular
and $F(z)$ will be understood in the principal-value sense. In the present context,
this will mean that

$$F(\zeta_0) = \mathcal{PV}[F(\zeta_0)] = \lim_{r \to 0} \frac{1}{2\pi i} \int_{C_r} \frac{f(\zeta)d\zeta}{\zeta - \zeta_0}, \tag{2}$$

where C_r is a curve obtained from contour C by removing its part contained in a
disc of radius $r \to 0$ and centered at ζ_0 (see Fig. 6.4.1).

Observe that, for any function f continuous on contour, function $F(z)$ is well
defined everywhere.

Formula (2) generalizes the concept of the principal value of a singular integral
on the real axis and we will study it for a function $f(z)$ analytic inside contour C
and continous on it. To that end substitute an identity

$$f(\zeta) = [f(\zeta) - f(\zeta_0)] + f(\zeta_0),$$

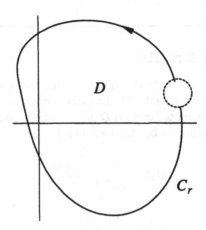

FIGURE 6.4.1
A schematic illustration of countour C_r.

into (2) and split the integral into two parts:

$$PV[F(\zeta_0)] = \frac{1}{2\pi i} \int_C \frac{[f(\zeta) - f(\zeta_0)]d\zeta}{\zeta - \zeta_0} + \lim_{r \to 0} \frac{f(\zeta_0)}{2\pi i} \int_{C_r} \frac{d\zeta}{\zeta - \zeta_0}.$$

We deliberately replaced C_r by C in the first integral since for $f(z)$ satisfying the Lipschitz condition (6.1.3) the integral is no longer singular and one can integrate over the whole closed contour C. Since its integrand is analytic inside the contour and continous on the contour, the first integral vanishes by Cauchy's theorem. Thus

$$PV[F(\zeta_0)] = \lim_{r \to 0} \frac{f(\zeta_0)}{2\pi i} \int_{C_r} \frac{d\zeta}{\zeta - \zeta_0}. \tag{3}$$

The last integral can be evaluated assuming that contour C is smooth in the vicinity of point ζ_0, which is called the *regular point of the contour.* Let us add and subtract from (3) the integral over portion c_r of the located in D circle of radius r with center at ζ_0. Since the integral over the closed contour $C_r + c_r$ is equal to zero, equality (3) can be rewritten in the form

$$PV[F(\zeta_0)] = -\lim_{r \to 0} \frac{f(\zeta_0)}{2\pi i} \int_{c_r} \frac{d\zeta}{\zeta - \zeta_0}.$$

The latter integral is easy to evaluate:

$$\int\limits_{C_r} \frac{d\zeta}{\zeta - \zeta_0} = \int\limits_{\pi}^{0} \frac{ire^{i\varphi}d\varphi}{re^{i\varphi}} = -i\pi.$$

Hence, we obtain that

$$\mathcal{PV}[F(\zeta_0)] = f(\zeta_0)/2,$$

and the principal value of the analytic function is equal to

$$\mathcal{PV} \int\limits_{C} \frac{f(\zeta)d\zeta}{\zeta - \zeta_0} = i\pi f(\zeta_0), \qquad (4)$$

that is, the value of function f at the singular point of the integrand multiplied by $i\pi$. Thus, the behavior of the Cauchy integral in the neighborhood of a regular point ζ_0 of contour C can be summarized as follows:

$$\int\limits_{C} \frac{f(\zeta)d\zeta}{\zeta - z} = 2\pi i \begin{cases} f(\zeta_0), & z \to \zeta_0+, \\ \frac{1}{2}f(\zeta_0), & z = \zeta_0, \\ 0, & z \to \zeta_0-, \end{cases} \qquad (5)$$

where the plus sign corresponds to the limit value of the integral while ζ_0 is approached from the inside of contour C, and the minus sign corresponds to the approach from outside.

Example 1. Let us consider a contour C consisting of the interval $[-R, R]$ on the real axis and the semicircle C_R of radius R with the center at point $z = 0$ located in the upper half-plane. If $f(z)$ is analytic in the upper half-plane and such that the integral over C_R uniformly converges to zero as $R \to \infty$, then the Cauchy formula is transformed into

$$\int \frac{f(s)}{s - z} ds = 2\pi i f(z), \qquad y > 0, \qquad (6)$$

and equality (4) assumes the form

$$\mathcal{PV} \int \frac{f(s)}{s - x} ds = i\pi f(x). \qquad (7)$$

Recall that, by the well known *Jordan Lemma* in the complex functions theory, function

$$f(z) = \exp(i\lambda z), \qquad \lambda > 0, \qquad (8)$$

satisfies the conditions mentioned earlier so that the above formula implies that

$$\mathcal{PV} \int \frac{e^{i\lambda s}}{s - x} ds = \pi i e^{i\lambda x}. \tag{9}$$

In particular, it follows that for $x = 0$

$$-i\mathcal{PV} \int \frac{e^{i\lambda s}}{s} ds = \int \frac{\sin(\lambda s)}{s} ds = \pi. \tag{10}$$

Notice that the Cauchy formula (6) interpreted in the spirit of the distribution theory defines a new distribution

$$\hat{T}(s - z) = \frac{1}{2\pi i} \frac{1}{s - z}, \tag{11}$$

which is called the *analytic representation of the Dirac delta*. Its functional action assigns to a function $f(s)$, analytic in the upper half-plane and rapidly decaying at infinity, its value at the point z. Applied to any usual "well-behaved" function of the real variable s, it defines a new function of complex variable z which is analytic everywhere with the possible exception of the real axis. Crossing the real axis at the point $z = x$ the functional $\hat{T}(s - z)[f]$ has a jump of size $f(x)$; this follows from formulas of Section 2. We will extract this jump by introducing a new distribution

$$\hat{\delta}(s - z) = \hat{T}(s - z) + \hat{T}^*(s - z) = \frac{1}{\pi} \frac{y}{(s - x)^2 + y^2},$$

which is harmonic for $y \neq 0$ and which converges to the usual Dirac delta $\delta(s-x)$ as $y \to 0+$. As we will see in Volume 2, the corresponding functional $\hat{\delta}(s - z)[f(s)]$ solves the Dirichlet problem for the 2-D Laplace's equation in the upper half-plane $y > 0$.   ∎

6.5 The Hilbert transform

The *integral Hilbert transform*

$$\psi(t) = \frac{1}{\pi} \mathcal{PV} \int \frac{\varphi(s)}{s - t} ds \tag{1}$$

of function $\varphi(t)$ is also defined in terms of the principal value of the integral involved. We shall find the inversion formula for the Hilbert transform by applying the Fourier transform to both sides of definition (1). The left-hand side is transformed into

$$\tilde{\psi}(\omega) = \frac{1}{2\pi} \int \psi(t) e^{-i\omega t} dt,$$

and the right-hand side, after a change of the integration order and other simple manipulations, assumes the form

$$\frac{1}{2\pi^2} \int \varphi(s) e^{-i\omega s} \left[\mathcal{PV} \int \frac{e^{i\omega(s-t)}}{s-t} dt \right] ds.$$

Since

$$\mathcal{PV} \int \frac{e^{i\omega(s-t)}}{s-t} dt = i \int \frac{\sin(\omega t)}{t} dt = i\pi \text{ sign } (\omega),$$

we get that

$$\tilde{\psi}(\omega) = i\tilde{\varphi}(\omega) \text{ sign } (\omega). \tag{2}$$

Now let $g(t)$ be the Hilbert transform of function $\psi(t)$. Then, according to the above formula, its Fourier transform is

$$\tilde{g}(\omega) = i\tilde{\psi}(\omega) \text{ sign } (\omega) = -\tilde{\varphi}(\omega)$$

so that $g(t) = -\varphi(t)$. Thus, the inverse Hilbert transform has the form

$$\varphi(t) = -\frac{1}{\pi} \mathcal{PV} \int \frac{\psi(s)}{s-t} ds. \tag{3}$$

One of the most important applications of the Hilbert transform is related to the causality principle. We shall illustrate it in the example of an absolutely integrable function $h(t)$ describing a response of a linear physical system to the Dirac delta impulse $\delta(t)$. The Fourier transform $\tilde{h}(\omega)$ appears in many physical applications. Let us write it with the real frequency ω replaced by a complex number $\lambda = \omega + i\alpha$:

$$\tilde{h}(\lambda) = \frac{1}{2\pi} \int_0^\infty h(t) e^{-i\lambda t} dt. \tag{4}$$

The zero lower limit takes into account the causality principle which requires that the system's response cannot appear before the action of the impulse: $h(t < 0) \equiv 0$. It is obvious from (4) that function $\tilde{h}(\lambda)$ is analytic in the lower half-plane and continuous on the real axis $\alpha = 0$. This, in turn, means that $\tilde{h}(\omega)$ satisfies the relation

$$-i\pi\tilde{h}(\omega) = \mathcal{PV} \int \frac{\tilde{h}(\kappa)}{\kappa - \omega} d\kappa,$$

similar to (6.4.7). The complex function $\tilde{h}(\omega)$ can be represented as a sum of its real and imaginary parts: $\tilde{h}(\omega) = \varphi(\omega) + i\psi(\omega)$. Substituting them into the last equality and comparing separately the real and the imaginary parts, we discover that φ and ψ are related to each other by the Hilbert transform. The connecting formulas (1) and (3) applied to the real and imaginary parts of the Fourier transform of response function are called in physics the *dispersion relations* and are widely used in the theory of wave propagation in dispersing media.

6.6   Analytic signals

Another physical application of the Hilbert transform is related to the notion of *analytic signal*, which appears in various areas of physical sciences, from electrical engineering to quantum optics. It can be introduced in the following way. To each real process $\xi(t)$, which will be assumed to be absolutely integrable, we will assign a complex signal $\zeta(t)$ with the real part equal to $\xi(t)$ and the imaginary part $\eta(t)$ defined by the condition that $\zeta(z)$ is an analytic function of the complex variable $z = t + i\beta$ in the upper half-plane. The analyticity can be achieved by making the Fourier transform of the signal $\xi(t)$ vanish for negative frequencies, that is, by replacing $\tilde{\xi}(\omega)$ by $2\chi(\omega)\tilde{\xi}(\omega)$. The last expression can be also written in the form

$$\tilde{\zeta}(\omega) = 2\chi(\omega)\tilde{\xi}(\omega) = \tilde{\xi}(\omega) + \tilde{\xi}(\omega)\,\text{sign}\,(\omega). \tag{1}$$

The first component on the right-hand side is the Fourier transform of the original signal and the second is the Fourier transform of the imaginary part of the analytic signal $\zeta(t)$. Formula (6.5.2), relating the Fourier transforms of a function and its Hilbert transform, implies that the imaginary part of the analytic signal of real variable is expressed through its real part by

$$\eta(t) = -\frac{1}{\pi}\mathcal{PV} \int \frac{\xi(s)}{s - t} ds. \tag{2}$$

The concept of analytic signal helps to solve the crucial electrical engineering problem of definition of amplitude and phase of the *narrow band signal* whose Fourier transform is concentrated in the small neighborhood of the carrying frequency Ω. Such a signal is usually represented in the form

$$\xi(t) = A(t)\cos[\Omega t + \psi(t)], \tag{3}$$

where $A(t)$ and $\psi(t)$ are slowly varying within the period $T = 2\pi/\Omega$. The standard engineering problem of finding $A(t)$ and $\psi(t)$ from the known form of $\xi(t)$ is not well posed mathematically, since it reduces to solving one equation for two unknowns A and φ. That fundamental difficulty makes, for example, comparing accuracy for phase measurements by different phase detectors questionable. From the theoretical viewpoint the best prescription is to uniquely define the amplitude and the phase via the concept of the analytic signal, the imaginary part providing the missing second equation

$$\eta(t) = A(t)\sin[\Omega t + \psi(t)]. \tag{4}$$

6.7 Fourier transform of Heaviside function

Having armed ourselves with the notion of the principal value of singular integral, we are now in a position to explore Fourier transforms of the Heaviside function and related distributions. This topic was only briefly mentioned in Chapter 4. We shall begin by rewriting the Hilbert transform (6.5.1) in the language of convolutions:

$$\psi(t) = \varphi(t) * \left[-\frac{1}{\pi} \mathcal{PV}\left(\frac{1}{t}\right) \right].$$

Comparing the Fourier transform (6.5.2) of this function with the formula $f(t) * \varphi(t) \longmapsto 2\pi \tilde{f}(\omega)\tilde{\varphi}(\omega)$ (3.2.8), after simple transformations we obtain that

$$\mathcal{PV}\left(\frac{1}{t}\right) \longmapsto -\frac{i}{2}\,\mathrm{sign}\,(\omega).$$

Inverting this expression with the help of formula (3.2.5) we get that

$$\mathrm{sign}\,(t) \longmapsto \frac{1}{\pi}\mathcal{PV}\left(\frac{1}{i\omega}\right). \tag{1}$$

Now, we are ready to recover the Fourier transform of the Heaviside function. To do that we will represent the latter in the form

$$\chi(t) = \frac{1}{2} + \frac{1}{2}\,\mathrm{sign}\,(t).$$

The Fourier transform of the first summand is equal to $\delta(\omega)/2$ so that for the Fourier transform of the Heaviside function we have

$$\chi(t) \longmapsto \frac{1}{2}\delta(\omega) + \frac{1}{2\pi}\mathcal{PV}\left(\frac{1}{i\omega}\right).$$

It is convenient to write this relation with the help of the distributions (6.2.3):

$$\chi(t) \longmapsto \frac{1}{2\pi i} \cdot \frac{1}{\omega - i0}. \tag{2}$$

Once the Fourier transform of the Heaviside function has been calculated we can evaluate Fourier transforms of a wide class of functions

$$F(t) = \int_{-\infty}^{t} f(\tau)d\tau \tag{3}$$

representable by integrals with variable upper limit of absolutely integrable functions $f(t)$. Indeed, if we represent $F(t)$ as a sum

$$F(t) = F(\infty)\chi(t) + G(t),$$

where

$$G(t) = F(t) - F(\infty)\chi(t),$$

then, according to (3), the Fourier transform of $F(t)$ can be expressed in terms of the Fourier transform of $G(t)$ by

$$\tilde{F}(\omega) = F(\infty)\frac{1}{2\pi i}\frac{1}{\omega - i0} + \tilde{G}(\omega). \tag{4}$$

Example 1. Consider an absolutely integrable function $f(t) = \chi(t)e^{-pt}$, $p > 0$. Then $G(t) = -\chi(t)e^{-pt}/p$. Its Fourier transform exists in the classical sense. Thus, equation (4) implies that

$$\tilde{F}(\omega) = \frac{1}{2\pi i p}\left[\frac{1}{\omega - i0} - \frac{1}{\omega - ip}\right]. \tag{5}$$

∎

Integration of (3) leads, in turn, to a new function which linearly increases as $t \to \infty$. We shall learn how to find the Fourier transform of such functions by first evaluating the Fourier transform of the absolute value function $|t|$.

Example 2. Let us write the absolute value function in the product form $|t| = t \operatorname{sign}(t)$. The Fourier transforms of each of the two factors are already known to us. Recall that $t \mapsto i\delta'(\omega)$ and that the Fourier transform of $\operatorname{sign}(t)$ is given by formula (6.7.1). In this fashion, with the help of formula (3.2.10) according to which the Fourier transform of a product is equal to convolution of the Fourier transforms of factors, we get that

$$|t| = t \operatorname{sign}(t) \longmapsto i\delta'(\omega) * \frac{1}{\pi} \mathcal{PV}\left(\frac{1}{i\omega}\right) = \frac{1}{\pi} \mathcal{PV}\frac{d}{d\omega}\left(\frac{1}{\omega}\right). \qquad (6)$$

Let us identify the new distribution arising on the right-hand side through its action as a functional on an arbitrary test function $\phi \in S$:

$$\int \phi(\omega)\frac{d}{d\omega}\mathcal{PV}\left(\frac{1}{\omega}\right) d\omega = -\mathcal{PV}\int \frac{\phi'(\omega)}{\omega}d\omega.$$

Recall that the principal value of the above integral is, by definition, equal to

$$\mathcal{PV}\int \frac{\phi'(\omega)}{\omega}d\omega = \lim_{\varepsilon \to o}\left[\int_{-\infty}^{-\varepsilon} \frac{\phi'(\omega)}{\omega}d\omega + \int_{\varepsilon}^{\infty} \frac{\phi'(\omega)}{\omega}d\omega\right].$$

For the first integral, integrating by parts expression in the brackets, we obtain that

$$\int_{-\infty}^{-\varepsilon} \frac{\phi'(\omega)}{\omega}d\omega = \int_{-\infty}^{-\varepsilon} \frac{[\phi(\omega) - \phi(-\varepsilon)]}{\omega^2}d\omega.$$

A similar transformation of the second integral yields eventually that

$$\mathcal{PV}\int \frac{\phi'(\omega)}{\omega}d\omega = \lim_{\varepsilon \to o}\left[\int_{-\infty}^{-\varepsilon} \frac{[\phi(\omega) - \phi(-\varepsilon)]}{\omega^2}d\omega + \int_{\varepsilon}^{\infty} \frac{[\phi(\omega) - \phi(\varepsilon)]}{\omega^2}d\omega\right].$$

The latter limit defines a new distribution

$$\mathcal{PV}\left(\frac{1}{\omega^2}\right),$$

which functionally acts via the formula

$$\mathcal{PV}\left(\frac{1}{\omega^2}\right)[\phi(\omega)]$$

$$= \lim_{\varepsilon \to 0} \left[\int_{-\infty}^{-\varepsilon} \frac{[\phi(\omega) - \phi(-\varepsilon)]}{\omega^2} d\omega + \int_{\varepsilon}^{\infty} \frac{[\phi(\omega) - \phi(\varepsilon)]}{\omega^2} d\omega \right].$$

Obviously, the last equality can be written in the form

$$\mathcal{PV} \left(\frac{1}{\omega^2} \right) [\phi(\omega)] = \mathcal{PV} \int \frac{[\phi(\omega) - \phi(0)]}{\omega^2} d\omega, \tag{7}$$

or in an equivalent regularized form

$$\mathcal{PV} \left(\frac{1}{\omega^2} \right) [\phi(\omega)] = \int_0^{\infty} \frac{\phi(\omega) + \phi(-\omega) - 2\phi(0)}{\omega^2} d\omega. \tag{8}$$

Thus, we derived a new distribution-theoretic formula

$$\frac{d}{d\omega} \mathcal{PV} \left(\frac{1}{\omega} \right) = -\mathcal{PV} \left(\frac{1}{\omega^2} \right),$$

which yields the Fourier transform of function $|t|$:

$$|t| \longmapsto -\frac{1}{\pi} \mathcal{PV} \left(\frac{1}{\omega^2} \right). \tag{9}$$

∎

6.8 Fractal integration

In the last three sections of this chapter we develop another class of important singular integrals which arise when one tries to extend the notion of n-tuple integrals and of n-th order derivatives of classical calculus to noninteger (or fractional) n. We begin with the concept of *fractal* or *fractional integration*. It is natural to introduce it as a generalization of the *Cauchy formula*

$$(I^n g)(t) = \int_{-\infty}^t dt_1 \int_{-\infty}^{t_1} dt_2 \ldots \int_{-\infty}^{t_{n-1}} dt_n \, g(t_n)$$

$$= \frac{1}{(n-1)!} \int_{-\infty}^t (t-s)^{n-1} g(s) \, ds, \tag{1}$$

which expresses the result of n-tuple integration of function $g(t)$ of a single variable via the single integration operator.

Before we move on to fractal integrals, let us take a closer look at the Cauchy formula (1). It is valid for absolutely integrable functions $g(t)$ which decay for $t \to -\infty$ sufficiently rapidly to guarantee the existence of the integral on the right-hand side of (1). Assuming that the integrand $g(t)$ vanishes for $t < 0$, the Cauchy formula can be rewritten in the form

$$(I^n g)(t) = \int_0^t dt_1 \int_0^{t_1} dt_2 \ldots \int_0^{t_{n-1}} dt_n \, g(t_n)$$

$$= \frac{1}{(n-1)!} \int_0^t (t-s)^{n-1} g(s) \, ds. \tag{2}$$

Remark 1. The Cauchy formula can be viewed as an illustration of the general *Riesz Theorem* about representation of any linear continuous (in a certain precise sense) operator L transforming function g of one real variable into another function $L[g]$ as an integral operator

$$L[g](t) = \int h(t, s) g(s) \, ds \tag{3}$$

with an appropriate kernel $h(t, s)$. In our case, the n-tuple integration linear operator in (1) has a representation via the single integral operator with kernel $h(t, s) = (t - s)^{n-1}/(n - 1)!$.

Let us check the validity of the Cauchy formula (1) by observing that the n-tuple integral in (1) is a solution of the differential equation

$$\frac{d^n}{dt^n} x(t) = g(t) \tag{4}$$

satisfying the causality principle. Such a solution, in view of (2.2.4), can be written as convolution

$$x(t) = \int k_n(t - s) g(s) \, ds, \tag{5}$$

where

$$k_n(t) = \chi(t) y(t),$$

and $y(t)$ is the solution of the corresponding homogeneous equation

$$\frac{d^n}{dt^n} x(t) = 0$$

with the initial conditions

$$y(0) = y'(0) = \ldots = y^{(n-2)}(0) = 0, \quad y^{(n-1)} = 1.$$

Solving the above initial-value problem we get

$$k_n(t) = \frac{1}{(n-1)!} t^{n-1} \chi(t),$$

the kernel that appears in the Cauchy formula (1).

This is a good point to introduce fractal integrals. Replacing integer n in kernel k_n by an arbitrary positive real number α and the factorial $(n-1)!$ by the gamma function $\Gamma(\alpha)$ (see (4.4.3)) we arrive at the generalized kernel

$$k_\alpha(t) = \frac{1}{\Gamma(\alpha)} t^{\alpha-1} \chi(t). \tag{6}$$

So it is natural to call the convolution operator

$$(I^\alpha g)(t) = k_\alpha(t) * g(t), \tag{7}$$

the *fractal integration operator of order* α.

In the case when function $g(t) \equiv 0$ for $t < 0$ the convolution (7) reduces to the integral

$$(I^\alpha g)(t) = \frac{1}{\Gamma(\alpha)} \int_0^t (t-s)^{\alpha-1} g(s)\, ds \tag{8}$$

where the upper integration limit reflects the causality property of the operator of fractal integration.

Let us establish some of the important properties of the fractal integration operator assuming, for simplicity, that function $g(t)$ in (8) is bounded and continuous.

Existence of fractal integrals. For $\alpha \geq 1$ the integrand in (8) is bounded and continuous, and the integral exists in the Riemann sense. For $0 < \alpha < 1$, the kernel $(t-s)^{\alpha-1}$ is singular but the singularity is integrable and the integral is an absolutely convergent improper integral.

Zero-order integration. As $\alpha \to 0+$ the operators I^α tend to the identity operator. Indeed, in view of the recurrent formula $\Gamma(\alpha+1) = \alpha\Gamma(\alpha)$ for the gamma function and the fact that $\Gamma(1) = 1$ we have the asymptotics $\Gamma(\alpha) \sim 1/\alpha$, $(\alpha \to 0+)$. Hence,

$$\lim_{\alpha \to 0+} (I^\alpha g)(t) = \lim_{\alpha \to 0+} \int g(s)\chi(t-s)\alpha(t-s)^{\alpha-1}\, ds.$$

Example 1 in Section 1.9 shows that the function $\chi(t-s)\alpha(t-s)^{\alpha-1}$ weakly converges to the Dirac delta$\delta(t-s-0)$ as $\alpha \to 0+$.[2] Consequently,

$$(I^0 g)(t) = \lim_{\alpha \to 0+} (I^\alpha g)(t) = g(t),$$

or, equivalently, in the distributional language,

$$k_0(t) = \delta(t). \tag{9}$$

Iteration of fractal integrals. As in the case of usual n-tuple integrals, repeated application of fractal integrals is subject to the rule

$$I^\alpha I^\beta = I^{\alpha+\beta}. \tag{10}$$

To see this it suffices to check that, for any $\alpha, \beta > 0$,

$$k_\alpha * k_\beta = k_{\alpha+\beta}. \tag{11}$$

Indeed, the left-hand side of (11), in view of (6), equals

$$k_\alpha * k_\beta = \frac{\chi(t)}{\Gamma(\alpha)\Gamma(\beta)} \int_0^t s^{\alpha-1}(t-s)^{\beta-1} ds,$$

so that, passing to the new dimensionless variable of integration $\tau = s/t$,

$$k_\alpha * k_\beta = \frac{\chi(t) t^{\alpha+\beta-1}}{\Gamma(\alpha)\Gamma(\beta)} B(\alpha, \beta), \tag{12}$$

where

$$B(\alpha, \beta) = \int_0^1 \tau^{\alpha-1}(1-\tau)^{\beta-1} d\tau$$

is the *beta function*. It can be expressed in terms of the gamma function by

$$B(\alpha, \beta) = \frac{\Gamma(\alpha)\Gamma(\beta)}{\Gamma(\alpha+\beta)}.$$

Substituting it into (12) we obtain equality (11).

---

[2]This notation emphasizes that the support of this Dirac delta lies inside the interval $(0, t)$ so that $\int_0^t \delta(t-s-0) ds = 1$ and not $1/2$ as in (2.9.6). A similar situation was encountered in formula (2.9.4) for $\alpha \to +\infty$

Fractal integrals as continuous operators. The following two inequalities show that integration of fractal order has some continuity properties as a linear operation. These properties will find an application in our construction of Brownian motion in Chapter 14 of Volume 2, and for those purposes it will suffice to assume that $0 < \alpha < 1$. First, observe that, for $\alpha > \frac{1}{2}$, I^α is a continuous operator from $L^2[0, 1]$ *into* $L^p[0, 1]$ for each $p < \infty$. [3] Indeed, by the *Schwartz Inequality*,

$$\|I^\alpha f\|_p^p = \int_0^1 |(I^\alpha f)(t)|^p dt$$

$$\leq \int_0^1 \left(\int_0^t f^2(s)ds \right)^{p/2} \cdot \left(\int_0^t k_\alpha^2(s)ds \right)^{p/2} dt \tag{13}$$

$$\leq c(\alpha, p) \left(\int_0^1 f^2(s)ds \right)^{p/2} = c(\alpha, p)\|f\|_2^p,$$

where $c(\alpha, p)$ is a constant depending only on α and p.

Additionally, by a similar argument, but using the *Hölder Inequality* with $1/p + 1/q = 1$,

$$\|I^\beta f\|_\infty = \sup_{0 \leq t \leq 1} \left| \int_0^t k_\beta(t - s)f(s)ds \right|$$

$$\leq \left(\int_0^1 |k_\beta(s)|^q ds \right)^{1/q} \left(\int_0^1 |f(s)|^p ds \right)^{1/p} = c(\beta, p)\|f\|_p, \tag{14}$$

so that for any $\beta > 1/p$, the operator I^β is *continuous* from $L^p[0, 1]$ into the space of continuous functions $C[0, 1]$.

6.9   Fractal differentiation

The operator D^α of *fractal* or *fractional differentiation* is defined as the inverse of the operator I^α of the fractal integration, that is, via the operator equation

$$D^\alpha I^\alpha = \text{Id}, \tag{1}$$

---

[3] Recall the $L^p[0, 1]$ denotes the *Lebesgue space* of functions f on the interval $[0, 1]$ which have pth powers integrable, i.e., for which the norm $\|f\|_p := (\int_0^1 |f(s)|^p \, ds)^{1/p} < \infty$.

where Id denotes the identity operator. Similarly to the fractal integration operator, the fractal differentiation operator has an integral representation

$$(D^\alpha g)(t) = r_\alpha(t) * g(t), \tag{2}$$

where $r_\alpha(t)$ is the convolution kernel which will be identified next. To do that notice that the operator equation (1) is equivalent to the convolution algebra equation

$$r_\alpha(t) * k_\alpha(t) = \delta(t), \tag{3}$$

where k_α is the fractal integration kernel (6.8.6). Denote by γ the solution of the equation $\alpha + \gamma = n$, where $n = \lceil \alpha \rceil$ is the smallest integer greater than or equal to α. In other words,

$$n - 1 < \alpha \le n, \qquad \gamma = n - \alpha, \qquad 0 \le \gamma < 1. \tag{4}$$

Applying the fractal integration operator I^γ to both sides of (3) we get

$$r_\alpha(t) * k_n(t) = k_\gamma(t).$$

The expression on the left-hand side represents the usual n-tuple integral of the fractal differentiation kernel r_α. Thus, if we differentiate it n times we arrive at the explicit formula

$$r_\alpha(t) = k_\gamma^{(n)}(t), \qquad \gamma = n - \alpha > 0. \tag{5}$$

for the fractal differentiation kernel. The corresponding fractal differentiation operator is then given by the convolution

$$(D^\alpha g)(t) = k_\gamma^{(n)}(t) * g(t). \tag{6}$$

For $t > 0$, the n-th derivative of the kernel $k_\gamma(t)$ appearing in (5) exists in the classical sense and

$$r_\alpha(t) = \frac{1}{\Gamma(-\alpha)} t^{-\alpha-1} = k_{-\alpha}(t), \qquad t > 0. \tag{7}$$

So, the fractal differentiation kernel is equal to the fractal integration kernel with the opposite index $-\alpha$. Here, the gamma function $\Gamma(-\alpha)$ of negative noninteger variable is defined via the above mentioned recurrence property as follows:

$$\Gamma(-\alpha) = \Gamma(n - \alpha)/(n - \alpha - 1) \cdot \ldots \cdot (-\alpha).$$

Therefore, the fractal differentiation operator can be treated as the fractal integration operator of the negative order:

$$D^\alpha = I^{-\alpha}, \tag{8}$$

which adds attractive symmetry to the fractal calculus.

Consequently, for a function $g(t)$ which vanishes on the negative half-axis, equation (6) can be rewritten in the following symbolic integral form:

$$(D^\alpha)(t) = (I^{-\alpha}g)(t) = \frac{1}{\Gamma(-\alpha)} \int_0^t (t-s)^{-\alpha-1} g(s)\,ds. \tag{9}$$

For $\alpha \geq 0$, the above improper integral diverges in view of the nonintegrable singularity of the integrand in the vicinity of the upper limit of integration. Therefore, its values have to be taken as the values of the corresponding regularized integral which can be found treating equality (6) as the convolution of distributions. In view of properties of the distributional convolution, the operation of n-tuple integration can be shifted from the first convolution factor to the second so that

$$(D^\alpha g)(t) = k_\gamma(t) * g^{(n)}(t), \tag{10}$$

with converging integral on the right-hand side. In particular, for integer $\alpha = n$, taking (6.8.9) into account, we obtain (as expected) that

$$(D^n g)(t) = k_0(t) * g^{(n)}(t) = g^{(n)}(t).$$

Example 1.   Let us now consider the special case of a function $g(t)$ which vanishes identically for $t < 0$ and is of the form

$$g(t) = \chi(t)\phi(t), \tag{11}$$

where $\phi(t)$ is an arbitrary infinitely differentiable function. Differentiating (11) n times and taking into account the multiplier probing property (1.5.3) of the Dirac delta, we obtain

$$g^{(n)}(t) = \chi(t)\phi^{(n)}(t) + \sum_{m=0}^{n-1} \delta^{(m)}(t)\phi^{(n-m-1)}(0).$$

Substituting the above formula in (10), and remembering that $\gamma = n - \alpha$ and

$$k_\gamma^{(m)}(t) = k_{\gamma-m}(t), \qquad t > 0,$$

we finally get that

$$(D^\alpha g)(t) = \frac{1}{\Gamma(n-\alpha)} \int_0^t (t-s)^{n-\alpha-1} \phi^{(n)}(s)\, ds$$

$$+ \sum_{m=0}^{n-1} \phi^{(n-m-1)}(0) k_{n-m-\alpha}(t), \qquad t > 0. \tag{12}$$

In particular,

$$D^\alpha \chi = k_{-\alpha-1}, \qquad D^\alpha k_\beta = k_{\beta-\alpha}.$$

Also

$$D^\alpha(\chi(s)s^\beta \log|s|) = \frac{\Gamma(\beta+1)}{\Gamma(\beta-\alpha+1)} \chi(s)s^{\beta-\alpha}\Big[\log|s| + C\Big],$$

where the constant (see the literature in the Bibliography for its derivation and other formulas of the fractal calculus)

$$C = (\log\Gamma)'(\beta+1) - (\log\Gamma)'(\beta-\alpha+1).$$

Furthermore, for $0 < \alpha < 1$, we have

$$(D^\alpha g)(t) = \frac{1}{\Gamma(1-\alpha)} \int_0^t \frac{\phi'(s)}{(t-s)^\alpha}\, ds + \phi(0)\frac{1}{\Gamma(1-\alpha)} t^{-\alpha}, \qquad t > 0. \tag{13}$$

Observe that, in contrast to (9), the singular integral on the right-hand sides of (12) and (13) converges absolutely, and that the regularizations (10), (12) define a new distribution—the principal of function $r_\alpha(t)\chi(t)$ (7):

$$\mathcal{PV}\,\chi(t)\frac{1}{\Gamma(-\alpha)} t^{-\alpha-1}, \qquad \alpha \geq 0. \tag{14}$$

Its convolution

$$(D^\alpha g)(t) = \frac{1}{\Gamma(-\alpha)}\, \mathcal{PV} \int_0^t (t-s)^{-\alpha-1} g(s)\, ds$$

with any function of the form (11) has a distributional interpretation via the right-hand side of formula (12). ∎

The following properties of the operation of differentiation of fractal order highlight its peculiarities.

Nonlocal character. Values $D^n g(t)$ of the usual derivatives of integer orders depend only on values of function $g(t)$ in the immediate and arbitrarily small (infinitesimal) vicinity of the point t. By contrast, the fractal (noninteger) derivatives are *nonlocal operators* since the value $D^\alpha g(t)$ depends on the values of $g(\tau)$ for all $\tau < t$. In particular, this fact explains why a function's discontinuity at a certain point ($t = 0$ for function (11)) generates slowly decaying "tails" in its fractal derivatives (the last sum on the right-hand side of formula (12)).

Causality. Fractal derivatives enjoy the causality property: If function $g(t)$ is identically equal to zero for $t < t_0$ then so does its fractal derivative.

Scale invariance. Like usual derivatives, fractal derivatives are scale invariant. This means that differentiation of the compressed ($\kappa > 1$) function $g_\kappa(t) = g(\kappa t)$ just requires multiplication of the compressed derivative by the compression factor:

$$(D^\alpha g_\kappa)(t) = \kappa^\alpha (D^\alpha g)(\kappa t). \tag{15}$$

Fourier transform. Under Fourier transformation, fractal derivatives behave just like the ordinary derivatives. In the distributional sense

$$(D^\alpha g)(t) \longmapsto (i\omega)^\alpha \tilde{g}(\omega). \tag{16}$$

This formula follows directly from (10) and from the results of Section 4.4.

Remark 1. Fractal Laplacians. The above definitions of fractal differential operators for functions of one variable can be extended to fractal partial differential operators for functions of several variables (see the literature at the end for further details). For example the *fractal Laplacian* can be defined through the Fourier transform approach as follows: For any $\phi \in \mathcal{D}(\mathbf{R}^d)$,

$$(\Delta_\alpha \phi)(x) \longmapsto -\|\omega\|^\alpha \tilde{\phi}(\omega)$$

The following integration by parts formula is then obtained via the Parseval equality

$$\int |\Delta_{\alpha/2} \phi(x)|^2 dx = -\int \phi(x) \Delta_\alpha \phi(x) dx$$

It is also clear that the fundamental solution (Green's function) G_α of the equation $-\Delta_\alpha u = \delta$ has the Fourier transform

$$\tilde{G}_\alpha(\omega) = \|\omega\|^{-\alpha}.$$

The explicit inversion depends on the dimension d of the space. If $\alpha = 2$, $d \geq 3$, or $0 < \alpha < 2$, $d \geq 2$, or $0 < \alpha < 1$, $d = 1$, then

$$G_\alpha(x) = \frac{\Gamma((d-\alpha)/2)}{\pi^d \Gamma(\alpha/2)} \|x\|^{\alpha-d}.$$

6.10 Fractal relaxation

Recently, more and more frequently, physicists find applications for the fractal calculus which permit construction of generalized mathematical models of such phenomena as relaxation, diffusion and wave propagation. In this section we will illustrate these possibilities in the example of relaxation processes.

Informally, one says that a physical system has the relaxation property if within finite time τ (called the *relaxation time*) of cessation of external perturbation the system "forgets" the perturbation and returns to its original state. Such systems are encountered in the wide spectrum of applied problems from physics and electrical engineering to biology and economics. In the simplest case, the mathematical model of a linear relaxing system is the first-order ordinary differential equation

$$x' + px = g(t), \tag{1}$$

where function $g(t)$ describes the external perturbation of the system and $p = 1/\tau > 0$ is called the *relaxation frequency*. The response $x(t)$ of the system (1) to the external perturbation which satisfies the causality principle is given by formula (2.2.4):

$$x(t) = H(t) * g(t). \tag{2}$$

The fundamental solution $H(t)$ entering (2) satisfies equation (2.2.5) which, in our particular case, is of the form

$$H' + pH = \delta(t), \qquad H(t) = 0 \text{ for } t < 0. \tag{3}$$

Its well known solution is

$$H(t) = \chi(t) \exp(-pt). \tag{4}$$

From the physicist's perspective the main feature of the relaxation model (1) is the presence of a unique (for this system) characteristic relaxation time $\tau = 1/p$. The model itself is just a special case of the whole family of justifiable models described by *fractal differential equations*

$$D^\alpha x + p^\alpha x = g(t), \qquad 0 < \alpha \le 1. \tag{5}$$

For obvious dimensional reasons, to preserve the frequency dimension of p, equation (5) contains the α's power of p. Equation (5) appears in the physical literature as an adequate model of relaxation processes in viscoelastic materials.

As was the case for equation (1), the solution of equation (5) satisfying the causality principle is given by the expression (2)

$$x(t) = H(t) * g(t),$$

where the fundamental solution $H(t)$ satisfies the fractal differential equation

$$D^\alpha H + p^\alpha H = \delta(t), \qquad H(t) = 0 \text{ for } t < 0. \tag{6}$$

We will find the needed fundamental solution by the *recursive method*. As the first step, let us apply the operator of fractal integration I^α to both sides of (6) to obtain equation

$$H + p^\alpha I^\alpha H = k_\alpha(t), \tag{7}$$

where $k_\alpha(t)$ (6.8.6) is the kernel of the fractal integration operator. We shall represent the solution of (7) as a power series

$$H(t) = \sum_{n=0}^{\infty} p^{n\alpha} H_n(t), \tag{8}$$

in parameter p, which will be substituted in (7) to compare coefficients for the same powers $p^{n\alpha}$. As a result we obtain an infinite system of recurrence relations

$$H_0(t) = k_\alpha(t), \qquad H_{n+1}(t) = -I^\alpha H_n, \qquad n = 0, 1, 2, \ldots.$$

Taking into account property (6.8.10) of the fractal integration kernels, we can solve the above system to get that

$$H_n(t) = (-1)^n k_{(n+1)\alpha}(t) = \frac{(-1)^n}{\Gamma(n\alpha + \alpha)} t^{n\alpha + \alpha - 1} \chi(t).$$

Substituting this expression into (8) we arrive at the final formula for the fundamental solution of the fractal differential equation (5):

$$H(t) = \chi(t) t^{\alpha - 1} R(p^\alpha t^\alpha, \alpha) / \Gamma(\alpha), \tag{9}$$

where

$$R(\mu, \alpha) = \Gamma(\alpha) \sum_{n=0}^{\infty} \frac{(-1)^n \mu^n}{\Gamma(n\alpha + \alpha)} \tag{10}$$

is a function of the dimensionless variable $\mu = (pt)^\alpha$. It describes the fundamental laws of *fractal relaxation*. The Heaviside function on the right-hand side of (9) ensures that the causality principle is satisfied for the fundamental solution and the power factor $t^{\alpha-1}$ secures the correct dimensionality of H.

One can easily prove that the series (10) converges absolutely for any μ. This, in particular, implies that the function (9) is indeed a solution of the equation (6). For $\alpha = 1$, series (10) becomes the Taylor series for the standard exponential relaxation

$$R(\mu, 1) = \exp(-\mu). \tag{11}$$

For other values of α, function $R(\mu, \alpha)$ is a special example of the so-called *Mittag-Leffler functions* which relatively seldom appear in applied problems. For certain particular values of parameter α, function $R(\mu, \alpha)$ can be expressed in terms of more familiar special functions.

Example 1: Relaxation of order 1/2. For $\alpha = 1/2$, series (10) can be split

$$R = R_{even} - R_{odd} \tag{12}$$

into its even and odd parts and the odd part series can be summed easily to get

$$R_{odd} = \sqrt{\pi} \mu \sum_{m=0}^{\infty} \frac{\mu^{2m}}{m!} = \sqrt{\pi} \mu \exp(\mu^2), \tag{13}$$

since $\Gamma(1/2) = \sqrt{\pi}$. The even part

$$R_{even} = 1 + \sqrt{\pi} \sum_{m=1}^{\infty} \frac{\mu^{2m}}{\Gamma(m + 1/2)}.$$

Using the recurrent property of the gamma function the denominator can be written in the form

$$\Gamma(m + 1/2) = \frac{1 \cdot 3 \cdot \ldots \cdot (2m - 1)}{2^m} \sqrt{\pi}. \tag{14}$$

Finally, changing the index of summation to $n = m - 1$, we get

$$R_{even} = 1 + 2\mu \sum_{n=0}^{\infty} \frac{\mu^{2n+1} 2^n}{1 \cdot 3 \cdot \ldots \cdot (2n + 1)}.$$

Browsing through mathematical tables [4] we run into expansion

$$\text{erf}\,(z) := \frac{2}{\sqrt{\pi}} \int_0^z e^{-t^2}\, dt = \frac{2}{\sqrt{\pi}} e^{-z^2} \sum_{n=0}^{\infty} \frac{z^{2n+1} 2^n}{1 \cdot 3 \cdot \ldots \cdot (2n+1)},$$

for the well-known *error function* so that the even part can now be written in the form

$$R_{even} = 1 + \sqrt{\pi} \mu e^{\mu^2}\, \text{erf}\,(\mu). \tag{16}$$

Substituting (16) and (13) into (12) we arrive at the final expression for the 1/2-fractal relaxation function:

$$R(\mu, 1/2) = 1 - \sqrt{\pi} \mu e^{\mu^2}\, \text{erfc}\,(\mu), \tag{17}$$

where

$$\text{erfc}\,(z) = 1 - \text{erf}\,(z) = \frac{2}{\sqrt{\pi}} \int_z^{\infty} e^{-t^2}\, dt \tag{18}$$

is the *complementary error function*. The graphs of the exponential and 1/2-fractal

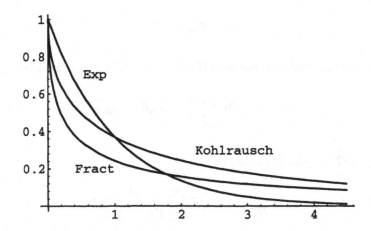

FIGURE 6.10.1
Comparison of exponential, fractal and Kohlrausch relaxation functions for $\alpha = 1/2$.

relaxation functions are compared in Fig. 6.10.1.   ∎

Remark 1. Asymptotics of fractal relaxation. For $\alpha < 1$, the fractal relaxation systems display slow, power-type long-range decay rates in response to Dirac delta

---

[4]e.g., *Handbook of Mathematical Functions*, M. Abramovitz and I.A. Stegun, Eds., formula 7.1.6

pulse perturbations—a contrast to the classical exponential relaxation function. We shall check this phenomenon for the particular case $\alpha = 1/2$ discussed in the above example. Since

$$\sqrt{\pi} z e^{z^2} \operatorname{erfc}(z) \sim 1 + \sum_{m=1}^{\infty} (-1)^m \frac{1 \cdot 3 \cdot \ldots \cdot (2m-1)}{(2z^2)^m}, \qquad (z \to \infty),$$

(see the asymptotic formula 7.1.23 in the above mentioned mathematical tables) the main asymptotics of 1/2-fractal relaxation function (18) is

$$R \sim \frac{1}{2\mu^2}, \qquad (\mu \to \infty).$$

The corresponding asymptotics of the fundamental solution $H(t)$ (9) for $\alpha = 1/2$ and large times $t \gg 1/p$ is described by the expression

$$H(t) \sim \frac{1}{2pt\sqrt{\pi t}}, \qquad (t \to \infty).$$

Remark 2. Kohlrausch relaxation function. Modelers of relaxation phenomena also often use the *Kohlrausch relaxation functions* (also called *stretched exponentials*)

$$\rho_\alpha(t) = \chi(t) \exp(-p^\alpha t^\alpha), \qquad 0 < \alpha < 1,$$

which, for large times t, decay slower than the classical exponential relaxation function but faster than the corresponding α-fractal relaxation functions (see Fig. 6.10.1). It is easy to see that function $\rho_\alpha(t)$ satisfies the differential equation

$$\rho_\alpha' + \alpha p^\alpha t^{\alpha-1} \rho_\alpha = \delta(t)$$

or equivalently (see (1.7.3))

$$\frac{d\rho_\alpha}{d(t^\alpha)} + p^\alpha \rho_\alpha = \delta(t^\alpha).$$

Thus, it can be seen as a version of the classical exponential relaxation function in α-fractally rescaled time or, to coin a new term, as an *α-fractal time relaxation function*. In this terminology, function H which was just called α-fractal relaxation function, should perhaps be called the *α-fractal frequency relaxation function* in view of its Fourier transform properties.

6.11   Exercises

Principal value

1. Find the solution $u(x, t)$ of the forced wave equation

$$\frac{\partial^2 u}{\partial t^2} = \frac{\partial^2 u}{\partial x^2} + \delta(x)\frac{df(t)}{dt},$$

where $f(t)$ is an absolutely integrable function. *Hint:* Pass to the Fourier image $\tilde{u}(k, \omega) = (1/4\pi^2)\int\int u(x, t)\exp(-ikx - i\omega t)\,dx\,dt$, of the solution, and then take the inverse Fourier transform taking into account the causality principle.

Hilbert transform

2. Find the Hilbert transform of the rectangular impulse $\varphi(t) = \chi(t) - \chi(t - \tau)$.

3. Find the Hilbert transform of the Lorentz-type function

$$\varphi(t) = \frac{1}{t^2 + \tau^2}.$$

4. What is the Hilbert transform of $\varphi(t) = e^{i\Omega t}$?

5. Find the Hilbert transform of $\varphi(t) = \sin(\nu t)/\nu t$.

6. Prove that if $\psi(t)$ is the Hilbert transform of $\varphi(t)$ (in brief, $\varphi(t) \overset{H}{\longmapsto} \psi(t)$), then:

(a) $\varphi(t + \tau) \overset{H}{\longmapsto} \psi(t + \tau)$;

(b) $\varphi(at) \overset{H}{\longmapsto} \psi(at),\ a > 0$;

(c) $\varphi(-t) \overset{H}{\longmapsto} -\psi(-t)$, and, in particular, if φ is an even function then ψ is odd and *vice versa;*

(d) If $\varphi(t)$ is an n-times differentiable function, then

$$\varphi^{(m)}(t) \overset{H}{\longmapsto} \psi^{(m)}(t), \qquad m = 1, 2, \dots, n;$$

(e) "Energies" (or, in other words, $\mathbf{L}^2$ norms) of $\varphi(t)$ and $\psi(t)$ are identical:

$$\int |\varphi(t)|^2\,dt = \int |\psi(t)|^2\,dt;$$

(f) The convolution $\varphi(t) * \varphi(t) \overset{H}{\longmapsto} -\psi(t) * \psi(t)$;

(g) If $\int \varphi(t)\,dt = 2\pi\tilde{\varphi}(0) \neq 0$, then $\psi(t) \sim -2\tilde{\varphi}(0)/t,\ (t \to \infty)$;

(h) If function $\varphi(t)$ has a jump $\lfloor\varphi\rfloor = \varphi(\tau + 0) - \varphi(\tau - 0)$ at the point $t = \tau$, then $\psi(t)$ has at that point a logarithmic singularity:

$$\psi(t) \sim -\frac{\lfloor\varphi\rfloor}{\pi}\ln|t - \tau|, \qquad (t \to \tau);$$

(i) If $\varphi(t)$ is even, smooth and bounded, then

$$\psi(t) \sim -ct, \qquad (t \to 0)$$

where

$$c = \frac{2}{\pi} \int_0^\infty \frac{\varphi(0) - \varphi(t)}{t^2} \, dt.$$

Analytic signals

7. Find the dependence of an analytic signal $\zeta(t) = \xi(t) + i\eta(t)$ on the real time variable if its real part

$$\xi(t) = \frac{\cos \Omega t}{t^2 + \tau^2}.$$

8. Let $\xi(t) = (\sin vt / vt) \cos \Omega t$. Find $\zeta(t)$.

9. Let

$$\xi(t) = f(t) \cos \Omega t + g(t) \sin \Omega t, \qquad \Omega > 0, \tag{1}$$

where functions $f(t)$ and $g(t)$ have finite support Fourier images, identically equal to zero for $|\omega| \geq \Omega$. Then, the corresponding analytic signal is equal to

$$\zeta(t) = [f(t) - ig(t)]e^{i\Omega t}. \tag{2}$$

Prove it.

Remark 1. Recall that the imaginary part of the analytic signal (6.6.2) is equal to minus the Hilbert transform (6.5.1) of $\xi(t)$. This implies the following corollary to the above result: If $\varphi = f \cos \Omega t + g \sin \Omega t$ then its Hilbert transform is $\psi = g \cos \Omega t - f \sin \Omega t$.

Remark 2. Signals with finite-support Fourier image seldom appear in electrical engineering applications. However, for narrow-band signals the replacement of the actual analytic signal by the expression (6.9.3) gives a rather good approximation. For example, the signal from Exercise 6 has an unbounded Fourier image but is narrow-band if $\Omega \tau \gg 1$. In this case, it is easy to see that the approximate expression

$$\zeta_a(t) = \frac{1}{t^2 + \tau^2} e^{i\Omega t}$$

is very close to $\zeta(t)$.

10. Let $\xi(t) = t \sin \Omega t$, $\Omega > 0$. Find $\zeta(t)$.

11. Use the concept of analytic signal to find the instantaneous amplitude, phase and frequency of

$$\xi(t) = A_1 \cos \Omega_1 t + A_2 \sin \Omega_2 t, , \qquad \Omega_1, \Omega_2 > 0.$$

12. Let $\xi(t) = \chi(t) \sin \Omega t$. Find the imaginary part $\eta(t)$ of the corresponding analytic signal.

13. Let $\xi(t) = \chi(t) t^{\alpha-1}$, $0 < \alpha < 1$. Find $\zeta(t)$.

14. Find the analytic signal $\zeta_e(t)$ which corresponds to the even function $\xi_e(t) = |t|^{\alpha-1}$.

15. Let $\xi(t) = \delta'(t)$. Find the imaginary part $\eta(t)$ of the corresponding analytic signal.

16. Let $\hat{H}$ be the Hilbert transform operator

$$\hat{H}\phi(t) = \frac{1}{\pi} \mathcal{PV} \int \frac{\phi(s)}{s - t} \, ds.$$

Find the set of functions $\phi(t)$ on which $\hat{H}$ acts as just the shift operator, i.e., $\hat{H}\phi(t) = \phi(t+T)$, for a certain T.

Fractal calculus

17. Extend the Cauchy formula to the case of the n-tuple integral

$$x(t) = \int_{-\infty}^{t} dt_1 a_1(t_1) \int_{-\infty}^{t_1} dt_2 a_2(t_2) \ldots \int_{-\infty}^{t_{n-1}} dt_n a_n(t_n) g(t_n),$$

where $a_1(t), a_2(t), \ldots, a_n(t)$, are known functions such that the above n-tuple integral converges absolutely for any integrable function $g(t)$.

18. Let $A \subset \mathbf{R}^n$. Express the multiple integral

$$I = \int \overset{n}{\ldots} \int_A a(x_1, x_2, \ldots, x_n) g(b(x_1, x_2, \ldots, x_n)) \, dx_1 \ldots dx_n$$

via a single integral of function $g(u)$.

Chapter 7

Uncertainty Principle and Wavelet Transforms

The method of *wavelet transforms*, which provides a decomposition of functions in terms of a fixed family of functions of constant shape but varying scales and locations, recently acquired broad significance in the analysis of signals and of experimental data from various physical phenomena. It is clear that the potential of this method has not yet been fully tapped. Nevertheless, its value for the whole spectrum of problems in many areas of science and engineering, including the study of electromagnetic and turbulent hydrodynamic fields, image reconstruction algorithms, prediction of earthquakes and tsunami waves, and statistical analysis of economic data, is by now quite obvious.

Although the systematic ideas of wavelet transforms have been developed only since the early 80s, to get the proper intuitions about sources of their effectiveness it is necessary to become familiar with a few more traditional ideas, tools and methods. One of those is the celebrated *uncertainty principle* for the Fourier transforms which will be given special attention in this chapter. A close relative of the wavelet transform—the *windowed Fourier transform*, will also be studied in this context. We begin though with a brief sketch of the notion of the *functional Hilbert space* which provide a convenient framework for our analysis.

7.1 Functional Hilbert spaces

The extension of classic 3-D Euclidean geometry concepts such as the space, vector, composition, multiplication of vectors by scalars, inner product of vectors, angle, orthogonality and parallelness, to a broad class of mathematical objects, was one of the success stories of twentieth century mathematics. As a result, a multitude of *functional spaces* were introduced, studied and added to our permanent arsenal. The *linear topological spaces of distributions* briefly described in Section 1.9 are one such example. In this section, we discuss another class of functional spaces called *Hilbert spaces*.

© Springer Nature Switzerland AG 2018
A. I. Saichev and W. Woyczynski, *Distributions in the Physical
and Engineering Sciences, Volume 1*, Applied and Numerical
Harmonic Analysis, https://doi.org/10.1007/978-3-319-97958-8_7

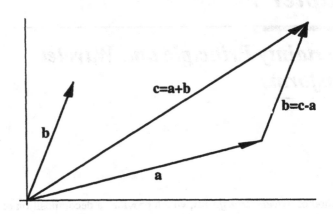

FIGURE 7.1.1
Composition of 2-D vectors.

At first, recall the basic notions of the usual 3-D geometry where each point of the space is identified, in a fixed Cartesian coordinate system, with a vector a anchored at the origin and with a tip at a given point. Each such vector is uniquely described by its coordinates (a_1, a_2, a_3)—an ordered triple of real numbers. If $b = (b_1, b_2, b_3)$ is another 3-D vector then the *inner* or *scalar product* of these two vectors is defined by the equality

$$(a, b) := a_1 b_1 + a_2 b_2 + a_3 b_3. \tag{1}$$

Since, alternatively,

$$(a, b) = \|a\| \|b\| \cos \alpha,$$

where, by Pythagoras' theorem,

$$\|a\|^2 = (a, a) \tag{2}$$

is the square of the vector's *norm (length, magnitude)* and α is the angle between the two vectors, the inner product of two vectors clearly depends on their mutual orientation. In particular,

$$a \perp b \quad \textit{if and only if} \quad (a, b) = 0.$$

The geometric composition of vector a with vector b (see, Fig. 7.1.1) corresponds to the algebraic operation

$$c = a + b \tag{3}$$

of vector addition. Vector $c = (c_1, c_2, c_3)$ is called the sum of vectors a and b if its coordinates are sums of corresponding coordinates of the summand vectors: $c_n = a_n + b_n$, $n = 1, 2, 3$. Such addition operation is obviously *commutative*, that is

$$a + b = b + a. \tag{4}$$

Besides addition, one introduces the operation of multiplication of a vector by a scalar:

$$d = (d_1, d_2, d_3) := rd = (rd_1, rd_2, rd_3), \tag{5}$$

which geometrically represents vector contraction for $|r| < 1$, vector dilation if $|r| > 1$ and vector reflection in case of $r = -1$.

The following three properties of the inner product and the norm are fundamental for the geometric properties of the Euclidean space:

(i) The norm is *homogeneous*, that is, for any scalar r and vector a,

$$\|ra\| = |r|\|a\|. \tag{6}$$

(ii) The norm and the addition operation are related via the *triangle inequality*, that is, for any two vectors a and b,

$$\|a + b\| \le \|a\| + \|b\|. \tag{7}$$

(iii) The norms and the inner product are related by the *Schwartz inequality*

$$|(a, b)| \le \|a\|\|b\|. \tag{8}$$

The first step in the generalization of the geometry of 3-D Euclidean spaces to abstract functional Hilbert spaces are two observations:

(a) The concept of the inner product (and related geometry of the space) can be immediately extended to d-dimensional Euclidean spaces by defining for any $a = (a_1, \dots, a_d)$ and $b = (b_1, \dots, b_d)$

$$(a, b) = a_1 b_1 + \dots + a_d b_d.$$

(b) If one wants to operate with vectors with complex (rather than real) coordinates and preserve the positivity of the norm the only adjustment in the definition of the inner product is as follows:

$$(a, b) = a_1^* b_1 + \dots + a_d^* b_d, \tag{9}$$

where, as usual, the asterisk denotes the complex conjugation. Then, the square

$$\|a\|^2 = \sum_{n=1}^{d} a_n^* a_n = \sum_{n=1}^{d} |a_n|^2 \geq 0 \qquad (10)$$

defines a positive norm

$$\|a\| = \sqrt{(a, a)}. \qquad (11)$$

In the complex case, the symmetry of the inner product is replaced by the *Hermitian property*

$$(a, b) = (b, a)^*. \qquad (12)$$

The linearity with respect to the second variable in the inner product is preserved, though, as for any complex number z,

$$(a, zb) = z(a, b). \qquad (13)$$

Example 1. Inner product space of polynomials. Let us consider the set l_N^2 of all polynomials

$$P_{N-1}(x) = a_0 + a_1 x + \ldots + a_{N-1} x^{N-1}$$

of degree at most $N - 1$ with complex coefficients. Each of these polynomials is uniquely determined by its coefficients for different powers of x, that is by the (complex) vector $a = (a_0, \ldots, a_{N-1})$. The sum of such polynomials is again a polynomial of the above type and the same is true for a product of a complex number and such polynomial. Moreover, the summation and multiplication by scalars in the family l_N^2 of such polynomials corresponds to the analogous operations on the coefficient vectors and identifies the vector space structure of the family l_N^2 of all polynomials of degree at most $N - 1$ as that of an N-dimensional complex vector space. The natural inner product leads to the notion of the "distance" $l = \|a - b\|$ between polynomials with coefficient vectors a and b.   ∎

Example 2. Inner product space of complex exponentials. Consider the set of all infinite sums of complex exponentials

$$E(x) = \frac{1}{\sqrt{2\pi}} \sum_{n=-\infty}^{\infty} a_n e^{inx}, \qquad x \in [-\pi, \pi] \qquad (14)$$

with complex coefficients. This set forms a natural vector space under termwise addition and multiplication by scalars and can be identified with the (infinite-dimensional) vector space of coefficient vectors (sequences) $a = (\ldots, a_{-1}, a_0, a_1,$

$a_2, \ldots$). However, if we wanted to introduce the inner product in such space associated with the norm

$$\|E\| = \sqrt{\sum_{n=-\infty}^{\infty} |a_n|^2}$$

we immediately run into the question of convergence of the above series and to proceed we have to assume additionally that

$$\sum_{n=-\infty}^{\infty} |a_n|^2 < \infty. \tag{15}$$

This is the first fundamental difference with the finite dimensional spaces. The subset l_∞^2 of sums E satisfying condition (15) remains closed under operations of termwise addition and multiplication by scalars since $\|zE\| = |z|\|E\|$ and

$$\|E + F\| \le \|E\| + \|F\|,$$

for any sums E, F and scalar z. ∎

Attempts to generalize the above examples immediately lead us to the idea of the *functional Hilbert space* $L^2(\mathbf{R})$ of complex-valued functions $f(x)$ defined on the entire real axis $\mathbf{R}$ and such that

$$\int |f(x)|^2 dx < \infty. \tag{16}$$

Now, we can introduce the inner product in $L^2(\mathbf{R})$ via the formula

$$(f, g) = \int f^*(x)g(x)\, dx \tag{17}$$

and the related norm

$$\|f\| = \sqrt{(f, f)} = \left(\int |f(x)|^2 dx \right)^{1/2}. \tag{18}$$

In view of the classical *integral Schwartz inequality*

$$|(f, g)| = \left| \int f^*(x)g(x)\, dx \right|$$

$$\leq \left(\int |f(x)|^2 dx \right)^{1/2} \left(\int |g(x)|^2 dx \right)^{1/2} = \|f\| \|g\|, \qquad (19)$$

condition (16) assures that the inner product (17) is well defined (i.e., that f^*g is integrable). The Schwartz inequality also immediately leads to (see Exercises) the triangle inequality

$$\|f + g\| \leq \|f\| + \|g\|, \qquad f, g \in L^2(\mathbf{R}), \qquad (20)$$

which, incidentally implies that the Hilbert space $L^2(\mathbf{R})$ is closed under the usual pointwise addition of functions. It is also closed under multiplication by scalars since, by (18),

$$\|zf\|^2 = |z|^2 \|f\|^2.$$

The above inner product (17) is Hermitian, that is

$$(f, g) = (g, f)^*$$

and homogeneous in the second variable, as

$$(f, zg) = z(f, g)$$

for any complex constant z and $f, g \in L^2(\mathbf{R})$.

Of course, the norm in $L^2(\mathbf{R})$ is nonnegative, that is for any $f \in L^2(\mathbf{R})$

$$\|f\| \geq 0$$

and if $f(x) \equiv 0$ then $\|f\| = 0$.

A mathematical aside: functions with vanishing norm and the Lebesgue integral.[1] It is quite clear that beside the function equal to zero identically there are other functions $f(x) \neq 0$ such that $\|f\| = 0$. In other words, $\|f\| = 0$ does not necessarily imply $f(x) \equiv 0$. For example, any function different from zero at a finite or countable number of points would have norm zero. This creates somewhat unpleasant situation of having two different functions $f(x) \neq g(x)$ for which the distance $\|f - g\| = 0$. The satisfactory resolution of this problem is not possible within the framework of the Riemann integral which we implicitly used throughout the preceding chapters (and which is sufficient for our other purposes). It requires introduction of the more general Lebesgue integral (hence letter L in the notation of the functional Hilbert space) which permits integration of a much broader class of functions than the Riemann integral (see the bibliographical notes at the end). For example the Dirichlet function

$$D(x) = \begin{cases} 0, & \text{if } x \text{ is irrational;} \\ 1, & \text{if } x \text{ is rational;} \end{cases}$$

---

[1]This material may be skipped by the first time reader.

is not integrable on $[0, 1]$ in the Riemann sense since the upper approximating sums (always equal to 1) and the lower approximating sums (always equal to 0) do not converge to the same number. However, it is integrable in the Lebesgue sense and its Lebesgue integral is equal to 0.

Interpreting the integrals in (16-18) as Lebesgue integrals, it is customary to formally define the functional Hilbert space $L^2(\mathbf{R})$ not just as the space of square-integrable functions but as the space of *equivalence classes* of square-integrable functions where two functions f, g are understood as equivalent (written $f \equiv g$) if and only if $\|f - g\| = 0$. It is easy to see that, if we define the Lebesgue measure $|A|$ of the set $A \subset \mathbf{R}$ as the Lebesgue integral of its indicator function $I_A(x)$ (equal to 1 on A and to 0 off A), then two functions belong to the same equivalence class if and only if they differ on a set of Lebesgue measure zero (or in measure theory jargon, are equal almost everywhere).

Now, the norms and inner products are the same for all functions in a given equivalence class, so practically one always does computations on concrete functions, but if elements f of the Hilbert space $L^2(\mathbf{R})$ are meant as equivalence classes of functions equal almost everywhere then the desired strict positivity of the norm is achieved, that is

$$\|f\| = 0 \quad \text{if, and only if} \quad f \equiv 0.$$

The norm $\|f\|$ in the functional Hilbert space $L^2(\mathbf{R})$, and the related distance $\|f - g\|$ of (equivalence classes of) functions f, g permit us to introduce the notion of the limit of functions in $L^2(\mathbf{R})$. Namely, we say that

$$f_n \to f \quad \text{in } L^2(\mathbf{R}) \tag{21}$$

if

$$\lim_{n \to \infty} \|f_n - f\| = 0.$$

This notion permits us to study approximation problems in the functional Hilbert space.

Having introduced in the functional Hilbert space the algebraic structure (addition and multiplication by scalars), the compatible inner product structure and the related metric (norm) structure one could sensibly introduce in $L^2(\mathbf{R})$ and study the geometric concepts such as the angles between functions, their orthogonality, etc. This very fruitful approach is the essence of the branch of mathematics called *functional analysis*.

Remark 2. Completeness of the functional Hilbert space. The finite dimensional inner product spaces introduced at the beginning of this sections enjoyed the important property of being complete, that is, the Cauchy criterion of convergence remained valid for them. The same criterion happens to hold true for the space $L^2(\mathbf{R})$. In other words, the functional Hilbert space is a *complete inner product space*, that is for any sequence of functions $f_n \in L^2(\mathbf{R})$, $n = 1, 2, \ldots$, such that

$$\lim_{n,m \to \infty} \|f_n - f_m\| = 0$$

there exists a function $f \in L^2(\mathbf{R})$ such that

$$\lim_{n \to \infty} \| f_n - f \| = 0.$$

Remark 3. Other Hilbert and normed functional spaces. The above discussion applies, with obvious adjustments, to Hilbert spaces of functions over other subsets of the real axis such as $L^2([a, b])$, $L^2([0, \infty))$, etc., as well as to their natural analogues $L^2(\mathbf{R}^d)$ defined for functions of several variables.

On the other hand, it is often necessary to consider functional spaces where the introduction of the inner product structure is impossible and one has to be satisfied only with the norm structure. Examples of non-Hilbertian functional normed spaces, such as $L^p(\mathbf{R})$, $1 \le p < \infty$, $p \ne 2$, which consists of all (equivalence classes of) functions for which

$$\| f \|_p = \left(\int |f(x)|^p \, dx \right)^{1/p} < \infty,$$

have been mentioned before in Chapter 6 (also, see the Bibliographical Notes). In particular, space $L^1 = L^1(\mathbf{R})$ consists of all absolutely integrable functions on the real axis.

7.2   Time-frequency localization and the uncertainty principle

Consider a (perhaps complex-valued) signal $f(t)$ such that

$$\int |f(t)|^2 dt = 1. \tag{1}$$

The quantity $|f(t)|^2$ can be thought of as the signal's "mass" density and describes its distribution in time. If the signal $f(t)$ is square integrable but (1) is not satisfied then one can always normalize it by considering $f(t)/(\int |f(t)|^2 dt)^{1/2}$. In this context, the quantity

$$\int t |f(t)|^2 dt$$

can be interpreted as the location in time of the signal's "center of gravity," or its mean location. For the purposes of this section, and without loss of generality, we will assume that its mean location is at 0 or, in other words, that $\int t|f(t)|^2 dt = 0$. In this case, the quantity

$$\sigma^2[f] = \int t^2 |f(t)|^2 dt \tag{2}$$

measures the average square deviation from the mean time location, or the degree of *localization* of the signal around its mean in the time domain.

On the other hand, the Fourier transform

$$\tilde{f}(\omega) = \frac{1}{2\pi} \int f(t) e^{-i\omega t} dt$$

displays no direct information about the signal's time localization, but has explicit information about its frequency localization. The square of its modulus $|\tilde{f}(\omega)|^2$ is the frequency domain counterpart of the time density $|f(t)|^2$. Note that, by Parseval's formula (3.2.11),

$$\int |\tilde{f}(\omega)|^2 d\omega = \frac{1}{2\pi},$$

so that $2\pi |\tilde{f}(\omega)|^2$ can be viewed as the signal's normalized density in the frequency domain. Assume (again, without loss of generality) that the mean frequency

$$2\pi \int \omega |\tilde{f}(\omega)|^2 d\omega = 0.$$

Then the quantity

$$\sigma^2[\tilde{f}] = 2\pi \int \omega^2 |\tilde{f}(\omega)|^2 d\omega \tag{3}$$

measures the mean square deviation from the mean frequency location, or the degree of localization of the signal in the frequency domain.

The *uncertainty principle* asserts that there exists a lower bound on the simultaneous localization of the signal in time and frequency domains. More precisely, it states that

$$\sigma^2[f]\sigma^2[\tilde{f}] \geq 1/4, \tag{4}$$

whenever the variances $\sigma^2[f]$ and $\sigma^2[\tilde{f}]$ are well defined. Note the universal constant $1/4$.

To see why the uncertainty principle holds true, consider the integral

$$I(x) = \int |xtf(t) + f'(t)|^2 dt \geq 0 \tag{5}$$

where x is a real parameter. Then, since

$$|xtf(t) + f'(t)|^2 = (xtf + f')(xtf^* + (f')^*)$$

we get that

$$I(x) = x^2 \int t^2 |f|^2 dt + x \int t(f(f')^* + f'f^*) dt + \int |f'|^2 dt. \qquad (6)$$

The first integral in (6) is just $\sigma^2[f]$ (by definition (2)). The second integral is equal to

$$\int t(ff^*)' dt = t|f(t)|^2 \Big|_{-\infty}^{\infty} - \int |f|^2 dt = -1,$$

since $t|f|^2$ decays to zero at $\pm\infty$ in view of the assumption $\sigma^2[f] < \infty$. Finally, the third integral is equal to

$$2\pi \int \omega^2 |\tilde{f}(\omega)|^2 d\omega = \sigma^2[\tilde{f}]$$

because of Parseval's formula (3.2.11) and the fact that the Fourier transform of f' is $i\omega \tilde{f}(\omega)$. As a result, the integral

$$I(x) = x^2 \sigma^2(f) - x + \sigma^2(\tilde{f}). \qquad (7)$$

This is a quadratic polynomial in variable x and, in view of (5), it is nonnegative for all values of x. As such, it has a nonpositive discriminant

$$1 - 4\sigma^2(f)\sigma^2(\tilde{f}) \leq 0,$$

which immediately yields the uncertainty principle (4).

Remark 1. The Heisenberg uncertainty principle in quantum mechanics. The (3-D version of the) above uncertainty principle concerning time-frequency localization has a celebrated interpretation in quantum mechanics, where the principle asserts that the position and the momentum of a particle cannot be simultaneously measured with arbitrary accuracy. Indeed, in quantum mechanics the particle is represented by a complex wave function $f(x)$, where $|f(x)|^2$ is the probability density of its position in space. The observables are represented by operators A on wave functions; the mean value of the observable is

$$\int (Af)(x) f^*(x) \, dx.$$

The *position observable* is represented by a multiplication by variable (vector) x and the *momentum observable* is represented by the operation of differentiation $\partial/\partial x$. However, via the Fourier transform, the latter also becomes an operation of multiplication but by an independent variable (vector) ω in the frequency domain. Thus the uncertainty principle (4) gives the universal lower bound for the product of variances of the probability distributions of the position and of the momentum. In the three-dimensional space, and in the physical units, the lower bound 1/4 in (4) has to be replaced by a different mathematical constant multiplied by a universal physical constant called the Planck constant. The employed above probabilistic concepts of means and variances will be further studied in Chapter 13.

Remark 2. One can check that the equality in the uncertainty principle (4) obtains only for the Gaussian function $f(t) = \pi^{-1/4} \exp(-t^2/2)$. Thus the optimal simultaneous time and frequency localization is attained for a Gaussian-shaped signal.

7.3 Windowed Fourier transform

7.3.1. Forward windowed Fourier transform. The uncertainty principle discussed in Section 7.2 is a basic law of mathematics and it is impossible to fool nature by measuring the frequency of the incoming signal with an arbitrary precision in a finite time interval. Moreover, for most of the signals we have to deal with in practical problems, such as speech, musical sounds, radar signals, the situation is often much worse than the basic uncertainty inequality permits and

$$\sigma(f)\sigma(\tilde{f}) \gg 1. \tag{1}$$

Nevertheless, it is often possible to process these signals in such a way that, without violation of the uncertainty principle, one can obtain information about the signal's "current" frequency and its time evolution. These various practical signal processing methods are adapted to different kinds of signals and pursue different goals. In this section we will take a look at one of these methods called the *windowed Fourier transform* which is closest perhaps to the spirit of the usual Fourier transform.

In what follows, to better grasp the mechanisms behind the windowed Fourier transform, it will be instructive to test them on a sample signal that we will call the *simplest tune.* Mathematically, it is described by the real part of the complex function

$$f(t) = \exp(i\,\Phi(t)), \tag{2}$$

where

$$\Phi(t) = \omega_0 t + \frac{\Omega}{\nu}\sin(\nu t) \tag{3}$$

is the signal phase. The simplest tune is plotted on Fig. 7.3.1.

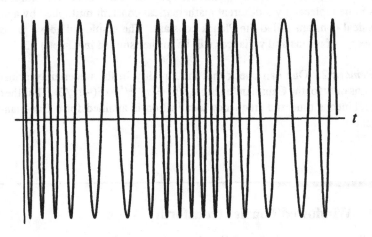

FIGURE 7.3.1
Plot of the simplest tune in case of $\omega_0 = 10\nu$ and $\beta = \Omega/\nu = 5$.

It is customary to say in the theoretical physics context that the simplest tune has the *instantaneous frequency* (admittedly, an oxymoron)

$$\omega_{inst}(t) = \frac{d\Phi(t)}{dt} = \omega_0 + \Omega\cos(\nu t), \tag{4}$$

which oscillates with period $T = 2\pi/\nu$ between its high value $\omega_0 + \Omega$ and low value $\omega_0 - \Omega$. By contrast with a theoretician, an experimenter has to deal not with mathematical formulas but with real signals and his job is to come up with a signal processing method that will discover the existence of frequency oscillations in the simplest tune.

The mathematical tool that is helpful in this situation is called the *windowed Fourier transform* which is just the usual Fourier transform

$$\tilde{f}(\omega, \tau) = \frac{1}{2\pi}\int f(t)g(t-\tau)e^{-i\omega t}\,dt \tag{5}$$

of the *time-windowed signal* $f(t)g(t-\tau)$, where $g(t)$ is the *windowing function* that usually is chosen to have value equal to 1 in a vicinity of the origin $t = 0$ (say, inside an interval of length λ), and that either vanishes or has values very close to 0 outside this neighborhood. This windowing function property will assure effective time-localization.

Usually, one defines the windowing function $g(t)$ via a *windowing shape function* $g_0(x)$ of a dimensionless variable x and the formula

$$g(t) = g_0(t/\lambda),$$ (6)

where λ is a scaling parameter. Some typical examples of normalized ($\|g_0\| = 1$) windowing shape functions are (see Fig. 7.3.2):

(a) *Finite memory window*

$$g_0(x) = \chi(x+1) - \chi(x);$$ (7a)

(b) *Relaxation window*

$$g_0(x) = 2\chi(-x)\exp(2x);$$ (7b)

(c) *Gaussian window*

$$g_0(x) = \pi^{-1/4}\exp(-x^2/2).$$ (7c)

Shift τ centers the window at different locations on the time-axis t. If $f(t)$ is a time-dependent signal and processing is performed in the *real-time* then τ is just the current time of the experiment and the time-window $g(t)$ has to satisfy the causality principle, i.e., $g(t) \equiv 0$ for $t > 0$. So, in this case, the finite memory and relaxation windows are appropriate but the Gaussian window is not. If the whole signal is recorded before processing, or the variable t has other interpretation (e.g., space, or angle variable) then the experimenter has more freedom in selecting the windowing shape function, and very often the Gaussian window is a good candidate.

7.3.2. Frequency localization. The time-window $g(t)$ was designed to separate well the time-localized pieces of duration λ of the incoming signal $f(t)$. Luckily, it turns out that the Fourier image of the time-window $g(t)$ can help in frequency localization. To see how this happens let us express the original signal $f(t)$ through its Fourier transform:

$$f(t) = \int \tilde{f}(\omega')e^{i\omega't}\,d\omega',$$ (8)

and substitute it into the right-hand side of (5). Note that, in the case of the simplest tune (2), $\tilde{f}(\omega)$ exists only in the distributional sense. The change of the integration

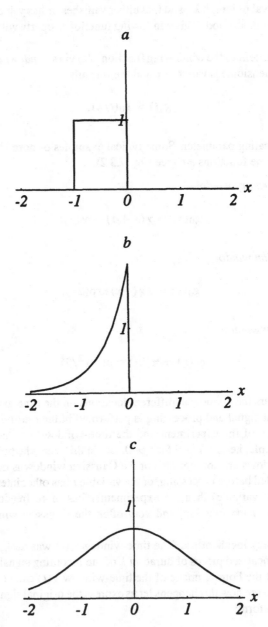

FIGURE 7.3.2
Examples of windowing shape functions. (a) Finite memory window; (b) Relaxation window; (c) Gaussian window.

order gives that

$$\tilde{f}(\omega, \tau) = e^{-i\omega\tau} \int \tilde{f}(\omega')\tilde{g}(\omega - \omega')e^{i\omega'\tau}\, d\omega'. \tag{9}$$

Remarkably, except for the nonessential factor in front of the integral, this expression looks like the symmetric counterpart of (5) in the frequency domain. Now, the role of the signal is played by its Fourier image $\tilde{f}(\omega)$ and the time-window has been replaced by the *frequency-window* $\tilde{g}(\omega)$.

The uncertainty principle (7.2.4) tells us that if the effective duration of the time-window is λ then one can expect the effective width of the frequency-window to be of order at least $1/\lambda$. In terms of the dimensionless window shapes $g_0(x)$ and $\tilde{g}_0(y)$ where, similarly to (6),

$$\tilde{g}(\omega) = \lambda\tilde{g}_0(\lambda\omega), \tag{10}$$

both $g_0(x)$ and $\tilde{g}_0(y)$ have to have a similar effective widths ~ 1.

However, the actual situation is a bit more complicated than the above juggling of the uncertainty principle may indicate. When the engineers talk about effectively localized frequency-window, they think about the compact support of the frequency-windowing shape function $\tilde{g}_0(y)$ or at least about its rapid decay outside a finite frequency band. However, we know from the properties of the Fourier transform that it is impossible for both the function and its Fourier transform to have compact supports. Furthermore, the frequency windowing shape function will decay rapidly for $|y| > 1$ only if the time windowing shape function is smooth. This fact eliminates time windowing shape functions (7a) and (7b), which have good time-localization properties, as good candidates for good frequency localization by their Fourier transforms. Abrupt truncations in them introduce discontinuities of the first kind which slow the decay of their Fourier transforms. For example, the modulus of Fourier image of the relaxation window (7b)

$$|\tilde{g}_0(y)| = \frac{1}{\pi\sqrt{4 + y^2}} \tag{11}$$

decays to zero slowly as $|\tilde{g}_0(y)| \sim 1/(\pi|y|)$, $(y \to \infty)$.

So, to achieve better frequency localization one has to take smoother windowing shape functions like, for example, functions described in Section 4.3.

Example 1. Compact time-window, power-law decay of the frequency window. Take the windowing shape function

$$g_0(x) = \frac{8}{3}\Big[\chi(x + 2) - \chi(x)\Big]\sin^2\left(\frac{\pi x}{2}\right). \tag{12}$$

corresponding to function (4.3.19), normalized appropriately and shifted to satisfy the causality principle. Its frequency counterpart

$$\tilde{g}_0(y) = e^{iy} \frac{4}{3} \pi \frac{\sin y}{y(\pi^2 - y^2)} \tag{13}$$

decays as $1/|y|^3$, faster than (11), which produces tolerable frequency localization while preserving perfect time localization. The power law of the frequency windowing shape (13) decay was caused by hidden discontinuities (in the second derivative) of the time windowing shape (12). ∎

Example 2. Gaussian time and frequency windows; Gabor transform. Since the Fourier image of a Gaussian time windowing shape gives a Gaussian frequency windowing shape, in this case we have excellent localization in both time and frequency domains. Indeed, if $g_0(x)$ is given by (7c) then

$$\tilde{g}_0(y) = \frac{1}{\sqrt{2\pi}} g_0(y) = \frac{\pi^{-1/4}}{\sqrt{2\pi}} \exp(-y^2/2). \tag{14}$$

The extra factor $1/\sqrt{2\pi}$ is the result of our asymmetric, but physical definition (3.1.1) of the Fourier transform. In mathematics, for esthetic reasons, one often prefers a *symmetric* definition of the Fourier transform and its inverse:

$$\tilde{f}(y) = \frac{1}{\sqrt{2\pi}} \int f(x) e^{-ixy} \, dx, \qquad f(x) = \frac{1}{\sqrt{2\pi}} \int \tilde{f}(x) e^{ixy} \, dy. \tag{15}$$

In this case, $\tilde{g}_0(y) \equiv g_0(y)$. The windowed Fourier transform

$$G(\omega, \tau) = \pi^{-1/4} \int f(t) e^{i\omega t - (t-\tau)^2/2} \, dt \tag{16}$$

based on the Gaussian window is called the *Gabor transform* in honor of the physicist who introduced it for studying quantum-mechanical problems. ∎

7.3.3. Energy density in the time-frequency domain. In applied problems the quantity of interest is usually not the complex function $\tilde{f}(\omega, \tau)$ itself but its squared modulus

$$E(\omega, \tau) = 2\pi |\tilde{f}(\omega, \tau)|^2. \tag{17}$$

It follows from (5) that

$$E(\omega, \tau) = \frac{1}{2\pi} \int dt \int d\theta \, e^{-i\omega\theta} f^*(t) g^*(t - \tau) f(t + \theta) g(t + \theta - \tau). \tag{18}$$

For the sake of symmetry between the time and frequency domains we permit both functions $f(t)$ and $g(t)$ to be complex-valued. Integrating the above equality over all ω, we get

$$\int E(\omega, \tau)\, d\omega = \int |f(t)|^2 |g(t - \tau)|^2\, dt. \tag{19}$$

Observe that the three unwieldy integrals on the right-hand side were reduced to an elegant single integral by noticing first that

$$\frac{1}{2\pi} \int e^{-i\omega\theta}\, d\omega = \delta(\theta), \tag{20}$$

and then using the probing property of the Dirac delta to get rid of another integral. After integration of (19) over all τ we have

$$\int d\omega \int d\tau\, E(\omega, \tau) = \|g\|^2 \|f\|^2, \tag{21}$$

where the norms on the right-hand side are the Hilbertian L^2-norms introduced in (7.1.18).

Since we assumed at the beginning of this section that the time windowing shape function g_0 is normalized, i.e., $\|g_0\| = 1$, the squared L^2-norm of the time windowing function itself

$$\|g\|^2 = \lambda, \tag{22}$$

i.e., it is equal to the duration of the time-window. Remembering that the squared norm

$$\|f\|^2 = \int |f(t)|^2\, dt$$

represents the energy E_f of the original signal $f(t)$, formula (21) implies the following energetic relation:

$$E_f = \frac{1}{\lambda} \int d\omega \int d\tau\, E(\omega, \tau).$$

Thus the function $E(\omega, \tau)/\lambda$ has a physical interpretation as the *joint frequency-time density* of the signal's energy.

7.3.4. Mean frequency and standard deviation of the windowed Fourier transform.
Recall that the windowed Fourier transform was introduced earlier in this section to track (with a precision determined by duration λ of the time window) the time τ evolution of the "instantaneous frequency" $\omega_{inst}(\tau)$ of signal $f(t)$. The latter was sufficiently clearly defined for the simplest tune signal, but for the general

situation we need a more rigorous definition. A good, and analytically convenient definition is the mean value

$$\bar{\omega}(\tau) = c(\tau) \int \omega E(\omega, \tau)\, d\omega, \tag{23}$$

where

$$c(\tau) = 1/ \int |f(t)|^2 |g(t-\tau)|^2\, dt \tag{24}$$

is the normalizing constant, although some physicists would perhaps prefer to use, as the definition of the instantaneous frequency at time τ, the value $\omega_{max} = \omega_{max}(\tau)$ which maximizes the joint energy density $E(\omega, \tau)$.

However, before advising the reader to go ahead and apply the above definition in research problems, let us step back and see what happens in a typical situation where $f(t)$ and $g(t)$ are real functions. Then, in accordance with the Fourier transform properties,

$$\tilde{f}(-\omega, \tau) = \tilde{f}^*(\omega, \tau),$$

so the energy density $E(\omega, \tau)$ is an even function of ω for each τ and, necessarily, $\omega(\tau) \equiv 0$. In this context, the above notion of the "instantaneous frequency" is useless.

The situation is different and more promising for analytic signals

$$f(t) = A(t) \exp(i\Phi(t)), \tag{25}$$

where $A(t)$ and $\Phi(t)$ are, respectively, the signal's real-valued amplitude and phase. In physical and engineering applications, the real signal $f(t)$ is often a narrow-band process which can be written in the form

$$f(t) = A(t) \cos(\omega_0 t + \varphi(t)), \tag{26}$$

where $A(t)$ and $\varphi(t)$ are slowly varying in comparison to $\cos(\omega_0 t)$. The "simplest tune" signal introduced earlier in this section is narrow-band if

$$\omega_0 \gg \Omega, \quad \text{and} \quad \omega_0 \gg \nu. \tag{27}$$

In such cases we will consider "approximately analytic" signals, replacing in (25) the exact amplitude and phase by the amplitude $A(t)$ and phase

$$\Phi(t) = \omega_0 t + \varphi(t) \tag{28}$$

of a narrow-band process.

Let us calculate the current frequency $\bar{\omega}(\tau)$ (23) of an analytic signal (25). Multiplying (18) by ω, integrating, and keeping in mind that the differentiation of (20) with respect to θ gives

$$\frac{1}{2\pi} \int \omega e^{-i\omega\theta} \, d\omega = i\delta'(\theta),$$

we obtain

$$\bar{\omega}(\tau) = -ic(\tau) \int f^*(t) g^*(t - \tau) \frac{d}{dt} [f(t) g(t - \tau)] \, dt.$$

Substitute in this formula the expression (25) for the analytic signal and take into account that the window $g(t)$ is a real-valued function to arrive at the formula

$$\bar{\omega}(\tau) = \bar{\omega}_{inst}(\tau) = \int \omega_{inst}(t) P(t; \tau) \, dt, \tag{29}$$

where

$$\omega_{inst}(t) = \frac{d\Phi(t)}{dt} \tag{30}$$

is the "instantaneous frequency" of the analytic signal, and

$$P(t; \tau) = c(\tau) |f(t)|^2 |g(t - \tau)|^2 \tag{31}$$

is the normalized power of the signal taking into account the window's weight $|g(t - \tau)|^2$. If $|f(t)|^2$ is constant, as in the case of the simplest tune signal (2-4), then the power

$$P(t; \tau) = \frac{1}{\lambda} g^2(t - \tau) = \frac{1}{\lambda} g_0^2 \left(\frac{t - \tau}{\lambda} \right).$$

For $\lambda \to 0$, the power function $P(t; \tau)$ weakly converges to $\delta(t - \tau)$ and $\bar{\omega}_{inst}(\tau)$ converges to the instantaneous frequency $\omega_{inst}(\tau)$.

Unfortunately, the above conclusion does not imply that the windowed Fourier transform permits, in the limit $\lambda \to 0$, the precise measurement of the signal's instantaneous frequency. Actually, the accuracy of such measurement is determined by the *frequency deviation* $\sigma(\tau) = \sqrt{D(\tau)}$, where

$$D(\tau) = c(\tau) \int (\omega - \bar{\omega}(\tau))^2 E(\omega, \tau) \, d\omega. \tag{32}$$

Simple algebra shows that

$$D(\tau) = \overline{\omega^2}(\tau) - (\bar{\omega}(\tau))^2, \tag{33}$$

where the second frequency moment

$$\overline{\omega^2}(\tau) = c(\tau) \int \omega^2 E(\omega, \tau) \, d\omega. \tag{34}$$

Calculations similar to those that brought us to the expression (29) for the current frequency give

$$\overline{\omega^2}(\tau) = -\int dt \, f^*(t) g(t - \tau) \frac{d^2}{dt^2} [f(t) g(t - \tau)],$$

or, after substitution of the analytic signal (25),

$$\overline{\omega^2}(\tau) = \overline{\omega^2_{inst}}(\tau) + D_0(\tau). \tag{35}$$

Here, in agreement with notation from formula (29)

$$\overline{\omega^2_{inst}}(\tau) = \int \omega^2_{inst}(t) P(t; \tau) \, dt, \tag{36}$$

$$D_0(\tau) = c(\tau) \int \left[\frac{d}{dt} \left| f(t) g(t - \tau) \right| \right]^2 dt. \tag{37}$$

Finally, substituting (35) into (33) we obtain

$$D(\tau) = D_{inst}(\tau) + D_0(\tau), \tag{38}$$

where

$$D_{inst}(\tau) = \overline{\omega^2_{inst}}(\tau) - \left(\tilde{\omega}_{inst}(\tau) \right)^2. \tag{39}$$

Both terms on the right-hand side of (38) have an obvious physical interpretation. The first term (39) takes into account the error in determining the instantaneous frequency caused by averaging over the time-window of duration λ, and it converges to 0 as $\lambda \to 0$. On the other hand, the second term, (37), blows up to ∞ as $\lambda \to 0$ and has a more fundamental nature related to the uncertainty principle considered in Section 7.2.

Example 3. Current frequency for the simplest tune seen through a Gaussian window. Let us consider in some detail the behavior of the current frequency $\tilde{\omega}_{inst}(\tau)$ (29), and the competition between two components of the current frequency deviation $D(\tau)$ (38), in the case of the simplest tune signal (4) and the Gaussian windowing shape (7c).

Taking into account that $|f(t)|^2 = 1$ for the simplest tune, elementary calculations give

$$\bar{\omega}_{inst}(s) = \omega_0 + \Omega e^{-\alpha^2/4} \cos s, \tag{40a}$$

$$D_{inst}(s) = \frac{1}{2}\Omega^2 \left(1 - e^{-\alpha^2/2}\right)\left(1 - e^{-\alpha^2/2}\cos(2s)\right), \tag{40b}$$

and

$$D_0 = \frac{1}{2\lambda^2}, \tag{40c}$$

with the dimensionless parameters

$$\alpha = \nu\lambda, \qquad s = \nu\tau. \tag{41}$$

Fig. 7.3.3 shows, in cases when $\alpha = 1$ and $\alpha = 2$, the (dimensionless) time

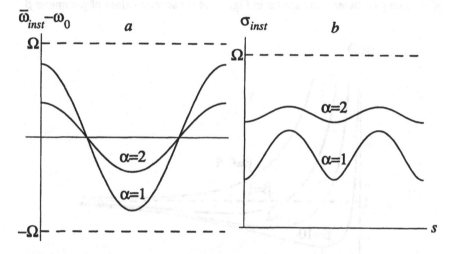

FIGURE 7.3.3
Time dependence of (a) the current frequency, and (b) the frequency deviation, for the simplest tune signal for two values, 1 and 2, of the dimensionless parameter α.

dependence of the current frequency $\bar{\omega}_{inst}(\tau)$ measured by the windowed Fourier transform (9), and the instantaneous frequency deviation $\sigma_{inst}(\tau) = (D_{inst}(\tau))^{1/2}$ due to time-window averaging.

Recall that for $\alpha = 0$, the current frequency $\bar{\omega}_{inst}(\tau)$ coincides with the instantaneous frequency $\omega_{inst}(\tau)$. The amplitude of $\bar{\omega}_{inst}(\tau)$ is smaller in the case $\alpha = 2$ than in the case $\alpha = 1$ which is a consequence of the smoothing action of time-averaging. Fig. 7.3.3(b) shows that the instantaneous frequency deviation $\sigma_{inst}(\tau)$ has peaks at the times when the instantaneous frequency changes quickly, and valleys when it changes slowly.

To evaluate the efficiency of instantaneous frequency measurement via the windowed Fourier transform we will consider the ratio of the full frequency dispersion (38) at $s = 0$ to its limit value $D^\infty = \Omega^2/2$ for $\lambda \to \infty$:

$$\rho^2(0) = \frac{D(0)}{D^\infty} = \frac{1}{4\alpha^2\beta^2} + \left(1 - e^{-\alpha^2/2}\right)^2, \tag{42}$$

where the new dimensionless parameter

$$\beta = \Omega/\nu. \tag{43}$$

Note that $s = 0$ is in a sense the "best" case since at that time the instantaneous frequency changes slowly and $D(\tau)$ has a minimum. The graphs of dependence of $\rho(0)$ on parameter α are shown in Fig. 7.3.4 for several values of parameter β.

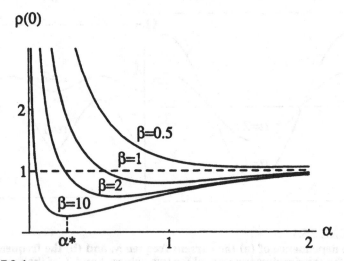

FIGURE 7.3.4
The graphs of dependence of $\rho(0)$ on parameter α for several values of parameter β.

As we explained before, the blow-up of the graphs to infinity as $\alpha \to 0$ is due to the uncertainty principle effects which guarantee that the window of a shorter duration will lead to greater indeterminacy of the measured frequency. As α increases, the uncertainty principle effects become negligibly small, but the instantaneous frequency measurement error due to the time-averaging, increases.

The value α^* for which ρ attains its minimal value ρ_{min} determines the optimal duration $\lambda^* = 2\alpha^*/\nu$ of the time-window. It is clear from Fig. 7.3.4 that only if $\beta \gg 1$ does $\rho_{min} \ll 1$ and, as a result, the windowed Fourier transform algorithm is capable of accurately tracking the instantaneous frequency. The latter values correspond to slow and/or large changes of the instantaneous frequency. ∎

To conclude our windowed Fourier analysis of the simplest tune, a word of caution is necessary. Conclusions based on simple integral characteristics of the joint frequency-time density $E(\omega, \tau)$ (17), such as the mean frequency (23) and deviation (32), can be sometimes misleading. They are quite coarse and lose a lot of information contained in the joint density. So, it is useful to indicate their area of applicability which is luckily possible for the simplest tune signal in view of the relatively simple structure of its windowed Fourier transform $\tilde{f}(\omega, \tau)$.

First, note the formula

$$e^{ia \sin b} = \sum_{n=-\infty}^{\infty} J_n(a)e^{inb}, \tag{44}$$

where

$$J_n(a) = \frac{1}{\pi} \int_0^{\pi} \cos(a \sin x - nx)\,dx, \tag{45}$$

are *Bessel functions of integer order* n which will be encountered often throughout the remainder of this book. Substituting $a = \Omega/\nu$ and $b = \nu t$ in (44) and multiplying it by $e^{i\omega_0 t}$ we will obtain for the simplest tune signal (2) the formula

$$f(t) = \sum_{n=-\infty}^{\infty} J_n(\Omega/\nu)e^{i(\omega_0+n\nu)t}. \tag{46}$$

Hence, $f(t)$ has the following distributional Fourier image:

$$\tilde{f}(\omega) = \sum_{n=-\infty}^{\infty} J_n(\Omega/\nu)\delta(\omega - \omega_0 - n\nu). \tag{47}$$

Applying it to the definition (9) of the windowed Fourier transform we get, in view of (10), that

$$\tilde{f}(\omega, \tau) = \frac{\alpha}{\nu} \sum_{n=-\infty}^{\infty} J_n(\beta)\tilde{g}_0(\alpha(\gamma - n))e^{-i(\gamma-n)s}, \tag{48}$$

where $\gamma = (\omega - \omega_0)/\nu$. It follows from (48) that our integral characteristics-based analysis of the joint density $E(\omega, \tau)$ is certainly not applicable for $\alpha \gg 1$ when the right-hand side of (48) collapses to a sum of separate peaks and $E(\omega, \tau)$ (see (17)) becomes a polymodal function of variable ω.

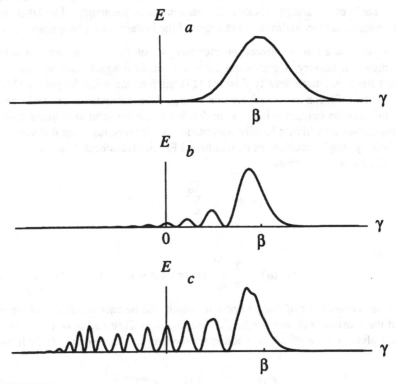

FIGURE 7.3.5
Plots of simplest tune's joint density $E(\omega, \tau)$ in case of $\beta = 10$ and (a) $\alpha = \alpha^* = 0.27$, (b) $\alpha = 1$, (c) $\alpha = 2$. The unimodality disappears and multimodality sets in at $\alpha \approx 1$.

7.3.5. Inverse windowed Fourier transform. As for any other integral transform, the question whether there is enough information contained in the windowed Fourier transform $\tilde{f}(\omega, \tau)$ to recover from it the original function $f(t)$ is of paramount importance. To answer this question let us keep in mind that the windowed Fourier transform of $f(t)$ is nothing but the ordinary Fourier transform of the windowed function $f(t)g(t - \tau)$. Hence, applying the usual formula (3.2.4) for the inverse Fourier transform we immediately get

$$f(t)g(t - \tau) = \int \tilde{f}(\omega, \tau)e^{i\omega t}\, d\omega. \tag{49}$$

This identity permits the recovery of values of function $f(t)$ only inside the window $g(t - \tau)$, or in practical terms, only where $g(t - \tau)$ is not too small.

To remove this limitation let us select a function $h^*(t)$, multiply both sides of (49) by its shift by τ and integrate them over all τ. The result is

$$Af(t) = \int d\tau \int d\omega \tilde{f}(\omega, \tau) h^*(t - \tau) e^{i\omega t}, \tag{50}$$

where

$$A = \int h^*(\theta) g(\theta)\, d\theta. \tag{51}$$

If the auxiliary function $h(\theta)$ is chosen so that the constant $A \neq 0$, then the desired inverse windowed Fourier transform formula takes the form

$$f(t) = \frac{1}{A} \int d\tau \int d\omega \tilde{f}(\omega, \tau) h^*(t - \tau) e^{i\omega t}. \tag{52}$$

Clearly, the above inverse formula is not unique as it depends on the choice of function $h(\theta)$. For example, if we take

$$h(t - \tau) = \delta(t - \tau - \tau_{max}),$$

where τ_{max} is the time when $g(t)$ has its maximum value, i.e.,

$$g(t) \leq g_{max} = g(\tau_{max}),$$

then $A = g_{max}$ and the inverse formula takes the form

$$f(t) = \frac{1}{g_{max}} \int \tilde{f}(\omega, t - \tau_{max}) e^{i\omega t}\, d\omega. \tag{53}$$

The above nonuniqueness of the inverse formulas is caused by the obvious *redundancy* of the windowed Fourier transform, which maps function $f(t)$ of a single variable into function $\tilde{f}(\omega, \tau)$ of two variables. Sometimes, however, such an overdetermination is useful. For instance, if one has to reconstruct the entire signal $f(t)$ from an incomplete information about its transform, the overdetermination present in the windowed Fourier transform can be helpful.

Among all the possible inverse formulas (52) one can try to find the optimal one in the sense that it would maximize the value of coefficient A (important in numerical computations) among all the auxiliary functions $h(\theta)$, such that

$$\|h\| = \|g\|. \tag{54}$$

We assume that the window $g(t)$ is given and is of finite energy, i.e., both $g, h \in L^2$. In terms of Section 7.1, A is just the inner product (h, g) and, by the Schwartz inequality

$$A \leq \|h\|\|g\| = \|g\|^2.$$

Thus, obviously, the greatest possible value of $A = \|g\|^2$ is attained if $h(t) = g(t)$, and the optimal (in the above sense) inverse formula is

$$f(t) = \frac{1}{\|g\|^2} \int d\tau \int d\omega \tilde{f}(\omega, \tau) g^*(t - \tau) e^{i\omega t},$$

or, in view of (6) and (22),

$$f(t) = \int d\tau \frac{1}{\lambda} g_0 \left(\frac{t - \tau}{\lambda}\right) \int d\omega \, \tilde{f}(\omega, \tau) e^{i\omega t}. \tag{55}$$

The asterisk was dropped because the windowing shape function $g_0(x)$ was real-valued.

7.3.6. From windows to wavelets. Although windowing was adopted in this section as our favored method, it is only fair to take now the last parting look at the windowed Fourier transform

$$\tilde{f}(\omega, \tau) = \frac{1}{2\pi} \int f(t) g_0 \left(\frac{t - \tau}{\lambda}\right) e^{-i\omega t} \, dt, \tag{56}$$

to assess impartially its merits and shortcomings.

First of all, note that the right-hand side of (56) contains three free parameters τ, ω, and λ, but only two of them—the current time τ and the frequency ω—are variables. The remaining parameter, the window duration λ, is usually assumed to be constant. This line of thinking is tied to the intuition that the windowed Fourier transform is just a parameterized version of the regular Fourier transform introduced for the purpose of tracking instantaneous frequencies of the signal. Such motivation is, however, also the source of limitations of the windowed Fourier transform and keeping λ constant, restricts our ability to simultaneously analyze the time-frequency properties of the signal. To see more precisely what we mean, let us consider the signal

$$f(t) = f_0(t/a)$$

of a given shape $f_0(x)$ but with variable width governed by the parameter a. Suppose that both $f_0(x)$ and $g_0(x)$ are well localized in the vicinity of $x = 0$. Then, for $a \ll \lambda$, approximately

$$\tilde{f}(\omega, \tau) \approx \tilde{f}(\omega) g_0(-\tau/\lambda).$$

This means that, for $a \ll \lambda$, the windowed Fourier transform really measures the signal's ordinary Fourier image and is not very good at doing its localization-in-time-and-frequency job. The window simply becomes too broad to be of any value. In the opposite case, when the window becomes very narrow ($\lambda \ll a$),

$$\tilde{f}(\omega, \tau) \approx f(\tau)\tilde{g}(\omega)e^{-\omega\tau},$$

the windowed Fourier transform does an excellent job at time-frequency localization but fails to provide any information about its spectral properties.

So, the windowed Fourier transform has a limited applicability field and seems to be most suitable for time-frequency analysis of narrow-band signals with phase $\Phi(t)$ subject to strong but slow nonlinear time-evolution. The example here is the simplest tune (2-4) for $\Omega \gg \nu$ ($\omega_0 \gg \Omega$). However, most of the signals of interest to scientists and engineers, from cardiograms and seismograms to stock market quotations and turbulent velocity fields, do not resemble simple tunes (Fig. 7.3.1) and have a much richer structure which often includes appearance of the wide range of scales. Tools more flexible than windowed Fourier transform are necessary for their satisfactory analysis, and it is clear from the very beginning that the *scale parameter* λ has to be treated as one of the primary variables. This leads to a suggestion of the new signal processing algorithm described by

$$\hat{f}(\lambda, \tau) = A(\lambda) \int f(t)\psi^* \left(\frac{t - \tau}{\lambda}\right) dt$$

which is called the *continuous wavelet transform* and which takes into account both the location and the scale properties via parameters τ and λ, respectively. The shape function $\psi(x)$ of the wavelet transform kernel is usually called the *mother wavelet*. In contrast with the windowed Fourier transform where the window's shape g_0 plays a minor role, the choice of the mother wavelet is of utmost importance and we will devote a lot of attention to it in the following sections.

The reader has probably noticed already that the notion of frequency has been lost in the process. This is not accidental and abandoning the frequency paradigm (closely tied to selecting the mother wavelet containing trigonometric functions $e^{-i\omega t}$) in favor of the scale paradigm (which permits full flexibility in selecting the mother wavelet) turns out to be not a weakness but the main strength of the wavelet transforms. The wavelet transforms can be tuned to the peculiarities of a signal we have to work with. If the signal is narrow-band then we can take an oscillating wavelet

$$\psi(x) = e^{iQx}g_0(x)$$

giving rise to a wavelet transform similar to the windowed Fourier transform but with different emphasis. The wavelet transform acts like a microscope, narrowing the visible area of the signal with the growth of the wavelets frequency $\omega = Q/\lambda$.

7.4 Continuous wavelet transforms

7.4.1. Definition and properties of continuous wavelet transform. In this section we take a general look at the continuous wavelet transform both theoretically and as it relates to physical and engineering problems. Mathematical questions concerning particular wavelet systems will be dealt with in the last three sections of this chapter.

The *continuous wavelet image* of signal $f(t)$ is defined by

$$\hat{f}(\lambda, \tau) = A(\lambda) \int f(t)\psi^* \left(\frac{t - \tau}{\lambda} \right) dt, \qquad \lambda > 0, \tag{1}$$

where $\psi(x)$ is a certain function called the *mother wavelet* and λ and τ are called, respectively, the *scale variable* and the *location variable*. Function $A(\lambda)$ will be specified later. Note that, to distinguish it from the Fourier transform

$$\tilde{f}(\omega) = \frac{1}{2\pi} \int f(t)e^{-i\omega t}\, dt, \tag{2}$$

the continuous wavelet transform will be denoted by applying a "hat" $\hat{f}$ to the original function f.

Observe that the mother wavelet $\psi(x)$ plays the role of complex exponentials $\exp(iz)$ in the Fourier transform (then, also $A = 1/2\pi$). Varying in (2) the frequency ω by compressing or dilating the complex exponential function, we obtain, after integration, a new function $\tilde{f}(\omega)$ which represents the complex amplitude of the corresponding harmonic component of the original signal:

$$f(t) = \int \tilde{f}(\omega)e^{i\omega t}\, dt. \tag{3}$$

Consequently, the Fourier image $\tilde{f}(\omega)$ measures the contribution of different harmonics to the in general nonharmonic signal $f(t)$.

A similar compression and dilation of the mother wavelet is accomplished for the continuous wavelet transform by the scaling parameter λ. Its exact analog for the Fourier transform is the period $T = 2\pi/|\omega|$. In a sense, one can interpret the value of the continuous wavelet transform $\hat{f}(\lambda, \tau)$ as a measure of the contribution of the rescaled by λ mother wavelet $\psi((t - \tau)/\lambda)$ to the signal $f(t)$.

The coefficient $A(\lambda)$ can be selected arbitrarily as to magnify or reduce sensitivity of the transform to different scales. However, very often it is simply selected as

$$A(\lambda) = 1/\sqrt{\lambda}, \tag{4}$$

so that

$$\hat{f}(\lambda, \tau) = \frac{1}{\sqrt{\lambda}} \int f(t) \psi^* \left(\frac{t - \tau}{\lambda} \right) dt. \tag{5}$$

This choice guarantees that the arbitrary rescaling of the mother wavelet preserves the mother wavelet's L^2-norm. Indeed,

$$\left\| \frac{1}{\sqrt{\lambda}} \psi^* \left(\frac{t - \tau}{\lambda} \right) \right\|^2 = \frac{1}{\lambda} \int \left| \psi^2 \left(\frac{t - \tau}{\lambda} \right) \right| dt = \int |\psi^2(x)| \, dx = \| \psi(x) \|^2. \tag{6}$$

One could say that with this choice of $A(\lambda)$ all the scales carry equal weight.

As we already mentioned in the previous section, for all its great features discussed at length in Chapter 3, the Fourier transform has from the point of view of a physicist one essential shortcoming: its "mother wavelet" $\exp(iz)$ has unbounded support. As a result, based on information contained in the Fourier image $\tilde{f}(\omega)$ it is difficult to assess where signal $f(t)$ (or its special features) is located on the t axis and where it is equal to 0. In particular, this type of information is totally lost in the "spectral density" $|\tilde{f}(\omega)|^2$ of the distribution of harmonic components over the frequency ω axis. That drawback will be removed in the continuous wavelet transform by selecting a *localized* mother wavelet $\psi(z)$ which decays rapidly to zero as $z \to \pm\infty$. Consequently, in the continuous wavelet transform, in addition to the scale parameter λ, there appears another primary parameter—the location shift τ. Varying it we can track the time t evolution of the "events."

Example 1. Wavelet transform expressed via the Fourier transform. To complete the general picture one should note that the continuous wavelet transform can be, obviously, expressed in terms of the Fourier images of the original function and the mother wavelet:

$$\hat{f}(\lambda, \tau) = 2\pi \lambda A(\lambda) \int \tilde{f}(\omega) \tilde{\psi}^*(\omega\lambda) e^{i\omega\tau} \, d\omega. \tag{7}$$

In particular, substituting the distributional Fourier image of function $\psi(z) = \exp(iz)$, we immediately get

$$\tilde{\psi}(\omega\lambda) = \delta(\omega\lambda - 1) = \delta(\omega - 1/\lambda)/\lambda,$$

and setting $A(\lambda) = 1/2\pi$ we obtain

$$\hat{f}(\lambda, \tau) = \tilde{f}(1/\lambda) \exp(i\tau/\lambda). \tag{8}$$

∎

Example 2. Morlet wavelets. The often encountered in practical application mother wavelet

$$\psi(z) = e^{iQz} \varphi(z), \tag{9}$$

with the Gaussian windowing function

$$\varphi(z) = \exp(-z^2/2), \tag{10}$$

is traditionally called the *complex-valued Morlet wavelet* (the plot of its real part, for $Q = 10$, is shown in Fig. 7.4.1). As a result, the Fourier image of $\psi(z)$ is also Gaussian:

$$\tilde{\psi}(\omega) = \frac{1}{\sqrt{2\pi}} \exp\left(-\frac{(\omega - Q)^2}{2}\right). \tag{11}$$

Recall that the Gaussian shape of the windowing function is the minimizer in the uncertainty principle (7.2.4) and, consequently, it optimizes the joint resolution in time t and frequency ω. Indeed, the continuous wavelet image

$$\hat{f}(\lambda, \tau) = A(\lambda) \int f(t)\varphi\left(\frac{t-\tau}{\lambda}\right) \exp\left(-i\frac{Q}{\lambda}(t-\tau)\right) dt$$

$$= A(\lambda) \int f(t) \exp\left(-i\frac{Q}{\lambda}(t-\tau) - \frac{(t-\tau)^2}{2\lambda^2}\right) dt \tag{12}$$

contains information about the original (not too fast increasing) function $f(t)$ in

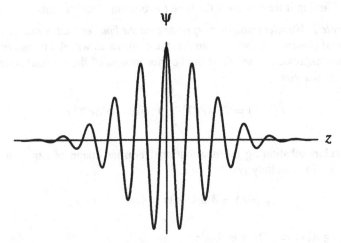

ψ

FIGURE 7.4.1
The plot of the real part of the Morlet mother wavelet for Q=10.

the window of effective length $\sim \sigma[\varphi] = \lambda/\sqrt{2}$. Expressing, as in (7), $\hat{f}(\lambda, \tau)$ through the Fourier images of the analyzed functions we get

$$\hat{f}(\lambda, \tau) = 2\pi\lambda A(\lambda) \int \tilde{f}(\omega)\varphi(\lambda\omega - Q)e^{i\omega\tau} d\omega$$

$$= \sqrt{2\pi} \lambda A(\lambda) \int \tilde{f}(\omega) \exp\left(-\frac{\lambda^2}{2}\left(\omega - \frac{Q}{\lambda}\right)^2\right) e^{i\omega\tau} \, d\omega. \tag{13}$$

This means that $\hat{f}(\lambda, \tau)$ depends on the values of the Fourier image $\tilde{f}(\omega)$ in the frequency band of width $\sigma[\tilde{\psi}] = 1/\lambda\sqrt{2}$ centered at the frequency

$$\Omega = Q/\lambda. \tag{14}$$

In other words, $\hat{f}(\lambda, \tau)$ supplies information about the spectral properties of the original function with resolution $1/\lambda\sqrt{2}$. The arbitrary parameter Q entering in the definition (9) of the Morlet wavelet could be called the *efficiency factor* of the Morlet wavelet since the quantity $Q/2\pi$ is of the order of the number of Morlet wavelet's periods contained in its window.   ∎

Just as the first automobiles of the last century took inspiration from and mimicked the horse-drawn carriages, and only later developed their own identity, the wavelets underwent a similar evolution which started with their identity as "improved" versions of the Fourier transform and only gradually developed into being recognized for their own outstanding capabilities. These capabilities, still far from being fully tapped, are related to the fact that the mathematical theory of wavelets, as we will see later on, imposes very few restrictions on the choice of the mother wavelet's shape. We will illustrate them on concrete applications in the rest of this section.

One of the powerful applications of the continuous wavelet transform is the study of open and hidden singularities in the incoming signal $f(t)$. Usually, the singularities are caused by physical (biological, economic, etc.) laws, whose validity the experimenter is trying to confirm, or come from the existence of the sharp boundaries between the regions where the process $f(t)$ evolves smoothly. The mother wavelets that are useful in this context are quite unlike the Morlet wavelet (9-10).

Example 3. Mexican hat wavelet. Differentiation can bring to the surface function's hidden singularities. For this reason one often selects mother wavelets so that the corresponding continuous wavelet transform converges, for $\lambda \to 0$, to a desired derivative of the function being analyzed. One of such examples is the *Mexican hat*

$$\psi(z) = -\frac{d^2}{dz^2}\varphi(z) = (1 - z^2)\exp\left(-\frac{z^2}{2}\right), \tag{15}$$

which is just the second derivative of the Gaussian function (10). Its Fourier image is

$$\tilde{\psi}(\omega) = \frac{\omega^2}{\sqrt{2\pi}}\exp\left(-\frac{\omega^2}{2}\right). \tag{16}$$

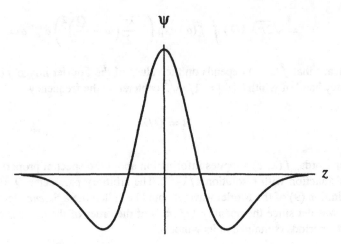

FIGURE 7.4.2
The Mexican hat mother wavelet.

Substituting (15) into (1), and integrating by parts twice, we get

$$\hat{f}(\lambda, \tau) = -A(\lambda)\lambda^2 \int \varphi\left(\frac{t-\tau}{\lambda}\right) \frac{d^2}{dt^2} f(t) \, dt. \qquad (17)$$

It is customary to select

$$A(\lambda) = 1/\lambda^3 \sqrt{2\pi}, \qquad (18)$$

so that $\hat{f}(\lambda, \tau)$ converges, for $\lambda \to 0$, to exactly $f''(\tau)$. ∎

Another property of the continuous wavelet transform essential to understand
its mechanism is based on the Schwartz inequality (7.1.19)

$$\left|\int f(t)g^*(t) \, dt\right|^2 \leq \int |f(t)|^2 \, dt \int |g(t)|^2 \, dt, \qquad (19)$$

which applied to the function

$$g(t) = \frac{1}{\sqrt{\lambda}} \psi\left(\frac{t-\tau}{\lambda}\right) \qquad (20)$$

yields the inequality

$$|\hat{f}(\lambda, \tau)|^2 \leq \|f\|^2 \|\psi\|^2. \qquad (21)$$

The inequality provides an upper bound on possible values of the modulus of the
continuous wavelet transform (5) of $f(t)$. Let us assume, without loss of generality,

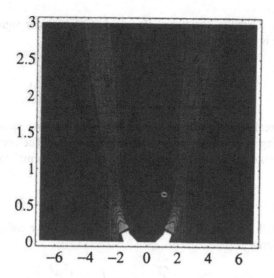

FIGURE 7.4.3
The grey-scale plot of the wavelet image of $f(t) = \exp(-|t|)$ in the case of the Mexican hat mother wavelet. The horizontal axis represents the τ-variable and the vertical—λ-variable. The grey-scale level changes from black to white as the values of the wavelet image increase. The black oval spot in the lower middle portion of the plot is a consequence of the singularity of the original function's second derivative at $t = 0$.

that both the signal and the mother wavelet are normalized so that $\|f\| = \|\psi\| = 1$. It is clear that the maximum values are achieved, and the inequality (21) becomes an equality, if the original function $f(t)$ is equal, for certain $\lambda = \lambda_0$ and $\tau = \tau_0$, to the wavelet

$$f(t) = \frac{1}{\sqrt{\lambda_0}} \psi\left(\frac{t - \tau_0}{\lambda_0}\right). \tag{22}$$

Informally, we can say that the continuous wavelet transform is best tuned to, or resonates with signals that have shapes similar to that of the mother wavelet.

Note that the more *complex-structured* the mother wavelet (20) and the resonating signal (22) are, the more pronounced the above resonance property of the corresponding continuous wavelet transform is. To make things a bit more formal let us define the signal as *complex-structured* if its time (7.2.2) and frequency (7.2.3) localizations satisfy the *"strong" uncertainty principle*:

$$\sigma^2[f]\sigma^2[\tilde{f}] \gg 1/4. \tag{23}$$

Example 4. Complex-structured signal. Let us consider signal $f(t)$ whose Fourier image is the familiar Gaussian function

$$\tilde{f}(\omega) = \frac{\pi^{-1/4}}{\sqrt{2\pi}\,\mu} \exp\left(-\frac{\omega^2}{2\mu^2}(1+i\gamma)\right), \qquad (24)$$

where γ is a real number and the constant μ (with the dimension of frequency) has the meaning of effective width of the Fourier image. The coefficient in front of the exponential function has been selected so that the normalization condition

$$\|f\|^2 = 2\pi \|\tilde{f}\|^2 = 1$$

is satisfied. With the help of the integral formula (3.2.3), we compute the inverse Fourier transform

$$f(t) = \frac{\sqrt{\mu}}{\pi^{1/4}\sqrt{1+i\gamma}} \exp\left(-\frac{t^2\mu^2}{2(1+i\gamma)}\right). \qquad (25)$$

In turn, using the integral formula

$$\int x^2 \exp(-r^2 x^2)\, dx = \frac{\sqrt{\pi}}{2r^3},$$

we find the frequency (7.2.3) and time (7.2.2) localizations of the complex-valued signal (25):

$$\sigma^2[\tilde{f}] = \frac{\mu^2}{2}, \qquad \sigma^2[f] = \frac{1}{2\mu^2}(1+\gamma^2). \qquad (26)$$

Substituting these expressions into (25) we get the following condition for the signal $f(t)$ (25) to be complex-structured:

$$\gamma \gg 1. \qquad (27)$$

∎

Remark 1. To better see reasons why signal (25) turned out to be complex-structured let us write the complex-valued Fourier image $\tilde{f}(\omega)$ of an arbitrary signal $f(t)$ in the exponential form

$$\tilde{f}(\omega) = A(\omega) \exp(-i\Phi(\omega)), \qquad (28)$$

where $A(\omega) = |\tilde{f}(\omega)|$ is the nonnegative amplitude and $\Phi(\omega)$—the real phase of the complex Fourier image $\tilde{f}(\omega)$. The amplitude and phase of the Fourier image

(24) of signal (25) are

$$A(\omega) = \frac{\pi^{-1/4}}{\sqrt{2\pi\mu}} \exp\left(-\frac{\omega^2}{2\mu^2}\right), \qquad \Phi(\omega) = \frac{\gamma\omega^2}{2\mu^2}. \tag{29}$$

The complex structure of signal (25) was conditioned on the fast nonlinear variation of the phase of the Fourier image (24) as a function of ω. Indeed, according to (3), the signal can be written in the form

$$f(t) = \int A(\omega) \exp\left(i\left(\omega t - \gamma\frac{\omega^2}{2\mu^2}\right)\right) d\omega.$$

Employing the stationary phase method, asymptotically ($\gamma \to \infty$), the value of signal $f(t)$ at a given instant t is determined by the integral contribution in the small neighborhood of the stationary point, in our case $\Omega = 2t\mu^2/\gamma$. Substituting here, instead of Ω, the effective width μ of the Fourier image we shall find the effective duration of the complex-structured signal:

$$T \approx \gamma/\mu, \qquad (T\mu \gg 1).$$

Remark 2. The approximate estimate of the signal (25) duration obtained above via the stationary phase method may seem unnecessary at first sight since we already know the exact form of the signal and the exact formula for its time localization:

$$\sigma[f] = \sqrt{(1+\gamma^2)/2\mu^2}. \tag{30}$$

Nevertheless, the above argument has a heuristic value, emphasizing the principal role of the phase in complex-structure signal formation. It also shows a universal method of calculation of its form and duration.

Example 5. Complex-structured mother wavelet. As another example of mother wavelet let us take function $\psi(z)$ coinciding with the complex-structured signal $f(z)$ (25). The continuous wavelet image $\check{f}(\lambda, \tau)$ of function $f(t)$ to which the mother wavelet is perfectly tuned is

$$K(\lambda, \tau) = \frac{1}{\sqrt{\lambda}} \int f(t) f^*\left(\frac{t-\tau}{\lambda}\right) dt. \tag{31}$$

Recall that the form (5) of the continuous wavelet transform selected here guarantees that, for any λ, the normalization condition

$$\left\|\frac{1}{\sqrt{\lambda}} f\left(\frac{t-\tau}{\lambda}\right)\right\| = 1$$

is satisfied. Notice that we also introduced special notation $K(\lambda, \tau)$ for the special continuous wavelet image of the mother wavelet itself. Function $K(\lambda, \tau)$ is sometimes called the *wideband ambiguity function* of the mother wavelet and it plays an important role in wavelet theory. In terms of the Fourier images

$$K(\lambda, \tau) = 2\pi \sqrt{\lambda} \int \tilde{f}(\omega) \tilde{f}^*(\omega\lambda) e^{i\omega\tau} \, d\omega, \tag{32}$$

so that substituting (24) we obtain

$$K(\lambda, \tau) = \frac{1}{\mu} \sqrt{\frac{\lambda}{\pi}} \int \exp\left(-\frac{1}{2}\rho \frac{\omega^2}{\mu^2} + i\omega\tau\right) d\omega, \tag{33}$$

where

$$\rho = (1 + \lambda^2) + i\gamma(1 - \lambda^2). \tag{34}$$

Finally, evaluation of the integral (33) gives

$$K(\lambda, \tau) = \sqrt{\frac{2\lambda}{\rho}} \exp\left(-\frac{\tau^2\mu^2}{2\rho}\right).$$

The above function has a maximum at $\tau = 0$, and its modulus square has the following dependence on λ :

$$I(\lambda) = |K(\lambda, 0)|^2 = \frac{2\lambda}{|\rho|} = \frac{2\lambda}{\sqrt{(1 + \lambda^2)^2 + \gamma^2(1 - \lambda^2)^2}}.$$

It is natural to interpret function $I(\lambda)$ as a sort of resonance curve which characterizes the response efficiency of the continuous wavelet transform as a function of the scale parameter λ. Fig. 7.4.4 shows graphs of function $I(\lambda)$ for signals of different complexity, as measured by parameter γ. It is clear from the illustrations that the resonance is best emphasized for large values of γ, that is for signals of large complexity.

The maximal value of $I(\lambda)$ is achieved for $\lambda = 1$. It is related to the fact that for $\lambda = 1$ function (31) becomes the *autocorrelation function*

$$K(\tau) = \int f(t) f^*(t - \tau) \, dt$$

of the original signal. The autocorrelation function has some remarkable properties. In particular, it transforms any signal, however complex, into a simple signal whose Fourier image,

$$\tilde{K}(\omega) = 2\pi |\tilde{f}(\omega)|^2, \tag{36}$$

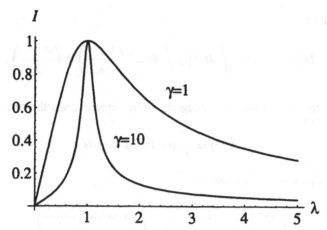

FIGURE 7.4.4
Graphs of function $I(\lambda)$ for different complexity-structure of signals as measured by γ.

is real and nonnegative, with the phase $\Phi(\omega) \equiv 0$. In electrical engineering one often says that all the harmonics of the autocorrelation function $K(t)$ have identical phases.

The autocorrelation function achieves its maximum at $\tau = 0$ and decays relatively rapidly as $|\tau|$ increases. In particular, it is easy to see that its localization properties are determined by

$$\sigma[K] = \frac{1}{\mu\sqrt{2}},$$

so that, in view of (30), it is clear that for $\gamma \gg 1$ the autocorrelation function $K(\tau) = K(\lambda = 1, \tau)$ is much better localized on the τ-axis than the original signal $f(t)$ (25) on the t-axis. ∎

7.4.2. Inversion of the continuous wavelet transform. As for any other integral transform the basic question is: Does the continuous wavelet image $\hat{f}(\lambda, \tau)$ contain sufficient information permitting recovery of the original function $f(t)$? In more practical terms: Does there exist an inversion formula for the continuous wavelet transform?

To answer these questions let us multiply equality (1) by $\psi((\theta - \tau)/\lambda)$ and integrate over all τ. The result is the auxiliary integral

$$I(\lambda, \theta) = \int \hat{f}(\lambda, \tau)\psi\left(\frac{\theta - \tau}{\lambda}\right) d\tau. \tag{37}$$

Equivalently,

$$I(\lambda, \theta) = A(\lambda) \int dt \, f(t) \int d\tau \, \psi^* \left(\frac{t - \tau}{\lambda}\right) \psi \left(\frac{\theta - \tau}{\lambda}\right). \tag{38}$$

It is easy to see that the inner integral can be expressed via the autocorrelation function (35)

$$K(z) = \int \psi(s)\psi^*(s - z) \, ds \tag{39}$$

of the mother wavelet as follows:

$$\int \psi^* \left(\frac{t - \tau}{\lambda}\right) \psi \left(\frac{\theta - \tau}{\lambda}\right) d\tau = \lambda K \left(\frac{\theta - t}{\lambda}\right). \tag{40}$$

As a result,

$$I(\lambda, \theta) = \lambda A(\lambda) \int f(t) K \left(\frac{\theta - t}{\lambda}\right) dt. \tag{41}$$

To solve this integral equation for $f(t)$ let us multiply (41) by a function $B(\lambda)$, to be selected later, and integrate over all λ:

$$\int_0^\infty I(\lambda, \theta)B(\lambda) \, d\lambda = \int f(t)g(\theta - t) \, dt, \tag{42}$$

where

$$g(s) = \int_0^\infty K(s/\lambda)C(\lambda) \, d\lambda, \tag{43}$$

and

$$C(\lambda) = \lambda A(\lambda)B(\lambda). \tag{44}$$

Clearly, the right-hand side of (42) would be reduced to $f(\theta)$, thus solving equation (41) for function $f(t)$ if

$$g(s) = \int_0^\infty K(s/\lambda)C(\lambda) \, d\lambda = \delta(s). \tag{45}$$

Let us find $C(\lambda)$ for which the distributional equation (45) is satisfied. Remembering that the Fourier image of the autocorrelation function $K(z)$ is $2\pi|\tilde{\psi}(\omega)|^2$, we get the equation

$$\tilde{g}(\omega) = 2\pi \int_0^\infty |\tilde{\psi}(\omega\lambda)|^2 \lambda C(\lambda) \, d\lambda = 1/2\pi, \tag{46}$$

equivalent to equation (45). To eliminate the dependence of the above integral on ω we shall select $C(\lambda)$ so that

$$\lambda C(\lambda) = 1/D\lambda. \tag{47}$$

In this case, (46) becomes

$$\frac{4\pi^2}{D} \int_0^\infty |\tilde{\psi}(\omega\lambda)|^2 \frac{d\lambda}{\lambda} = 1, \tag{48}$$

where

$$D = 4\pi^2 \int_0^\infty |\tilde{\psi}(\kappa)|^2 \frac{d\kappa}{\kappa} \tag{49}$$

is the normalizing constant that can be calculated from (48) by introducing the new variable of integration $\kappa = \omega\lambda$ to get

$$\frac{4\pi^2}{D} \int_0^\infty |\tilde{\psi}(\kappa)|^2 \frac{d\kappa}{\kappa} = 1, \tag{50}$$

Putting together (44), (47) and (49) we get that

$$B(\lambda) = \frac{1}{D\lambda^3 A(\lambda)}, \tag{51}$$

so that, from (51) and (42),

$$\frac{1}{D} \int_0^\infty \frac{I(\lambda, \theta) \, d\lambda}{\lambda^3 A(\lambda)} = f(\theta).$$

Substituting expression (37) for $I(\lambda, t)$ we finally obtain the inverse continuous wavelet transform

$$f(t) = \frac{1}{D} \int_0^\infty \frac{d\lambda}{\lambda^3 A(\lambda)} \int d\tau \hat{f}(\lambda, \tau) \psi\left(\frac{t-\tau}{\lambda}\right). \tag{52}$$

In particular, if the continuous wavelet transform is defined by (4-5), then the inversion formula takes the form

$$f(t) = \frac{1}{D} \int_0^\infty \frac{d\lambda}{\lambda^2 \sqrt{\lambda}} \int d\tau \hat{f}(\lambda, \tau) \psi\left(\frac{t-\tau}{\lambda}\right). \tag{53}$$

However, the above inversion formulas require several caveats.

Remark 3.   First of all we have to admit that in the process of making our calculations transparent we cheated quite a bit. The observant reader would have noticed that the passage from (48) to (50) is justified only if $|\tilde{\psi}(\omega)|^2$ is an even function. For that reason formulas (52-53) are valid only for *two-sided* mother wavelets, as mother wavelets with even square modulus Fourier image are called. To this class belong all the purely real-valued mother wavelets such as the Mexican hat (15-16). On the other hand, the complex-valued Morlet wavelet (9-10) is not of this type. For that reason mathematicians often work with *one-sided* mother wavelets whose Fourier image is

$$\tilde{\psi}(\omega) \equiv 0, \qquad \omega \le 0. \tag{54}$$

For such mother wavelets, instead of (46) we have the equality

$$\tilde{g}(\omega) = \frac{2\pi}{D} \int_0^\infty |\tilde{\psi}(\omega\lambda)|^2 \frac{d\lambda}{\lambda} = \frac{1}{2\pi}\chi(\omega). \tag{55}$$

To explain its consequences let us express the right-hand side of (42) in terms of the Fourier images $\tilde{f}(\omega)$ and $\tilde{g}(\omega)$:

$$\int I(\lambda, \theta)B(\lambda)\,d\lambda = 2\pi \int \tilde{f}(\omega)\tilde{g}(\omega)e^{i\omega\theta}\,d\omega.$$

Substituting here (37), (51) and (55), we arrive at the relation that replaces equality (52) for one-sided mother wavelets:

$$\int \tilde{f}(\omega)\chi(\omega)e^{i\omega\theta}\,d\omega = \frac{1}{D}\int_0^\infty \frac{d\lambda}{\lambda^3 A(\lambda)} \int d\tau \hat{f}(\lambda, \tau)\psi\left(\frac{\theta - \tau}{\lambda}\right).$$

As we have shown in Section 6.6, the Fourier integral on the left-hand side is, up to coefficient 1/2, equal to the analytic signal

$$F(t) = \frac{2}{D}\int_0^\infty \frac{d\lambda}{\lambda^3} A(\lambda) \int d\tau \hat{f}(\lambda, \tau)\psi\left(\frac{\theta - \tau}{\lambda}\right) \tag{56}$$

corresponding to the original signal $f(t)$. Remembering that the real part of the analytic signal coincides with $f(t)$, we arrive at the inversion formula for the continuous wavelet transform for one-sided mother wavelets:

$$f(t) = \frac{2}{D}\,\text{Re}\,\int_0^\infty \frac{d\lambda}{\lambda^3} A(\lambda) \int d\tau \hat{f}(\lambda, \tau)\psi\left(\frac{\theta - \tau}{\lambda}\right). \tag{57}$$

Example 6. Poisson wavelets. As an example of one-sided mother wavelets consider

$$\psi_m(z) = (1 - iz)^{-m-1}, \qquad m > 0, \tag{58}$$

which are called *Poisson wavelets*. Their Fourier images

$$\tilde{\psi}_m(\omega) = \frac{1}{2\pi} \int \frac{e^{-i\omega z}\, dz}{(1 - iz)^{m+1}} \tag{59}$$

can be calculated by means of the residues method to be

$$\tilde{\psi}_m(\omega) = \frac{1}{\Gamma(m + 1)} \omega^m e^{-\omega} \chi(\omega). \tag{60}$$

Poisson wavelets can be used to identify open and hidden singularities of signal $f(t)$ and, for $m = 2$, like the Mexican hat, in the search for edges between different regimes of the original function $f(t)$. Indeed, for $m = 2$, the Poisson wavelet

$$\psi_2(z) = -\frac{1}{2}\frac{d^2}{dz^2}\frac{1}{1 - iz}. \tag{61}$$

Its real part

$$\operatorname{Re}\psi_2(z) = -\frac{1}{2}\frac{d^2}{dz^2}\frac{1}{1 + z^2}$$

has a shape similar to that of the Mexican hat and possesses, for $\lambda \to 0$, the same differentiating properties. ∎

Remark 4. The above derivation indicates that the necessary condition for the existence of the inversion formulas is finiteness of coefficient D (49), or equivalently, the inequality

$$\int |\tilde{\psi}(\kappa)|^2 \frac{d\kappa}{|\kappa|} < \infty. \tag{62}$$

Mother wavelets satisfying condition (62) are called *admissible wavelets*. The complex-valued Morlet wavelet (9-10) is not admissible since its Fourier image does not vanish for $\omega = 0$ and, consequently, the integral (62) diverges. Nevertheless, in practice this is not a serious obstacle. First of all, the inversion formula is not always needed, and, secondly, for sufficiently large values of the "goodness" parameter Q, the Fourier image of the Morlet wavelet takes a very small value at $\omega = 0$, and it is not difficult to adjust it a little bit to make it admissible.

Remark 5. It follows from condition (62) that the Fourier images of admissible wavelets satisfy condition $\tilde{\psi}(\omega = 0) = 0$, which is equivalent to the condition

$$\int \psi(z)\, dz = 0. \tag{63}$$

This in turn implies that any admissible wavelet has to have a oscillatory (sign-changing) nature—this provides a partial explanation of the term *wavelet*.

Remark 6. The fact that the continuous wavelet image $\tilde{f}(\lambda, \tau)$ of the function $f(t)$ of a single variable depends on two variables indicates that the continuous wavelet transform contains redundant information and is overdetermined. One of the consequences of this fact is that the inversion formula is not unique. This is easily seen by multiplying (1) not by the mother wavelet, as was done earlier, but by another function

$$\varphi\left(\frac{\theta - \tau}{\lambda}\right),$$

and integrating the resulting equation over all τ. For the sake of simplicity of the argument let us assume that $\varphi(z)$ has a one-sided Fourier image, that is $\tilde{\varphi}(\omega) \equiv 0$ for $\omega \leq 0$. As a result we get an analog of expression (41), the only difference being that instead of the autocorrelation function (39) one enters the cross-correlation function

$$K(z) = \int \varphi(s)\psi^*(s - z)\,ds \tag{64}$$

whose Fourier image is $2\pi\tilde{\varphi}(\omega)\tilde{\psi}^*(\omega)$. Replacing $|\tilde{\psi}(\omega)|^2$ by $\tilde{\varphi}(\omega)\tilde{\psi}^*(\omega)$ in all the preceding formulas, we arrive at an infinite variety of continuous wavelet inversion formulas:

$$f(t) = \operatorname{Re}\frac{2}{D}\int_0^\infty \frac{d\lambda}{\lambda^3 A(\lambda)}\int d\tau\, \hat{f}(\lambda, \tau)\varphi\left(\frac{\theta - \tau}{\lambda}\right), \tag{65}$$

which are all well defined as long as

$$\left|\int_0^\infty \varphi(\kappa)\tilde{\psi}^*(\kappa)\frac{d\kappa}{\kappa}\right| < \infty. \tag{66}$$

The, complex in general, coefficient D entering in formula (65) is equal to

$$D = 4\pi^2 \int_0^\infty \varphi(\kappa)\tilde{\psi}^*(\kappa)\frac{d\kappa}{\kappa}.$$

Notice that the condition (66) can be fulfilled not only for admissible mother wavelets but also if $\tilde{\varphi}(\omega)$ converges sufficiently fast to zero as $\omega \to 0+$. This means that the inversion formula (65) remains valid also in cases when the formulas (52-53) do not make sense. For example, formula (65) permits recovery of the original function $f(t)$ from the continuous Morlet wavelet image $\hat{f}(\lambda, \tau)$ (9-10) if one takes $\varphi(z)$ to be the Poisson wavelet (58).

The question of how to make wavelet transforms more economical and less redundant, while preserving their good scale and time localization properties is a subtle mathematical problem. For the usual Fourier transform and the Fourier series the lack of overdetermination, and the uniqueness of the inverse Fourier transform (or Fourier coefficient sequences) is guaranteed by the Hilbert space L^2 orthogonality of complex exponentials (or trigonometric functions) on the interval $[0, 2\pi]$, that is, by the condition that

$$\int_0^{2\pi} e^{imt}(e^{int})^* dt = 0, \qquad \text{if} \qquad m \neq n.$$

The mathematically difficult task of constructing orthogonal wavelet systems will be discussed at some length in the next three sections.

7.5 Haar wavelets and multiresolution analysis

In this section we will take a look at a special (one can say digital) series representation for real-valued signals in terms of the so-called *Haar wavelets*. This idealized system provides a good easy introduction to the concepts of *wavelet transforms* and *multiresolution analysis*. Each term of the expansion will provide information about both the time and the frequency localization of the signal. The Haar wavelets will be obtained from a single prototype—a *mother wavelet*—by translations in time and frequency, although the explicit shift in frequency will be

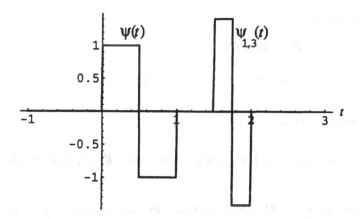

FIGURE 7.5.1
The Haar mother wavelet, and a wavelet of order (1,3).

replaced by a more natural in this case *dilation* (rescaling, stretching) in time. This will guarantee that all the wavelets have the same shape. To eliminate redundancy and overdetermination, we will make the wavelet system orthogonal.

The *Haar mother wavelet* is defined as follows

$$\psi(t) = \begin{cases} 1, & \text{for } 0 \le t < 1/2; \\ -1, & \text{for } 1/2 \le x < 1; \\ 0, & \text{otherwise.} \end{cases} \tag{1}$$

The *Haar wavelet*

$$\psi_{m,n}(t) := 2^{m/2}\psi(2^m t - n) = 2^{m/2}\psi(2^m(t - 2^{-m}n)), \tag{2}$$

of order (m, n), $m, n = \ldots, -1, 0, 1, \ldots$, is obtained by rescaling (dilating or compressing) the time in the mother wavelet $\psi(t)$ by a factor of 2^m and then translating the resulting wavelet by an integer n multiplicity of 2^{-m}. The dilation makes the wavelet $\psi_{m,n}(t)$ fit in the interval of length 2^{-m}, and the translation places its support finally in the interval $[2^{-m}n, 2^{-m}(n + 1)]$ (see Fig. 7.5.1). We will call parameter m—the *level of resolution* of the wavelet, and parameter n—the *location* parameter of the wavelet. Then the number 2^{-m} can be seen as its *resolution*, and $2^{-m}n$—as its *location*.

The coefficient $2^{m/2}$ in the definition (2) was selected to make all the Haar wavelets normalized in $L^2(\mathbf{R})$, that is, to compensate for the dilation operation to guarantee that

$$\|\psi_{m,n}\|^2 = \int \psi_{m,n}^2(t)\, dt = 1. \tag{3}$$

It turns out that:

The system of Haar wavelets

$$\psi_{m,n}(t), \qquad m, n = \ldots -2, -1, 0, 1, 2, \ldots \tag{4}$$

is orthogonal, that is

$$(\psi_{j,k}, \psi_{m,n}) = \int \psi_{j,k}(t)\psi_{m,n}(t)\, dt = 0, \quad \text{if} \quad (j, k) \ne (m, n), \tag{5}$$

and complete in $L^2(\mathbf{R})$. The latter means that any function $f \in L^2(\mathbf{R})$ has an L^2-convergent representation

$$f = \sum_{m=-\infty}^{\infty} \sum_{n=-\infty}^{\infty} w_{m,n}\psi_{m,n} \tag{6}$$

where, in view of the orthonormality, the expansion coefficients

$$w_{m,n} = w_{m,n}[f] = (f, \psi_{m,n}) = \int f(t)\psi_{m,n}(t)\, dt. \tag{7}$$

The above properties of orthogonality and completeness parallel properties of the trigonometric system of functions on a finite interval (say, $[0, 2\pi]$) which give rise to the usual Fourier series expansions.

The orthogonality (5) can be shown as follows. For a fixed resolution level $j = m$, if location parameters k, n are different, then wavelets $\psi_{j,k}(t)$ and $\psi_{j,n}(t)$ have disjoint supports, and the integral of their product is clearly zero. At different resolution levels, say $j < m$, either the supports of $\psi_{j,k}(t)$ and $\psi_{m,n}(t)$ are disjoint and the previous argument applies, or the support of $\psi_{m,n}(t)$ sits entirely within the interval where $\psi_{j,k}(t)$ is constant (either $+2^{j/2}$ or $-2^{j/2}$), and again the integral of their product vanishes because

$$\int \psi_{j,k}(t)\psi_{m,n}(t)\, dt = \pm 2^{j/2} \int \psi_{m,n}(t)\, dt = 0. \qquad \blacksquare$$

The completeness of the Haar wavelet system (4) is more difficult to establish and the proof relies on demonstrating that if all the wavelet coefficients $w_{m,n}[f] = 0$, then function f is necessarily 0 in L^2. We will give a flavor of the proof by showing that this is indeed the case if $f \in L^1 \cap L^2$. So, assume that $w_{m,n} = 0$, $m, n = \ldots - 1, 0, 1, \ldots$.

Since $w_{0,0} = 0$ then

$$\int_0^{1/2} f(t)\, dt = \int_{1/2}^1 f(t)\, dt = \frac{1}{2}\int_0^1 f(t)\, dt.$$

However, since $w_{-1,0} = 0$,

$$\int_0^1 f(t)\, dt = \int_1^2 f(t)\, dt = \frac{1}{2}\int_0^2 f(t)\, dt$$

and, by induction, for any n

$$\int_0^{1/2} f(t)\, dt = \frac{1}{2^{n+1}}\int_0^{2^n} f(t)\, dt = \lim_{n\to\infty} \frac{1}{2^{n+1}}\int_0^{2^n} f(t)\, dt = 0$$

since we assumed the finiteness of the integral $\int |f(t)|\, dt$ ($f \in L_1$). Clearly, the same argument can be repeated for any dyadic interval of the form $[2^{-m}n, 2^{-m}(n+$

1)], so that, by approximation, for any interval $[a, b]$

$$\int_{[a,b]} f(t)\, dt = 0.$$

This implies that $f = 0$ in $L^1 \cap L^2$, and the proof of completeness of the Haar wavelets is done. ∎

Remark 1. An alert reader would observe a seemingly paradoxical nature of expansion (6), where an arbitrary square integrable function in L^1, for which in general $\int f(t)\, dt \neq 0$, has an expansion into a series of Haar wavelets for which $\int \psi_{m,n}(t)\, dt = 0$. The explanation is that the convergence of the series (6) is in L^2 (that is in the mean square sense) so that the integrals themselves need not be preserved in the limit. To avoid this phenomenon one sometimes considers functions $I_{[0,1]}(t - n), n = \ldots, -1, 0, 1, \ldots$ in combination with the Haar wavelet subsystem $\psi_{m,n}$ for $m \geq 0$. We will return to this theme later.

Note that the inner series in the expansion (6) consists of wavelets of fixed *resolution* 2^{-m}, that is, it represents a function with constant values on dyadic intervals $[2^{-m}n, 2^{-m}(n + 1)]$, $n = \ldots, -1, 0, 1, \ldots$, and gives the *contents of function f at fixed resolution level m* (see Fig. 7.5.2) Then the partial sum

$$f_{R,S}(t) = \sum_{m=R}^{S} \sum_{n=-\infty}^{\infty} w_{m,n} \psi_{m,n}(t) \tag{8}$$

of the expansion (6) gives an approximation of function $f(t)$ at resolutions finer than 2^{-R} and coarser than 2^{-S} (see Fig. 7.5.3).

Thus expansion (6) may be interpreted as a *multiresolution analysis* of the function space $L^2(\mathbf{R})$.

Remark 2. Scaling function. We have already observed in Remark 1 that the multiresolution analysis of functions in $L^2(\mathbf{R})$ can be accomplished by means of a slightly different system that starts out with the *scaling function*

$$\varphi(t) = I_{[0,1]}(t)$$

and its integer translates

$$\varphi_n(t) = \varphi(t - n), \quad n = \ldots, -1, 0, 1, \ldots,$$

and supplements them with Haar functions $\psi_{m,n}(t)$ with nonnegative resolution levels $m = 0, 1, 2, \ldots$ and arbitrary integer location parameter n. Note that the

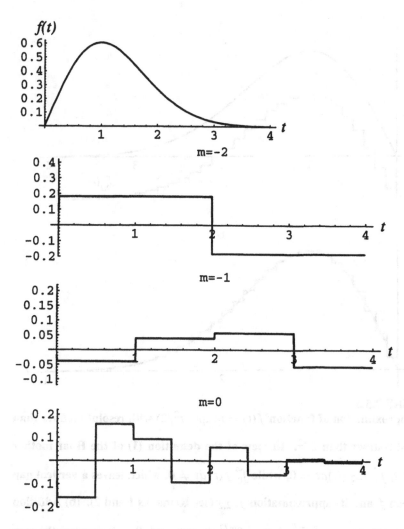

FIGURE 7.5.2
**Function $f(t) = t \exp(-t^2/2)$ and its contents at resolution levels $m =$
$-2, -1, 0.$**

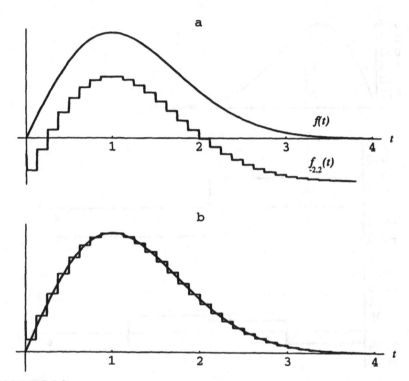

FIGURE 7.5.3
(a) Approximation of function $f(t) = t \exp(-t^2/2)$ with resolution finer than 2^2 and coarser than 2^{-2}. In view of the definition (1) of the Haar mother wavelet, $\int_0^4 f_{-2,2}(t)dt = 0$, while $\int_0^4 f(t)\,dt \neq 0$, which leaves a vertical gap between f and its approximation $f_{-2,2}$ (see Remarks 1 and 2). (b) Addition of the constant $c_R = \sum_{m=-\infty}^{-R-1} w_{m,0} 2^{m/2}$ (in our case, $R = 2$) removes the gap.

resulting system is still orthonormal and complete, and gives a multiresolution expansion of a function $f \in L^2(\mathbf{R})$ of the form

$$f = \sum_{n=-\infty}^{\infty} w_n \varphi_n(t) + \sum_{m=0}^{\infty} \sum_{n=-\infty}^{\infty} w_{m,n} \psi_{m,n}, \tag{9}$$

with coefficients

$$w_n = w_n(f) = \int f(t)\varphi_n(t)\,dt, \tag{10}$$

and $w_{m,n}$ as in formula (7).

Remark 3. Self-similar (fractal) properties of Haar wavelets. The crucial observation for the general theory of wavelets (to be discussed in the next section) is that the scaling function $\varphi(t)$ (the indicator function of the interval $[0,1]$) is *self-similar* in the sense that it satisfies the *scaling relation*

$$\varphi(x) = \varphi(2x) + \varphi(2x - 1), \tag{11}$$

and that the mother wavelet $\psi(t)$ can be obtained from the scaling function via the formula

$$\psi(x) = \varphi(2x) - \varphi(2x - 1). \tag{12}$$

The scaling relation (11) asserts that the scaling function is a certain linear combination of its own dilations and translations. It completely characterizes the indicator function $\varphi(t)$ up to a constant multiplier. Indeed, given values of $\varphi(t)$ at $t = 0$ and 1, the scaling relation (11) permits computation of values of φ at all dyadic rationals, i.e., real numbers of the form $2^{-m}n$.

7.6 Continuous Daubechies' wavelets

The Haar wavelets discussed in the previous section enjoyed many useful properties such as orthonormality, completeness, compact support and self-similarity but, as elegant as was their construction, they were anything but smooth. As a matter of fact they were not even continuous—a property important in many applications. So, in the present section we will explore the possibility of constructing smoother wavelets.

Since the scaling relation (7.5.11) characterizes the indicator scaling function I and thus the Haar wavelets, more complex scaling relations will have to be allowed. It turns out that one can find smooth scaling functions which satisfy a scaling relation

$$\varphi(t) = \sum_{k=0}^{N} a_k \varphi(2t - k)$$

for some positive integer $N > 2$ and coefficients a_k (by (7.5.11), $N = 2$ was necessary and sufficient for Haar wavelets). Then the mother wavelet can be selected to be

$$\psi(t) = \sum_{k=0}^{N} (-1)^k a_{N-k} \varphi(2t - k),$$

and the corresponding wavelet system can be built with its help via formula (7.5.2). Such an approach was suggested by Ingrid DAUBECHIES in 1988, and the resulting wavelets are called *Daubechies wavelets*.

Conceptually, the above construction is a clearcut generalization of the construction of Haar wavelets from the scaling function I provided in the previous section. However, for $N > 2$, the selection of coefficients a_k becomes highly nontrivial. Also, as a rule, the smoother one wants the wavelets one wants, the larger the N one has to take.

Below, we provide a sketch of the relatively simple construction of continuous Daubechies wavelets which is due to David POLLEN (1992). Their scaling function $\varphi(t)$ satisfies the *scaling relation*

$$\varphi(t) = a\varphi(2t) + (\overline{1-a})\varphi(2t-1) + (1-a)\varphi(2t-2) + \bar{a}\varphi(2t-3), \qquad (1)$$

where

$$a = \frac{1+\sqrt{3}}{4}, \qquad (2)$$

and where, for real numbers of the form $\alpha + \beta\sqrt{3}$ with (dyadic) rational α, β, the overline indicates the "conjugation" operation

$$\overline{\alpha + \beta\sqrt{3}} = \alpha - \beta\sqrt{3}.$$

The support of the resulting $\varphi(t)$ is contained in the interval $[0, 3]$ and, additionally,

$$\sum_{k=-\infty}^{\infty} \varphi(k) = 1. \qquad (3)$$

Assume that there exists a scaling function $\varphi(t)$ supported by $[0,3]$ and satisfying (1) and (3) for integer values of the argument t. The scaling relation (1) written for $t = 0, 1, 2, 3$ becomes a matrix equation

$$
\begin{pmatrix} \varphi(0) \\ \varphi(1) \\ \varphi(2) \\ \varphi(3) \end{pmatrix} = \begin{pmatrix} a & 0 & 0 & 0 \\ (1-a) & \overline{1-a} & a & 0 \\ 0 & \bar{a} & (1-a) & \overline{1-a} \\ 0 & 0 & 0 & \bar{a} \end{pmatrix} \begin{pmatrix} \varphi(0) \\ \varphi(1) \\ \varphi(2) \\ \varphi(3) \end{pmatrix}
$$

which, in view of condition (3), has exactly one solution:

$$
\varphi(0) = 0, \quad \varphi(1) = \frac{1 + \sqrt{3}}{2}, \quad \varphi(2) = \frac{1 - \sqrt{3}}{2}, \quad \varphi(3) = 0.
$$

Starting with these prescribed values and using the scaling relation (1) one can produce values of the scaling function $\varphi(t)$ for any dyadic rational t. For example,

$$
\varphi(1/2) = \frac{2 + \sqrt{3}}{4}, \quad \varphi(3/2) = 0, \quad \varphi(5/2) = \frac{2 - \sqrt{3}}{4},
$$

and so on.

The values of $\varphi(t)$ for dyadic t are clearly of the form $\alpha + \beta\sqrt{3}$ with dyadic α and β. One can also prove (see the Bibliographical Notes) that they also satisfy two extended *partition of unity* (see also (3)) formulas

$$
\sum_{k=-\infty}^{\infty} \varphi(t - k) = 1
$$

and

$$
\sum_{k=-\infty}^{\infty} \left(\frac{3 - \sqrt{3}}{2} + k \right) \varphi(t - k) = t.
$$

Since the support of $\varphi(t)$ is contained in $[0, 3]$ the above properties also give the interval translation properties for dyadic $t \in [0, 1]$:

$$
2\varphi(t) + \varphi(t + 1) = t + \frac{1 + \sqrt{3}}{2},
$$

$$
2\varphi(t + 2) + \varphi(t + 1) = -t + \frac{3 - \sqrt{3}}{2},
$$

$$
\varphi(t) - \varphi(t + 2) = t + \frac{-1 + \sqrt{3}}{2}.
$$

Combining them with the scaling relation (1) gives the scaling relations for dyadic $t \in [0, 1]$:

$$\varphi\left(\frac{0+t}{2}\right) = a\varphi(t);$$

$$\varphi\left(\frac{1+t}{2}\right) = \bar{a}\varphi(t) + at + \frac{2+\sqrt{3}}{4};$$

$$\varphi\left(\frac{2+t}{2}\right) = a\varphi(1+t) + \bar{a}t + \frac{\sqrt{3}}{4};$$

$$\varphi\left(\frac{3+t}{2}\right) = \bar{a}\varphi(1+t) - at + \frac{1}{4};$$ (4)

$$\varphi\left(\frac{4+t}{2}\right) = a\varphi(2+t) - \bar{a}t + \frac{3-2\sqrt{3}}{4};$$

$$\varphi\left(\frac{5+t}{2}\right) = \bar{a}\varphi(2+t).$$

Compared with the original scaling relation (1), they have a clear advantage: the values of $\varphi(t)$ at the next resolution level depend only on one value at the previous resolution level (instead of four in (1)).

The above formulas form a basis for the following recursive construction of the continuous version of the scaling function on the whole interval [0,3]. Start with function $g_0(t)$ which is equal to $\varphi(t)$ at integers 0,1,2,3, and which linearly interpolates φ in-between these integers. Clearly, $g_0(t)$ is continuous. In the next step, form $g_1(t)$ at the second resolution level by applying the (right-hand sides of) scaling relations (4) to g_0. More precisely, for $t \in [0, 1]$, define

$$g_1\left(\frac{0+t}{2}\right) = ag_0(t);$$

$$g_1\left(\frac{1+t}{2}\right) = \bar{a}g_0(t) + at + \frac{2+\sqrt{3}}{4};$$

$$g_1\left(\frac{2+t}{2}\right) = ag_0(1+t) + \bar{a}t + \frac{\sqrt{3}}{4};$$

$$g_1\left(\frac{3+t}{2}\right) = \bar{a}g_0(1+t) - at + \frac{1}{4};$$

$$g_1\left(\frac{4+t}{2}\right) = ag_0(2+t) - \bar{a}t + \frac{3-2\sqrt{3}}{4};$$

$$g_1\left(\frac{5+t}{2}\right) = \bar{a}g_0(2+t).$$

Outside $[0, 3]$ set $g_1(t) = 0$. Function $g_1(t)$ is continuous and coincides with $\varphi(t)$ at dyadic points with resolution 2^{-1} (in-between, it again provides a linear interpolation). Continuing this procedure we obtain a sequence g_n of continuous, piecewise linear functions (zero outside $[0, 3]$) which agree with $\varphi(t)$ at dyadic points of the form $k2^{-n}$.

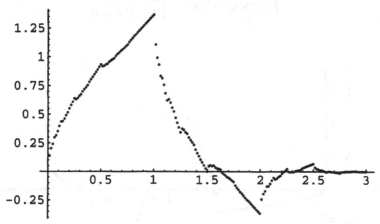

FIGURE 7.6.1
Values of the Daubechies' scaling function computed at dyadic points $t = n \cdot 2^{-6}, 0 \leq t \leq 3$, via the scaling relation (1).

Notice that functions $|g_n(t)| \leq 3$ for all $n = 1, 2, \ldots$, and since $0 \leq |\bar{a}| \leq a < 1$ (see (2)), we get that

$$\max_t |g_k(t) - g_{k+j}(t)| \leq a^k \max_t |g_0(t) - g_j(t)| \leq 6a^k.$$

Hence the sequence of functions $g_k(t)$ satisfies uniformly the Cauchy condition, and the limit

$$\varphi(t) = \lim_{n \to \infty} g_n(t).$$

is a continuous function. This is the scaling function we were searching for.

Remark 1. Note that the scaling function $\varphi(t)$ is not differentiable because

$$\lim_{j \to \infty} \frac{\varphi(2^{-j}) - \varphi(0)}{2^{-j}} = \lim_{j \to \infty} \frac{\varphi(2^{-j})}{2^{-j}} = \lim_{j \to \infty} \frac{a^j \varphi(1)}{2^{-j}} = \lim_{j \to \infty} (2a)^j \varphi(1) = \infty,$$

since $2a > 1$ and $\varphi(0) \neq 0$.

With some additional work one can now establish that

$$\int \varphi(t)\, dt = 1,$$

and that the integer translations of $\varphi(t)$ form an orthonormal system, that is

$$\int \varphi(t)\varphi(t-k)\, dt = \begin{cases} 0, & \text{if } k \neq 0; \\ 1, & \text{if } k = 0. \end{cases}$$

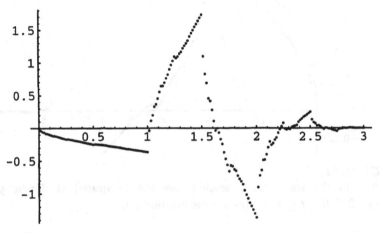

FIGURE 7.6.2
Values of the Daubechies' mother wavelet computed at dyadic points $t = n \cdot 2^{-6}, 0 \leq t \leq 3$, via the formula (5).

Following the general scheme explained in detail for the Haar wavelets in Section 7.5, we can now define the mother wavelet $\psi(t)$ via equality

$$\psi(t) = -\bar{a}\varphi(2t) + (1-a)\varphi(2x-1) - (\overline{1-a})\varphi(2x-2) + a\varphi(2x-3), \quad (5)$$

and check that

$$\int \psi(t)\, dt = 0.$$

The integer shifts of the mother wavelet are orthonormal, that is

$$\int \psi(t)\psi(t-k)\, dt = \begin{cases} 0, & \text{if } k \neq 0; \\ 1, & \text{if } k = 0. \end{cases}$$

Moreover, the scaling function φ and the mother wavelet ψ are orthogonal as well, that is

$$\int \varphi(t)\psi(t-k)\,dt = 0.$$

Thus, again, by an argument similar to that used for the Haar wavelets, the set of Daubechies wavelets

$$\psi_{m,n}(t) = 2^{m/2}\psi(2^m - n), \quad m,n = \ldots -1, 0, 1 \ldots,$$

forms an orthonormal complete basis in $L^2(\mathbf{R})$, and so does the set of functions

$$\varphi_n(t) = \varphi(t-n), \quad n = \ldots, -1, 0, 1, \ldots,$$

$$\psi_{m,n}(t) = 2^{m/2}\psi(2^m - n), \quad m = \ldots 0, 1, 2 \ldots, \quad n = \ldots, -1, 0, 1, \ldots.$$

7.7 Wavelets and distributions

The scaling relation

$$\Phi(t) = 2^{1/2} \sum_k a_k \Phi(2t - k) \tag{1}$$

can also have distributional, rather than just function solutions (following our convention we denote distributions by capital letters). In the most trivial case when (1) has only one nonzero term, say $a_0 = \sqrt{2}$, such a solution may be guessed immediately:

$$\Phi(t) = \delta(t),$$

since $\delta(t) = 2\delta(2t)$.

However, in the case of the scaling relation (1) with finitely many (but at least two) nonzero coefficients a_k, the scaling distribution Φ can no longer be a linear combination of Dirac deltas, that is, of the form

$$\Phi(t) = \sum_{k=0}^{n} c_k \delta(t - t_k). \tag{2}$$

This can be seen as follows. Suppose that the nonzero coefficients are a_0 and a_1, and perhaps some others. Since the support of Φ must be contained in the interval

[0, 1], we may take $0 \leq t_0 < t_1 < \ldots < t_n \leq 1$ in (2). The scaling relation (1) then forces equation

$$\sum_{k=0}^{n} c_k \delta(t - t_k) = 2^{1/2} \sum_{k=0}^{n} \left[a_0 \frac{c_k}{2} \delta \left(t - \frac{t_k}{2} \right) + a_1 \frac{c_k}{2} \delta \left(t - \frac{t_k + 1}{2} \right) + \ldots \right].$$

Comparing coefficients for the same Dirac deltas on either side of the above equation, we get in the case $t_0 = 0$ that

$$c_0 = 2^{1/2} a_0 \frac{c_0}{2},$$

and since $a_0 \neq 2^{1/2}$ we get that $c_0 = 0$. If $t_0 > 0$ then

$$2^{1/2} c_0 \frac{a_0}{2} = 0$$

so that again $c_0 = 0$. By the same argument we get that $c_1 = c_2 = \ldots = c_n = 0$ because $t_k/2 < t_k < (t_k + 1)/2$. So, there are no solutions of the form (2).

The situation is better if infinitely many nonzero coefficients are permitted in the scaling relation (1), and we will indicate some avenues that can be pursued in such a case.

One option is to seek a solution $\Phi \in S'$ of (1) which is a distribution with compact support. Then its Fourier transform $\tilde{\Phi}(\omega)$ is an analytic function in the entire complex plane C, and the scaling relation (1) translates into the following relation for $\tilde{\Phi}(\omega)$:

$$\tilde{\Phi}(\omega) = 2^{-1/2} \sum_{k} a_k \exp(-i\omega k/2) \tilde{\Phi}(\omega/2).$$

We, however, will not follow this route, and instead will construct the scaling distribution (and the corresponding wavelets) by demanding that its integer translates form an orthonormal basis at the zero resolution level.

A mathematical aside: orthogonality of distributions.[2] To make a rigorous discussion of the *orthogonality of distributions* possible we have to select an inner product of (at least some) distributions. One such possibility is the inner product

$$\langle T, S \rangle = \frac{1}{2\pi} \int \frac{\tilde{T}(\omega) \tilde{S}^*(\omega)}{\omega^2 + 1} d\omega, \tag{3}$$

---

[2]This material may be skipped by the first time reader.

which is well defined for all the distribution in the *Sobolev space*

$$\mathcal{H} := \left\{ f \in \mathcal{S}' : \|f\|^2 = \int \frac{|\tilde{f}(\omega)|^2}{\omega^2 + 1} d\omega < \infty \right\}, \tag{4}$$

which is a subspace of the space of tempered distributions $\mathcal{S}'$. Clearly, the Dirac delta $\delta(t) \in \mathcal{H}$ since its Fourier transform is identically equal to $1/2\pi$.

Now, our job is to construct a multiresolution analysis of the Sobolev space $\mathcal{H}$, and the first step is to find a scaling distribution Φ, the integer translates thereof would be a orthonormal basis at the zero resolution level, that is for

$$\mathcal{V}_0 = \{ f \in \mathcal{H} : \text{supp } f \subset \mathbf{Z} \}.$$

Distributions in $\mathcal{V}_0$ are of the form

$$T(t) = \sum_k c_k \delta(t - k), \qquad \sum |c_k|^2 < \infty. \tag{5}$$

Any distribution in $\mathcal{H}$ can be then approximated by partial sums of dilations of (5).

The first difficulty one encounters is that the usual dilations do not preserve the norm $\|.\|$ in $\mathcal{H}$. So to give ourselves some leeway, we will allow use of other inner products in $\mathcal{H}$, and introduce the family of inner products

$$\langle T, S \rangle_\alpha = \frac{1}{2\pi} \int \frac{\tilde{T}(\omega) \tilde{S}^*(\omega)}{\omega^2 + \alpha^2} d\omega, \tag{6}$$

parameterized by parameter $\alpha > 0$. They all generate norms $\|.\|_\alpha$ equivalent to the original norm $\|.\| = \|.\|_1$, and the orthogonality notions for all of them are equivalent. Their role will become clear later on.

If Φ is of the form (5), then the orthonormality condition gives that

$$\delta_{ok} = \langle \Phi(t), \Phi(t - k) \rangle_\alpha = \frac{1}{2\pi} \int \frac{|\tilde{\Phi}(\omega)|^2 e^{-i\omega k}}{\omega^2 + \alpha^2} d\omega.$$

$$= \frac{1}{2\pi} \sum_n \int_0^{2\pi} \frac{|\tilde{\Phi}(\omega + 2\pi n)|^2}{(\omega + 2\pi n)^2 + \alpha^2} e^{-i\omega k} d\omega$$

Thus δ_{ok} are Fourier coefficients (in $L^2(\mathbf{R})$) of a function appearing as a fraction in the last integral, which therefore must be equal (in $L^2(\mathbf{R})$) to its Fourier series

$$\sum_n \frac{|\tilde{\Phi}(\omega + 2\pi n)|^2}{(\omega + 2\pi n)^2 + \alpha^2} = \sum_k \delta_{0k} e^{i\omega k} = 1.$$

Since $\tilde{\Phi}$ has period 2π, the desired scaling function in $\mathcal{V}_0 \subset \mathcal{H}$ has the Fourier transform

$$\tilde{\Phi}(\omega, \alpha) = \left[\sum_n \frac{1}{(\omega + 2\pi n)^2 + \alpha^2} \right]^{-1/2} = \left[\frac{2\alpha(\cosh \alpha - \cos \omega)}{\sinh \alpha} \right]^{1/2}. \tag{7}$$

That the sum of the above infinite series is as indicated can be seen as follows (see also formula (8.7.7)).

By expanding $\exp(-ixt)$ on $[0,1]$ in the Fourier series, and evaluating it at $t = 0$, we get that

$$\sum_{n=-\infty}^{\infty} \frac{1}{x + 2\pi n} = \frac{1}{2} \frac{\cos(x/2)}{\sin(x/2)}$$

so that

$$a(\omega, \alpha) := \sum_n \frac{1}{(\omega + 4\pi n)^2 + \alpha^2}$$

$$= \frac{i}{4\alpha} \sum_n \frac{1}{(\omega + \alpha i)/2 + 2\pi n} - \sum_n \frac{1}{(\omega - \alpha i)/2 + 2\pi n}$$

$$= \frac{i}{8\alpha} \left[\frac{\cos(\omega + i\alpha)/4}{\sin(\omega + i\alpha)/4} - \frac{\cos(\omega - i\alpha)/4}{\sin(\omega - i\alpha)/4} \right]$$

$$= \frac{i}{8\alpha} \frac{\sin(-i\alpha/2)}{\sin^2 \omega/4 \cos^2 \alpha i/4 - \cos^2 \omega/4 \sin^2 \alpha i/4}$$

$$= \frac{1}{4\pi} \frac{\sinh \alpha/2}{\cos \alpha/2 - \cos \omega/2}.$$

The Fourier transform $\tilde{\Phi}$ given by (7) cannot be simply inverted. However, it turns out that the construction of the corresponding mother wavelet distribution Ψ leads to an invertible Fourier transform. Indeed, since

$$\Psi(t) = \sum_k d_k \sqrt{2} \Phi(2t - k) \tag{8}$$

for some coefficients d_k, and since the integer translations of the mother wavelet have to be orthonormal and orthogonal to the scaling function Φ (that is, orthogonal to all integer translations of the Dirac delta) we obtain the following conditions on the Fourier transform $\tilde{\Psi}$:

$$\sum_n \frac{|\tilde{\Psi}(\omega + 2\pi n)|^2}{(\omega + 2\pi n)^2 + \alpha^2} = \sum_k \delta_{0k} e^{i\omega k} = 1,$$

$$\sum_n \frac{\tilde{\Psi}(\omega + 2\pi n)}{(\omega + 2\pi n)^2 + \alpha^2} = \sum_k \delta_{0k} e^{i\omega k} = 0.$$

In view of the scaling relation (8), $\hat{\Psi}$ has period π, and both series can be simplified by separating odd and even terms, which leads to equations

$$|\hat{\Psi}(\omega)|^2 a(\omega, \alpha) + |\hat{\Psi}(\omega + \pi)|^2 a(\omega + 2\pi, \alpha) = 1,$$

$$\hat{\Psi}(\omega)a(\omega, \alpha) + \hat{\Psi}(\omega + \pi)a(\omega + 2\pi, \alpha) = 0.$$

Their easily identifiable solution is

$$\hat{\Psi}(\omega, \alpha) = e^{-i\omega/2} \left(\frac{a(\omega + 2\pi)}{a(\omega, \alpha)(a(\omega + 2\pi, \alpha) + a(\omega, \alpha))} \right)^{1/2}$$

$$= e^{-i\omega/2}(\cosh \alpha/2 - \cos \omega/2) \left(\frac{4\alpha}{\sinh \alpha} \right)^{1/2},$$

(taking just the positive root does not work here as it did in (7), since it does not satisfy the second of the above pair of equations).

For $\alpha = 2^{-m}$ (and this is just what we need for the construction of a wavelet basis in $\mathcal{H}$) the easily evaluated inverse transform of $\hat{\Psi}(\omega, \alpha)$ gives the following formula for the mother wavelet

$$\Psi(t, 2^{-m}) = C \left[2 \cosh(2^{-m-1})\delta(t - 1/2) - \delta(t) - \delta(t - 1) \right]$$

for a certain constant C.

Now, we are finally ready for the construction of the orthonormal wavelet basis in the Sobolev distribution space $\mathcal{H}$, using the mother wavelet Ψ as the starting point. Here, the advantage of using the family of norms $\|.\|_\alpha$ becomes obvious. As we observed earlier, the classical rescaling

$$\Psi_1(t) = 2^{1/2}\Psi(2t)$$

does not preserve the norm $\|.\|$, but

$$\|\Psi\|_\alpha = 2^{-1}\|\Psi_1\|_{2\alpha}.$$

Therefore, the correct rescaling in $\mathcal{H}$ which produces the first level resolution wavelet is

$$\Psi_1(t) = 2^{3/2}\Psi(2t, 2^{-1}),$$

and the complete multiresolution wavelet basis for $\mathcal{H}$ is provided by the integer translation of the scaling distribution

$$\Phi_n(t) = \Phi(t - n), \quad n = \ldots, -1, 0, 1, \ldots,$$

and the wavelets

$$\Psi_{m,n}(t) = 2^{3m/2}\Psi(2^m t - n, 2^{-m}), \quad n = \ldots, -1, 0, 1, \ldots, \quad m = 0, 1, \ldots,$$

or, alternatively, by the wavelet system

$$\Psi_{m,n}(t) = 2^{3m/2}\Psi(2^m t - n, 2^{-m}), \quad n, m = \ldots, -1, 0, 1, \ldots,$$

complemented by linear combinations of the Dirac delta δ.

Remark 1.   One can also expand distributions in weakly convergent series of smooth function wavelets. In this case, the typical result is that a tempered distribution which is the r-th derivative of a function (measure) of polynomial growth has a wavelet expansion with the coefficients $a_n = O(|n|^k)$, for some integer k.

More precisely, assume that the scaling function φ is in S_r, that is it has r continuous and rapidly decreasing derivatives. In other words,

$$|\varphi^{(k)}(t)| \le C_{p,k}(1 + |t|)^{-p}, \quad k = 0, 1, \ldots, r, \quad p \in \mathbf{N}, t \in \mathbf{R}.$$

The spaces S_r contain the space S of rapidly decreasing test functions. Then the mother wavelet ψ is also in S_r, and for any tempered distribution $T \in S_r'$ (of order r) the expansion with respect to $\varphi(t - n)$ and $\psi(t - n)$ exists and the coefficients $a_n = O(|n|^k)$ for some integer k.

As a result, for $T \in S_r'$, we have the usual wavelet expansion

$$T = \sum_{m,n} b_{m,n}\psi_{m,n}$$

where

$$\psi_{m,n}(t) = 2^{m/2}\psi(2^n t - n),$$

and where the convergence is in S_r' or, alternatively,

$$T(t) = \sum_{n} a_n\varphi(t - n) + \sum_{m \ge 0, n} b_{m,n}\psi_{m,n}(t).$$

7.8 Exercises

Function spaces:

1. Prove the inequality $P = \int |f(t)g(t)| \, dt \leq \|f\| \|g\|$ which was used, among others, in the proof of the uncertainty principle in Section 7.2. Compare with the Schwartz inequality (7.1.9).

2. Prove the triangle inequality $\|f + g\| \leq \|f\| + \|g\|$ for the functional Hilbert space L^2.

Windowed Fourier transform:

3. Let $\tilde{f}(\omega, \tau)$ be the windowed Fourier transform of the signal $f(t)$. Denote by $\tilde{f}'(\omega, \tau)$ the windowed Fourier transform of the derivative $f'(t)$. Express $\tilde{f}'$ in terms of $\tilde{f}$.

4. Signal $x(t)$ is a solution of the differential equation

$$\frac{dx(t)}{dt} + hx(t) = f(t),$$

where $f(t)$ is a signal with known windowed Fourier image $\tilde{f}(\omega, \tau)$ and h is a (real or complex) constant. Express the windowed Fourier image of $x(t)$ in terms of $\tilde{f}(\omega, \tau)$.

5. Let $\tilde{f}(\omega, \tau)$ be the windowed Fourier image of signal $f(t)$ (i.e. $f(t) \mapsto \tilde{f}(\omega, \tau)$). Find the windowed Fourier images of signals (a) $f(t)e^{i\omega_0 t}$, and (b) $f(t + \theta)$.

6. Find the windowed Fourier transform $\tilde{f}(\omega, \tau)$ of function $f(t) = e^{vt}$.

7. Utilizing results of the previous exercises, find the windowed Fourier image of the signal $f(t) = e^{vt} \cos \omega_0 t$.

8. Assume that the values of the windowed Fourier image $\tilde{f}(\omega, \tau)$ of signal $f(t)$ are known outside the interval $\tau \in [0, T]$ only. Is it possible, on the basis of this incomplete information, to recover values of signal $f(t)$ for all values of t?

Wavelets:

9. Provide the expression for coefficient D in (7.4.49) if one employs the more popular in mathematical literature definition (7.3.15) of the Fourier transform.

10. The distributional relation

$$\frac{1}{D} \int_0^\infty K\left(\frac{s}{\lambda}\right) \frac{d\lambda}{\lambda^2} = \delta(s) \tag{1}$$

played the principal role in derivation of formulas for the inverse continuous wavelet transform. In practice it is impossible to carry out the integration all the way to $\lambda = 0$. Show that the regularized integral

$$g_\epsilon(s) = \frac{1}{D} \int_\epsilon^\infty K\left(\frac{s}{\lambda}\right) \frac{d\lambda}{\lambda^2}$$

converges weakly to the Dirac delta as $\epsilon \to 0$.

11. Find an explicit formula for function $\Phi(x) = ((6 - x^2)/(8\sqrt{\pi}))\exp(-x^2/4)$ (see answer to the Exercise 10) in the case of the Mexican hat mother wavelet (7.4.15).

12. Obtain, in the case of two-sided mother wavelets, a formula connecting $|\hat{f}|^2$ and $|f|^2$, analogous to the Parseval formula for the ordinary Fourier transform.

13. Denote by $\hat{f}_a(\lambda, \tau)$ the continuous wavelet image of signal $f(at)$, $a > 0$, compressed ($a > 1$) or dilated ($a < 1$) in comparison with the original signal $f(t)$. Find out how $\hat{f}_a(\lambda, \tau)$ is related to the continuous wavelet image of signal $f(t)$ itself in the case of wavelet transform definition (7.4.5).

14. Find $\hat{f}(\lambda, \tau)$ (7.4.1) for the self-similar signal $f(t) = |t|^\alpha$.

15. Let
$$f(t) = (5 - 9t + 4t^2)/(5 - 12t + 8t^2)^3.$$
Find numerically and graphically the Haar wavelet expansion of $f(t)$ with resolution level coarser than 0 and finer than 6. Graph the resolution level n contents of $f(t)$ for $n = 0, 1, \ldots, 6$. Use your computer in order to estimate numerically the maximum error of your approximation.

16. Use your computer and the defining scaling relations to produce *numerical* values of the Daubechies scaling function and mother wavelet at the dyadic points up to resolution level 6.

Chapter 8

Summation of Divergent Series and Integrals

The theory of distributions has a flavor similar to the theory of summation of divergent series and integrals and, as we have seen in Chapter 6, is closely related to the theory of singular integrals. A generic alternating series

$$1 - 1 + 1 - 1 + \ldots = \sum \pm 1$$

is a good example here. It seems to make no sense to assign a specific value to this infinite sum. Nevertheless, mathematicians have produced certain reasonable rules of summation that assign to it value 1/2. Such an assignment is in complete agreement with the intuition of physicists who encounter similar series. In this chapter, we will see how one can sum this, or even more strange, divergent series and integrals. To gain a better insight into the essence of this problem, let us begin with elementary examples and recall basic notions and theorems of the ordinary theory of convergent infinite series.

8.1 Zeno's "paradox" and convergence of infinite series

8.1.1. Geometric series. Recall the celebrated 4th century B.S. Achilles-and-tortoise "paradox" due to the Greek philosopher Zeno. At time $t = 0$ Achilles is at point $x = 0$ and the tortoise at $x = 1$. Achilles begins to chase the tortoise with velocity +1, and the tortoise begins to run away with velocity $0 < v < 1$.

To catch the tortoise Achilles has to first reach point $x = 1$. This will happen at time $\tau_0 = 1$. By that time the tortoise will have moved by distance $v\tau_0 = v$. To cover that distance Achilles needs time $\tau_1 = v$, during which the tortoise will have moved further to the right by distance $v\tau_1 = v^2$ (see Fig 8.1.1).

The pattern continues *ad infinitum*, seemingly indefinitely delaying the moment

© Springer Nature Switzerland AG 2018
A. I. Saichev and W. Woyczynski, *Distributions in the Physical and Engineering Sciences, Volume 1*, Applied and Numerical Harmonic Analysis, https://doi.org/10.1007/978-3-319-97958-8_8

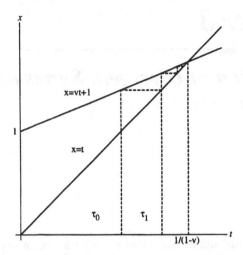

FIGURE 8.1.1
A graph representing paths of Achilles chasing the tortoise. They will meet at the point of intersection of the corresponding straight lines.

when Achilles catches the tortoise. Of course, time t needed by Achilles to catch the tortoise is the sum of an infinite series

$$t = \tau_0 + \tau_1 + \tau_2 + \ldots,$$

where $\tau_m = v^m$, $m = 0, 1, 2, \ldots$, so that

$$t = 1 + v + v^2 + v^3 + \ldots = \sum_{m=0}^{\infty} v^m, \tag{1}$$

is the sum of a *geometric series*.

Leaving aside philosophical significance of Zeno's "paradox" concerning the nature of space and time, let us underline the basic mathematical difficulty encountered in formula (1): we are trying to perform infinitely many mathematical operations, each of which takes a certain amount of time. At a naive level, this seems to be an impossible task if we want to do it exactly.

An insight into how this difficulty can be overcome is gained by a look at the problem from the physical viewpoint. To find time t when Achilles catches the tortoise, it suffices to solve the system of equations of motion for both of them:

$$x = vt + 1, \tag{2a}$$

$$x = t. \tag{2b}$$

The solution is

$$t = \frac{1}{1-v},$$
(3)

which, together with formula (1), gives

$$\sum_{m=0}^{\infty} v^m = \frac{1}{1-v}.$$
(4)

Equation (4) attaches to the infinite series on its left-hand side, a number from the analytic expression on its right-hand side. Note that the latter has a well defined mathematical and physical meaning, even if the left-hand side does not form a convergent series. For example, for $v = -1$, equation (4) states that

$$1 - 1 + 1 - 1 + \ldots = \frac{1}{2},$$
(5)

an equality anticipated in the preamble to this chapter. The question is: How can we provide a mathematical justification of this type of identities? The answer is given by the theory of *summability of divergent series* which formalizes procedures that are consistent with the physical principle of infinitesimal relaxation. On the other hand, we'll be able to elucidate rigorous mathematical constructs of summability theory by looking at their physical roots. Observe that series (1) which appears on the left-hand side of equality (4) arises in an attempt to solve equations of motion (2-3) by the method of consecutive approximations in parameter v.

8.1.2. Criteria for convergence. The above example illustrates the basic difference between finite sums

$$S_n = \sum_{m=0}^{n} a_m$$
(6)

and the infinite series

$$S = \sum_{m=0}^{\infty} a_m.$$
(7)

The value of a finite sum can always be explicitly computed (at least in principle), whereas it is not always possible to assign a numerical value to an infinite series. However, for some series, the sequence of *partial sums*

$$S_0, S_1, S_2, \ldots, S_n, \ldots$$
(8)

converges, as $n \to \infty$, to a finite limit

$$S = \lim_{n \to \infty} S_n,$$
(9)

and then it is natural to think about S as the sum of the infinite series, the value of which is being approximated by computable finite partial sums. Such series are called *convergent*.

One learns in calculus that it is possible to find out whether the series converges or not without actually computing the limit of its partial sums. One such approach is based on the so-called *Cauchy criterion* which states:

Series S converges if and only if for any given number $\varepsilon > 0$, one can find a integer $N = N(\varepsilon)$ such that, for any $n \geq N$ and arbitrary positive integer $k = 1, 2, \ldots$,

$$|S_{n+k} - S_n| < \varepsilon. \tag{10}$$

If the series converges, then its *remainder*

$$R_n = S - S_n = \sum_{m=n+1}^{\infty} a_m$$

is a well defined number, and its absolute value measures the error of approximation of the infinite series S by its finite partial sum S_n. In particular, if $n \geq N(\varepsilon)$ then $|R_n| < \varepsilon$.

Example 1. Geometric series. Let us return to the geometric series

$$S = \sum_{m=0}^{\infty} q^m = 1 + q + q^2 + \ldots.$$

Its partial sums S_n obviously satisfy identity

$$S_n = 1 + q + q^2 + \ldots + q^n = 1 + q(S_n - q^n),$$

from which we immediately get that

$$S_n = \frac{1 - q^{n+1}}{1 - q}.$$

Hence, the partial sums sequence S_n converges if and only if $|q| < 1$. If it does converge, its limit

$$S = \lim_{n \to \infty} S_n = \frac{1}{1 - q}.$$

The remainder R_n is also easily computable:

$$R_n = \frac{1}{1 - q} - S_n = \frac{q^{n+1}}{1 - q}. \qquad \blacksquare$$

A comparison of terms of an arbitrary series with corresponding terms of the geometric series gives a handy sufficient condition of convergence of general series.

If, for all sufficiently large m, we have that $|a_m|^{1/m} < q < 1$, then the series $\sum_{m=0}^{\infty} a_m$ converges.

If a series does not converge then we call it a *divergent series*. One obvious example is the geometric series with $|q| \geq 1$.

Example 2. Harmonic series. Consider the harmonic series

$$H = \sum_{m=0}^{\infty} \frac{1}{m+1} = 1 + \frac{1}{2} + \frac{1}{3} + \ldots.$$

As we have observed in Section 4.1, its partial sums H_n satisfy an asymptotic relation

$$H_n \sim \ln(n+1) + \gamma, \qquad (n \to \infty) \tag{11}$$

which immediately proves the divergence of the harmonic series. ∎

Note, however, that if we alternate signs of the harmonic series to obtain

$$L = \sum_{m=0}^{\infty} \frac{(-1)^m}{m+1} = 1 - \frac{1}{2} + \frac{1}{3} - \ldots,$$

then the latter series converges. This is due to the general phenomenon which makes all the alternating series of the form

$$D = \sum_{m=0}^{\infty} (-1)^m c_m, \qquad c_1 > c_2 > \ldots c_m \to 0_+,$$

converge. The corresponding formal statement is known in calculus as the *Leibniz Theorem*.

The alternating harmonic series is an example of a convergent series $\sum_m a_m$ for which the series of absolute values $\sum_m |a_m|$ diverges. Series $\sum_m a_m$ which converges together with the series $\sum_m |a_m|$ of absolute values is called *absolutely convergent*. Obviously, any geometric series with $|q| < 1$ converges absolutely.

8.1.3. Conditional and absolute convergence. Summation of convergent infinite series is an *associative operation*. This means that if we group the terms of a convergent series into finite blocks, then the sum of the whole series is equal to the sum of the series consisting of these blocks. More precisely, the associativity means that if we select an increasing sequence of integers

$$0 < k_0 < k_1 < k_2 < k_3 < \ldots < k_m < \ldots$$

and denote the finite sums

$$a_0' = a_0 + a_1 + \ldots + a_{k_0}, \quad \ldots \quad , a_m' = a_{k_{m-1}+1} + \ldots + a_{k_m}, \ldots,$$

sums in finite blocks generated by the above sequence, then the series

$$S' = \sum_{m=0}^{\infty} a_m'$$

also converges and $S = S'$. This immediately follows from the fact that the sequence of partial sums $\{S_0', S_1', \ldots, S_n', \ldots\}$ forms a subsequence of the original sequence of partial sums $\{S_0, S_1, \ldots, S_n, \ldots\}$.

In contrast, summation of convergent series is not necessarily a *commutative operation*. A series formed by a permutation (perhaps infinite) of terms of a convergent series may converge to a different limit, or may even diverge. If a series converges to the same limit after any permutation of its terms, then we call it *unconditionally convergent*. It turns out that the necessary and sufficient condition for the unconditional convergence of a series is its absolute convergence. If a series converges, but not absolutely, then in view of the above statement, one says that it converges *conditionally*. For a conditionally convergent series, the *Riemann Theorem* asserts that any number is the sum of a certain permutation of the original series. This striking theorem will not be proved here (see Bibliography for the proof), but we will illustrate it by a concrete example.

Example 3. A conditionally convergent series. Select positive integers p and q, and consider a permutation of the alternating harmonic series in which q negative terms follow p positive terms:

$$L' = 1 + \frac{1}{3} + \ldots + \frac{1}{2p-1} - \frac{1}{2} - \ldots - \frac{1}{2q} + \frac{1}{2p+1} + \ldots + \frac{1}{4p-1} - \ldots .$$

Consider partial sums $L'_{n(p+q)}$ containing the first $n(p+q)$ terms of the rearranged series. Since finite summation is commutative,

$$L'_{n(p+q)} = \sum_{m=0}^{np-1} \frac{1}{2m+1} - \sum_{m=0}^{nq-1} \frac{1}{2(m+1)}.$$

Now,

$$\sum_{m=0}^{np-1} \frac{1}{2m+1} = \sum_{m=0}^{2np-1} \frac{1}{m+1} - \frac{1}{2} \sum_{m=0}^{np-1} \frac{1}{m+1}$$

and

$$\sum_{m=0}^{nq-1} \frac{1}{2(m+1)} = \frac{1}{2} \sum_{m=0}^{nq-1} \frac{1}{m+1},$$

which implies that the sums in the above expression for $L'_{n(p+q)}$ can be expressed through partial sums of the harmonic series. In view of the asymptotic relation (11), as $n \to \infty$,

$$\sum_{m=0}^{2np-1} \frac{1}{m+1} \sim \ln(2np) + \gamma,$$

$$\sum_{m=0}^{np-1} \frac{1}{m+1} \sim \ln(np) + \gamma,$$

so that,

$$L'_{n(p+q)} \sim \ln(2np) - \frac{1}{2}\ln(np) - \frac{1}{2}\ln(nq), \quad (n \to \infty).$$

Hence, finally,

$$\lim_{n \to \infty} L'_{n(p+q)} = \ln\left(2\sqrt{\frac{p}{q}}\right). \tag{12}$$

The reader can complete the example by proving that not only subsequence $L'_{n(p+q)}$ but also the entire series L' converges to the same limit.

It is clear that, if we choose different integers p' and q', thus selecting a different rearrangement of the alternating harmonic series, then we'll get different limits as long as $p/q \neq p'/q'$. In particular, for $p = q = 1$, we get the well known formula

$$L = 1 - \frac{1}{2} + \frac{1}{3} - \ldots = \ln 2$$

but for $p = 2, q = 1$, we obtain that

$$L' = 1 + \frac{1}{3} - \frac{1}{2} + \frac{1}{5} + \frac{1}{7} - \frac{1}{4} + \ldots = \frac{3}{2} \ln 2. \qquad \blacksquare$$

The above example gives a taste of the proof of the Riemann Theorem. Although the latter seems to contradict common sense, it is sometimes used—although without crediting Riemann— by dishonest financiers taking new loans to repay growing old debts. The same argument explains illusory prosperity of declining nations which issue new paper money to cover inflationary budgets.

8.1.4. Functional series and uniform convergence. In what follows we will often work with infinite series

$$S(x) = \sum_{m=0}^{\infty} a_m(x), \qquad (13)$$

where the terms are not numbers but functions. For such series the notion of pointwise convergence, that is the notion of convergence for every point x separately, is not sufficient. For that reason one introduces a stronger notion of *uniform convergence* which will be discussed below.

Assume that series (13) converges for each $x \in [a, b]$. Then, by the definition of *pointwise convergence*, for any $x \in [a, b]$ and for any given number $\varepsilon > 0$, one can find an integer $N = N(\varepsilon, x)$ such that, for any $n \geq N(\varepsilon, x)$ and for an arbitrary positive integer $k = 1, 2, \ldots$, $|S_{n+k}(x) - S_n(x)| < \varepsilon$.

If we can make the selection of $N = N(\varepsilon, x)$ independent of $x \in [a, b]$ (one says "select N uniformly over $[a, b]$"), then the series is said to converge uniformly over $[a, b]$. Notice that the condition of uniform convergence is equivalent to the condition that

$$n(\varepsilon) = \max_{x \in [a,b]} N(\varepsilon, x) < \infty.$$

The importance of uniform convergence becomes obvious when we try to investigate properties of functions $a_m(x)$ that are inherited by the sum $S(x)$ of the whole series. It is easy to see that, in general, the continuity of $a_m(x)$ does not imply the continuity of $S(x)$ if the series (13) converges pointwise. Indeed, if $S_n(x) = x^n$, $x \in [0, 1]$, then although the corresponding terms $a_m(x) = S_m(x) - S_{m-1}(x) = x^m - x^{m-1}$ are continuous, the limit $S(x)$ which equals 0 for $0 \leq x < 1$, and 1 for $x = 1$, is a discontinuous function.

However, the uniform convergence of the series precludes the above situation, and we have the following theorem:

If functions $a_m(x)$, $m = 0, 1, 2, \ldots$, are continuous on $[a, b]$ and the series $\sum_m a_m(x)$ converges uniformly on $[a, b]$, then $S(x)$ is continuous on that interval.

Similarly, pointwise convergence is not sufficient to permit differentiation and integration of the functional series term by term or, in other words, interchanging the order of infinite summation with operations of differentiation and integration. However, in the presence of uniform convergence we have the following two useful theorems:

If functions $a_m(x)$, $m = 0, 1, 2, \ldots$, are integrable on $[a, b]$ and the series $\sum_m a_m(x)$ converges uniformly on $[a, b]$, then

$$\int_a^b \left[\sum_{m=0}^{\infty} a_m(x) \right] dx = \sum_{m=0}^{\infty} \int_a^b a_m(x)\, dx.$$

Assume that functions $a_m(x), m = 0, 1, 2, \ldots$, are defined on $[a, b]$ and are continuously differentiable on (a, b). Then, if the series $\sum_m a_m(x_0)$ converges for an $x_0 \in [a, b]$, and the series of derivatives $\sum_m a'_m(x)$ converges uniformly on (a, b), then

$$\left(\sum_{m=0}^{\infty} a_m(x) \right)' = \sum_{m=0}^{\infty} a'_m(x)$$

for any $x \in (a, b)$.

Checking the uniform convergence of a functional series is not always easy but there exist several criteria that can be helpful. One of the most useful is the following *Weierstrass criterion.*

Let $c_m > 0$ be a sequence of numbers such that the series $\sum_m c_m$ converges. If functions $a_m(x), m = 0, 1, 2, \ldots$, are defined on $[a, b]$, and satisfy, for all $m = 0, 1, 2, \ldots$, inequalities

$$\max_{x \in [a,b]} |a_m(x)| \leq c_m,$$

then the series $\sum_m a_m(x)$ converges uniformly on $[a, b]$.

   Example 4. An application of the Weierstrass criterion. Let $|q| < 1$. The above Weierstrass criterion, combined with properties of the geometric series, immediately yields the uniform convergence of the functional series

$$\sum_{m=0}^{\infty} q^m \cos(mx)$$

on the whole real line.

8.2   Summation of divergent series

In this section we will study the main topic of this chapter: methods of summability of divergent series

$$\sum_{m=0}^{\infty} a_m. \tag{1}$$

We have encountered divergent series before. One example was the series $\sum \pm 1$, which can be obtained by specifying $a_m = (-1)^m$ in expression (1). We also discussed physical situations, where divergent series arise. The solution of equations of motion (8.1.2), with $v = -1$, by the method of successive approximation, provides such an example. In mathematics, the study of divergent series is also motivated by other reasons as for example to deal with the fact that the product of two convergent series can be a divergent series. Quite reasonably, one would like to assign to this product a value equal to the product of sums of the two factor series.

The basic idea, underlying some of the most useful methods of summation of divergent series, is contained in the relatively innocent looking *Abel's Theorem*:

If series $\sum_{m=0}^{\infty} a_m$ converges to sum S then, for any $0 < q < 1$, the power series

$$S(q) = \sum_{m=0}^{\infty} a_m q^m \tag{2}$$

also converges and, additionally,

$$\lim_{q \to 1-} S(q) = S. \tag{3}$$

However, sometimes it happens that the series $\sum a_m$ diverges, but the power series (2) converges for any $0 < q < 1$, and the sums $S(q)$ have a finite limit as $q \to 1-$. In such a situation, S can be defined by formula (3), and is called the *generalized sum* of series (1) *in the Poisson-Abel sense*. Often, the above approach is just called the *Abel method of summation*, although Abel himself never worked on the summability of divergent series.

Example 1. Summing $\sum \pm 1$ by the Abel method. For the series $\sum \pm 1$, the auxiliary power series (2) has the sum

$$S(q) = \sum_{m=0}^{\infty} (-1)^m q^m = \frac{1}{1+q}, \quad |q| < 1.$$

As $q \to 1 - 0$, the above sum $S(q)$ has a finite limit 1/2. Thus, the Abel method leads to the familiar formula (8.1.5)

$$1 - 1 + 1 - 1 + \ldots = 1/2. \tag{4}$$

∎

Another approach, known as the *Cesaro method* of generalized summation, depends on the formation of *arithmetic averages*

$$A_0 = S_0, \quad A_1 = \frac{S_0 + S_1}{2}, \ldots, \quad A_n = \frac{S_0 + S_1 + \ldots + S_n}{n+1}, \ldots \tag{5}$$

of partial sums of the original series (1), and on the investigation of their convergence. If the sequence S_n itself converges, then the sequence of Cesaro averages A_n always converges to the same limit. But sometimes A_n converges even though S_n diverges. In this case, we shall say that the original series (1) is *Cesaro summable* and the limit of the sequence A_n is called the *generalized sum* of (1) *in the Cesaro sense*.

Example 2. Summing $\sum \pm 1$ by the Cesaro method. In the generic example of the series $\sum \pm 1$, the partial sums are $S_{2n} = 1$, $S_{2n+1} = 0$. Consequently,

$$A_{2k} = \frac{k+1}{2k+1}, \quad A_{2k+1} = \frac{1}{2}$$

so that $A_k \to 1/2$. Thus, the Cesaro method leads to the same generalized sum as the Abel method of summation. ∎

There exists a large number of other summability methods. To make them useful, mathematicians usually demand that they satisfy *regularity and linearity conditions*. By definition, a summability method is called *regular* if it assigns to an already convergent series a generalized sum equal to its usual sum S. The method is called *linear*, if a generalized sum for the series $\sum (pa_m + qb_m)$ is equal to $pS + qT$, provided series $\sum a_m$ and $\sum b_m$ have generalized sums S and T, respectively.

Note, that the regularity of the Abel method follows from the Abel Theorem. Its linearity is obvious. It is also not difficult to prove regularity and linearity of the Cesaro method.

Naturally, not all series have finite generalized sums. For example, series $1 + 1 + 1 + \ldots$ has infinite Abel and Cesaro generalized sums, that is, they are not Abel or Cesaro summable. Summable series with alternating signs are sometimes called *semiconvergent series*.

8.3 Tiring Achilles and the principle of infinitesimal relaxation

As we already noticed, properties of finite and infinite sums can be drastically different. Just recall the Riemann Theorem which implies that the addition in conditionally convergent infinite series is not commutative whereas in finite sums it is. For semiconvergent series addition is not associative either; such series are sensitive not only to rearrangements of their terms, but also to their grouping. For example, for the series $\sum \pm 1$, we have

$$(1 - 1) + (1 - 1) + (1 - 1) + \ldots = 0 + 0 + 0 \ldots = 0, \tag{1a}$$

but

$$1 - (1 - 1) - (1 - 1) - \ldots = 1 - 0 - 0 - \ldots = 1. \tag{1b}$$

A comparison of this result with the result of Abel (or Cesaro) summation of the same series (see (8.1.5)) may give you second thoughts about the summability methods introduced in the preceding section. One can argue that (1a) and (1b) taken together, although ambiguous, make more sense than the answer 1/2 obtained before, since one would like the sum of integers to be an integer as well. Similar misgivings can help motivate the search for additional summability methods based on physical arguments.

To illustrate what we have in mind we shall return to the Achilles and tortoise example. However, this time we will closely watch Achilles' consecutive steps along the x-axis, rather than just the overall time he needs to catch the tortoise. Since Achilles' speed is equal to 1, his total displacement is given by series (8.1.1) with t replaced by x:

$$x = \sum_{m=0}^{\infty} v^m = 1 + v + v^2 + \ldots \tag{2}$$

Let us imagine the chase as a series of physical stages, each consisting of Achilles reaching the previous position of the tortoise and resting for an instant (as Zeno prescribed originally), before embarking on the next stage.

In contrast to formula (8.1.1), series (2) has a physical meaning also for negative velocities v of the tortoise. If $v < 0$, the tortoise always runs towards Achilles. For example, if $v = -1$, Achilles reaches point $x = 1$ at the end of the first stage having met the tortoise on his way and, at the same time, the tortoise halts at $x = 0$. Then, Achilles turns around, and at the end of the second stage finds himself at point $x = 0$, while the tortoise has returned to point $x = 1$, and so on. The graph of Achilles' motion is presented in Fig. 8.3.1a.

Now let us strip Achilles of his semigod status and assume, realistically, that he is getting tired chasing the tortoise back and forth. As his, now human, strength is sapped, after each turn he slows down by a factor of q. The graph of Achilles' motion is presented on Fig. 8.3.1b, and his full displacement is given by the sum of an absolutely convergent series

$$x = \sum_{m=0}^{\infty} (-1)^m q^m = \frac{1}{1+q}.$$

Note that in the case $q \to 1 - 0$ of infinitesimally slow relaxation of Achilles' strength, the above sum is equal to 1/2; the infinitesimally tiring Achilles will end his chase in the middle of the interval originally separating him from the tortoise.

The calculation of Achilles' total displacement, provided above in the case of infinitesimal relaxation of his speed, can be applied to an arbitrary series (8.2.1)

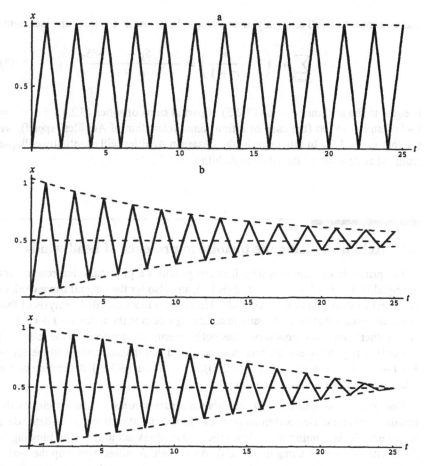

FIGURE 8.3.1
The graphs of motion of tiring Achilles.

in which case it coincides with the Abel generalized summation procedure (8.2.2) and (8.2.3).

The Cesaro method can also be "supported" by a similar physical argument. Suppose that the never tiring Achilles covers the distance a_m in each stage, but his tiring *alter ego* decreases his speed linearly (as opposed to the exponential decay seen in the above example related to the Abel method). In this case

$$v_m = 1 - \frac{m}{n+1}$$

and he comes to a full stop only after n steps. In this case, his total displacement

$$x_n = \sum_{m=0}^{n} a_m \left(1 - \frac{m}{n+1} \right) = \frac{S_0 + S_1 + \ldots + S_n}{n+1} \tag{3}$$

is equal to the arithmetic mean (8.2.5) of partial sums of series (8.2.1). For $a_m = (-1)^m$ and $n \to \infty$ (the case of infinitesimal relaxation of Achilles' speed), we see that $x_n \to 1/2$. In other words, the Cesarean Achilles will eventually collapse from exhaustion along the Abelian Achilles.

8.4 Achilles chasing the tortoise in presence of head winds

The principle of infinitesimal relaxation provides a physical interpretation for the equality $1 - 1 + 1 - 1 \ldots = 1/2$ (8.1.5), and also for the general framework of Abel and Cesaro summation methods. However, a more detailed analysis of that principle reveals that the ambiguity in assigning values to the series $1 - 1 + 1 - 1 \ldots$ and to other semiconvergent series, has not been entirely removed. In the following example, a regular linear method of generalized summation assigns to the series $1 - 1 + 1 - 1 \ldots$ values $1/2$ (as in (8.2.4)), 0, or 1 (as in (8.3.1)), depending on the selection of infinitesimal relaxation rates.

Example 1. Summation with variable relaxation rates. Achilles chases the tortoise in asymmetric conditions that slow him down at different rates depending on whether he is running up or down the x-axis (think about the wind blowing in the negative direction along the x-axis). As a result, Achilles' speed up the x-axis decreases by a factor of p at each stage, and his speed down decreases (increases) by a factor q, $p \neq q$. During the m-th stage his speed

$$v_{2k} = (qp)^k, \quad v_{2k+1} = p(qp)^k, \quad k = 0, 1, 2, \ldots. \tag{1}$$

Thus, the original divergent series $1 - 1 + 1 - 1 \ldots$ is replaced by a series

$$x = \sum_{m=0}^{\infty} (-1)^m v_m, \tag{2}$$

which, in view of the Cauchy criterion, converges absolutely for $0 < pq < 1$. Since the terms of an absolutely convergent series can be rearranged without changing their sum,

$$x = \sum_{k=0}^{\infty} (qp)^k - p \sum_{k=0}^{\infty} (qp)^k,$$

where we grouped together the terms of series (2) with even and odd indices, respectively. Formula (8.1.4) for the sum of a geometric series yields

$$x = \frac{1-p}{1-qp}.$$

Now, following the principle of infinitesimal relaxation, we let $qp \to 1 - 0$. That can be accomplished in different ways, and it turns out that the final answer is sensitive to the choice of rates at which q and p converge to 1.

If, for instance, $p = q^r$, then

$$x = \frac{1-q^r}{1-q^{r+1}}.$$

and in the limit $q \to 1-$, applying L'Hôpital's rule we get that

$$x = \frac{r}{r+1}. \tag{3}$$

For $r = 1$, that is when $p = q$ and the conditions of running to the right and to the left are the same, we get that $x = 1/2$—the result obtained by the Abel and Cesaro methods.

However, if we let $r \neq 1$ vary then the limiting x can take other values as well. To see this, it is useful to distinguish between the cases of two-sided and one-sided relaxations. If $r > 0$ and $q < 1$, then $p = q^r < 1$ and the Achilles' speed decreases independently of whether he moves to the right or to the left. In this case we say that a two-sided relaxation takes place, and the Achilles' total displacement x, given by formula (3), can take any value from the interval $(0,1)$. For $r \to 0$, we have $x \to 0$ which corresponds (see (8.3.1a)) to the rearrangement $(1-1)+(1-1)+ \ldots = 0$ of the series $1-1+1-1+\ldots$. For $r \to \infty$, we have $x \to 1$ which corresponds (see (8.3.1b)) to the rearrangement $1-(1-1)-(1-1)-\ldots = 1$.

Note that the condition $qp = q^{1+r} < 1$, equivalent to the absolute convergence of series (1), can also be fulfilled in the case when $p > 1$ and $p < 1/q$. This is the model of one-sided relaxation corresponding to the values $r < 0$ and Achilles running in presence of head winds blowing in the direction of the negative x-axis. Then, the limiting $x < 0$. The result has a transparent physical meaning. Running to the left Achilles goes further than running to the right. The accumulating drift of infinitesimal displacements leads to the full displacement which can be anywhere in the interval $(-\infty, 0)$. In this way the limit value $x = -\infty$ can be interpreted as achievable as a result of a perturbation of the rearrangement $1-(1-1)-(1-1)+\ldots$ of the series $1-1+1-1\ldots$, in which each expression in parentheses is replaced by a very small number ε, so that $\varepsilon + \varepsilon + \ldots = \infty$. ∎

8.5 Separation of scales condition

Examples provided in Section 8.4 demonstrated the existence of regular, linear methods of summation satisfying the principle of infinitesimal relaxation, which give sums different than those provided by the Abel and Cesaro methods. In other words, the principle of infinitesimal relaxation by itself does not produce a unique sum for a semiconvergent series. In this context it seems desirable to seek additional natural conditions which would eliminate the above ambiguities.

In this section we introduce the *separation of scales condition* which will restrict unwanted arbitrariness in the generalized methods of summation discussed so far. Roughly speaking, the condition requires that the relaxation of semiconvergent series terms should proceed at a much slower rate than the rate of internal oscillations around zero of the series itself. Without explicitly formulating the scales separation condition we will proceed to discuss it on a revealing example.

Consider the *φ-summation method* which associates with the series $\sum a_m$ an auxiliary series

$$S_\varphi(\delta) = \sum_{m=0}^{\infty} a_m \varphi(\delta m), \tag{1}$$

where $\varphi(x)$, $x > 0$, is a certain function. If series (1) converges for $\delta > 0$ and its sums have a finite limit

$$S_\varphi = \lim_{\delta \to 0_+} S_\varphi(\delta), \tag{2}$$

then S_φ is taken as a generalized sum of series $\sum a_m$.

To make the φ-summation method work, function $\varphi(x)$ will be assumed to be continuous and satisfy the following conditions:

• "Sufficiently" many derivatives $\varphi'(x)$, $\varphi''(x)$, $\ldots$, are continuous and integrable over the half-line $(0, \infty)$;

• $\varphi(0) = 1$;

• For "sufficiently" large $n = 1, 2, \ldots$,

$$\lim_{x \to \infty} x^n \varphi(x) = 0. \tag{3}$$

The conditions were deliberately stated in a somewhat vague form, but their specific role will become clear later on.

The method is obviously linear. Its regularity is assured by the continuity of $\varphi(x)$ at 0 which yields $\lim_{x \to 0} \varphi(x) = 1$. Function $\varphi(x)$, which completely determines the generalized φ-summation method, describes the relaxation law. Condition (3) demands that $\varphi(x)$ rapidly decreases to 0 at infinity, thus guaranteeing the relaxation principle. The separation of scales condition is fulfilled: inasmuch $\delta \to 0$, the multiplier $\varphi(\delta m)$ varies slower and slower as m increases.

Example 1. Abel and Cesaro methods as φ-summation. Formula (1) generates the Abel and the Cesaro methods as special cases. Indeed, taking $\varphi(x) = \exp(-x)$ we arrive at the auxiliary series

$$S_\varphi(\delta) = \sum_{m=0}^{\infty} a_m (e^{-\delta})^m, \tag{4}$$

which gives formula (8.2.2) of the Abel method after substituting $q = \exp(-\delta)$. On the other hand, if we select φ to be the triangular function

$$\varphi(x) = \begin{cases} 1 - x, & \text{for } 0 < x < 1; \\ 0, & \text{for } x \geq 1, \end{cases} \tag{5}$$

and replace $\delta \to 0$ by a sequence $\delta_n = 1/(n+1)$, $n \to \infty$, then we arrive at the Cesaro summation method. Either function fulfills the conditions imposed on function φ above and, as a result, Abel and Cesaro methods satisfy the condition of separation of scales. ∎

As promised at the beginning of this section, the condition of separation of scales guarantees uniqueness of the sum of series $\sum \pm 1$. More precisely, the generalized sum (3) of series $\sum \pm 1$ does not depend on the selection of the relaxation function $\varphi(x)$, as long as it satisfies the above listed conditions. To see this, substitute $a_m = (-1)^m$ in (1), and group the terms in pairs to get

$$S_\varphi(\delta) = \sum_{k=0}^{\infty} [\varphi(2k\delta) - \varphi(2k\delta + \delta)]. \tag{6}$$

In view of the Mean Value Theorem, for a continuously differentiable f,

$$f(b) - f(a) = f'(c)(b - a) \tag{7}$$

for a certain $c \in (a, b)$. Thus, sum (6) can be rewritten in the form

$$S_\varphi(\delta) = -\frac{1}{2} \sum_{k=0}^{\infty} \varphi'(2k\delta + c_k) 2\delta, \quad c_k \in (0, \delta).$$

The above series is an approximate sum, with the partition points $x_k = 2k\delta + c_k$, for the integral of function $\varphi'(x)$ over interval $(0, \infty)$. Hence, the assumed integrability of the derivative $\varphi'(x)$ implies that

$$\lim_{\delta \to 0} \delta \sum_{k=0}^{\infty} \varphi'(2k\delta + c_k) = \frac{1}{2} \int_0^{\infty} \varphi'(x) dx.$$

The factor 1/2 is due to the length of the partition intervals being 2δ. Finally,

$$S_\varphi = \lim_{\delta \to 0} S_\varphi(\delta) = -\frac{1}{2} \int_0^\infty \varphi'(x)dx = -\frac{1}{2}\varphi(x) \Big|_0^\infty = \frac{1}{2},$$

which proves our assertion that under the scales separation condition all generalized φ-summation methods (1-3) give the same sum $1 - 1 + 1 - 1 \ldots = 1/2$.

Remark 1. Divergent improper integrals. In a calculus course, the usual approach is to approximate integrals over finite intervals by finite sums. Extension of the φ-summation method of approximation to integrals over infinite intervals requires certain precautions, and it would be a good exercise for the reader to provide a rigorous proof here, and find assumptions that have to be imposed on functions $\varphi(x)$ and $\varphi'(x)$ to make the approximation work in this case.

Remark 2. Rapidly divergent semiconvergent series. Note that our proof of the validity of the generalized summation method (1-3) for the series $\sum \pm 1$ depended only on the integrability of the first derivative $\varphi'(x)$ over the interval $(0, \infty)$. This condition was obviously satisfied by the relaxation function $\varphi(x) = \exp(-x)$ of the Abel method, as well as the triangular function (5) of the Cesaro method. However, the stronger the growth of oscillations (as $m \to \infty$) of the terms of a semiconvergent series, the stronger smoothness conditions will have to be imposed on the relaxation function $\varphi(x)$. For that reason the Cesaro method does not work well for series diverging stronger than the series $\sum \pm 1$.

Example 2. A φ-summation for a rapidly diverging series. Consider series

$$\sum_{m=1}^\infty (-1)^m m = -1 + 2 - 3 + 4 - \ldots$$

The corresponding auxiliary series in (1) is

$$S_\varphi(\delta) = \sum_{m=1}^\infty (-1)^m m\varphi(\delta m) = \frac{1}{\delta} \sum_{m=0}^\infty (-1)^m g(\delta m),$$

where $g(x) = x\varphi(x)$. As before, we will pair up the terms of this series to get

$$S_\varphi(\delta) = \frac{1}{\delta} \sum_{m=0}^\infty [g(2k\delta) - g(2k\delta + \delta)]. \tag{8}$$

But in this case, the factor $1/\delta$ makes the Mean Value Theorem useless in determining the behavior of $S_\varphi(\delta)$ as $\delta \to 0$. What is needed is a more precise information contained in the *Taylor formula* which states that if function $f(x)$ has a continuous derivative of order $n + 1$ then, for some $c \in (a, b)$,

$$f(x) = \sum_{m=0}^{n} \frac{f^{(m)}(a)}{m!}(x - a)^m + \frac{f^{(n+1)}(c)}{(n + 1)!}(x - a)^{(n+1)}. \qquad (9)$$

For $n = 1$ we get, in particular, that

$$g(2k\delta) - g(2k\delta + \delta) = -g'(2k\delta)\delta - g''(c_k)\delta^2/2,$$

for a $c_k \in (2k\delta, 2k\delta + \delta)$ which in combination with (8) gives

$$S_\varphi(\delta) = -\sum_{k=0}^{\infty} g'(2k\delta) - \frac{\delta}{2} \sum_{k=0}^{\infty} g''(c_k). \qquad (10)$$

As a simple consequence of Taylor's formula (9)

$$\int_{a}^{a+\Delta} f(x)\,dx = f(a)\Delta + \frac{1}{2} \cdot f'(c)\Delta^2, \quad c \in (a, a + \Delta),$$

applied to function $f(x) = g'(x)$, $a = 2k\delta$ and $\Delta = 2\delta$, we get that

$$\int_{2k\delta}^{(2k+2)\delta} g'(x)dx = g'(2k\delta)2\delta + g''(e_k)2\delta^2, \qquad (11)$$

for some $e_k \in (2k\delta, (2k+2)\delta)$. This equality can now be used to eliminate $g'(2k\delta)$ from (10) and arrive at

$$S_\varphi(\delta) = -\frac{1}{2\delta} \int_{0}^{\infty} g'(x)dx + \delta \sum_{k=0}^{\infty} g''(e_k) - \frac{\delta}{2} \sum_{k=0}^{\infty} g''(c_k).$$

Observe that, for any δ and for a suitable selection of e_k's and c_k's, the above formula is exact. The integral

$$\int_{0}^{\infty} g'(x)dx = x\varphi(x)\Big|_{0}^{\infty} = 0$$

in view of assumptions on function φ. Integrability of $\varphi''(x)$ assures then the convergence, as $\delta \to 0$, of the remaining two sums to the corresponding integrals, so that finally

$$\lim_{\delta \to 0} S_\varphi(\delta) = \frac{1}{4} \int_0^\infty g''(x)dx = -\frac{1}{4}g'(0) = -\frac{1}{4}.$$

Thus the principle of infinitesimal relaxation combined with the separation of scales condition permitted us to arrive at a striking result

$$1 - 2 + 3 - 4 + \ldots = 1/4. \tag{12}$$

∎

In some physical situations the separation of scales condition is not satisfied. That was the case in the analysis of Achilles chasing tortoise in the presence of head winds. Another, and more convincing example of violation of the scales separation condition will be encountered later on when we study the summability of semiconvergent integrals. However, in physical applications, such cases are rare and the condition is widely applied.

8.6　Series of complex exponentials

Many problems of the theory of divergent (and also convergent) series can be solved by a study of the complex exponential series

$$\sum_{m=0}^\infty e^{imz}, \tag{1}$$

where $z = x + iy$. For $y > 0$ the series converges absolutely and

$$\sum_{m=0}^\infty e^{imz} = \frac{1}{1 - e^{iz}}.$$

In addition, for $y \to 0+$, the expression on the right gives the Abel summation formula of a divergent series

$$\sum_{m=0}^\infty e^{imx} = \frac{1}{1 - e^{ix}}, \tag{2}$$

which, for $x = \pi$, reduces to the familiar equality $1 - 1 + 1 - 1 \ldots = 1/2$. Separating the real and the imaginary parts in (2), we arrive at the well known formulas of generalized summation: if $x \neq \pm 2\pi n$, $n = 0, 1, 2, \ldots$,

$$\sum_{m=0}^{\infty} \cos mx = 1/2 \tag{3a}$$

and

$$\sum_{m=0}^{\infty} \sin mx = \frac{1}{2} \cot \frac{x}{2}. \tag{3b}$$

If $x = \pm 2\pi n$, $n = 0, 1, 2, \ldots$, then the series (2) becomes a real-valued series $1 + 1 + 1 + \ldots$ which diverges to $+\infty$, and formulas (3) can be complemented by formulas

$$\sum_{m=0}^{\infty} \cos mx = +\infty, \quad \sum_{m=0}^{\infty} \sin mx = 0, \quad x = \pm 2\pi n, n = 0, 1, 2, \ldots.$$

A more detailed investigation of the character of singularities of series (2) in the neighborhood of these points will be pursued at the beginning of the next section.

For $y > 0$, the absolute convergence of the series obtained by term-by-term differentiation of series (1) justifies formulas

$$(-1)^n \sum_{m=0}^{\infty} m^n e^{imz} = \frac{d^n}{dy^n} \frac{1}{1 - e^{iz}},$$

for $n = 1, 2, \ldots$. If, for $y \to 0+$, the right-hand sides of these equalities converge to finite limits, then we will take them as generalized sums of the corresponding divergent series on the left-hand side. Thus, putting $x = \pi$, we obtain that

$$\sum_{m=0}^{\infty} (-1)^m m^n = (-1)^n \frac{d^n}{dy^n} \frac{1}{1 + e^{-y}} \bigg|_{y=0} = (-1)^n f^{(n)}(0),$$

where

$$f(y) = \frac{1}{1 + e^{-y}} = \frac{1}{2} + \frac{1}{2} \tanh \left(\frac{y}{2} \right).$$

The Taylor expansion of the hyperbolic tangent function is of the form

$$\frac{1}{2} \tanh \left(\frac{y}{2} \right) = \sum_{k=1}^{\infty} \frac{2^{2k} - 1}{(2k)!} B_{2k} y^{2k-1}, \quad |y| < \pi,$$

where coefficients B_n entering in the above formula are called *Bernoulli numbers* and can be determined from the Taylor expansion

$$\frac{x}{e^x - 1} = -\frac{x}{2} + \frac{x}{2} \coth \frac{x}{2} = \sum_{k=1}^{\infty} \frac{B_k}{k!} x^n.$$

In particular,

$$B_0 = 1, \ B_1 = -\frac{1}{2}, \ B_2 = \frac{1}{6}, \ B_3 = B_5 = B_7 = \ldots = 0, \ B_4 = -\frac{1}{30},$$

$$B_6 = \frac{1}{42}, \ B_8 = -\frac{1}{30}, \ B_{10} = \frac{5}{66}, \ B_{12} = -\frac{691}{2730}, \ B_{14} = \frac{7}{6}, \ldots$$

Hence,

$$\sum_{m=0}^{\infty} (-1)^m m^{2k} = 0, \quad \sum_{m=0}^{\infty} (-1)^{m-1} m^{2k-1} = \frac{2^{2k} - 1}{2k} B_{2k}. \tag{4}$$

For $k = 1$, (4) gives equality $1 - 2 + 3 - 4 + \ldots = 1/4$ (8.5.12), and for $k = 2$ we get that

$$\sum_{m=1}^{\infty} (-1)^{m-1} m^3 = 1 - 8 + 27 - 64 + \ldots = -\frac{1}{8}.$$

Differentiation of series (1) led to the generalized summation formulas (4). The integration of the similar series

$$\sum_{m=1}^{\infty} e^{imz} = \frac{e^{iz}}{1 - e^{iz}}, \quad y > 0, \tag{5}$$

also gives rise to useful relations. Multiplying both sides of (5) by a function $f(y)$ and integrating them term-by-term with respect to y over $(0, \infty)$, we get

$$\sum_{m=1}^{\infty} F(m) e^{i(m-1)z} = \int_0^{\infty} \frac{f(y) dy}{e^y - e^{ix}},$$

where $F(t) = \int_0^{\infty} f(y) e^{-ty} dy$.

In particular, for $f(y) = y^{s-1}$, we get that $F(t) = \Gamma(s)t^{-s}$, where $\Gamma(s)$ is the gamma function (4.4.3). Hence, we obtain the equality

$$\sum_{m=1}^{\infty} \frac{1}{m^s} e^{i(m-1)x} = \frac{1}{\Gamma(s)} \int_0^{\infty} \frac{y^{s-1} dy}{e^y - e^{ix}}. \tag{6}$$

For $x = 0$, the above expression gives the well known formula for the *Riemann Zeta function*

$$\zeta(s) = \sum_{m=1}^{\infty} \frac{1}{m^s} = \frac{1}{\Gamma(s)} \int_0^{\infty} \frac{y^{s-1} dy}{e^y - 1}, \quad s > 1,$$

and, for $x = \pi$, we get

$$\eta(s) = \sum_{m=1}^{\infty} \frac{(-1)^{m-1}}{m^s} = \frac{1}{\Gamma(s)} \int_0^{\infty} \frac{y^{s-1} dy}{e^y + 1}, \quad s > 0.$$

If we substitute $s = 1$ in (6), then we get

$$\sum_{m=1}^{\infty} \frac{1}{m} e^{imx} = e^{ix} \int_0^{\infty} \frac{dy}{e^y - e^{ix}} = \int_{1-e^{iz}}^1 \frac{dz}{z},$$

where we introduced a new variable $z = 1 - e^{ix}e^{-y}$. After evaluation of the integral, we finally obtain that

$$\sum_{m=1}^{\infty} \frac{1}{m} e^{imx} = -\ln(1 - e^{ix}) = \frac{1}{2}\ln\frac{1}{2(1 - \cos x)} + \frac{i}{2}(\pi - x), \quad 0 < x < 2\pi.$$

To conclude this section, we derive another useful formula by integrating (5) with respect to y over (p, ∞), $p > 0$, and then substitute $x = 0$. This gives that

$$\sum_{m=1}^{\infty} \frac{e^{-mp}}{m} = \ln(1 + \coth(p/2)) - \ln 2, \quad p > 0.$$

For small p this series can be interpreted as "quasiharmonic" in which the contribution of large terms is damped by the exponential multiplier. In particular, for $p \to 0$ we get the asymptotic formula

$$\sum_{m=1}^{\infty} \frac{e^{-mp}}{m} \sim \ln(1/p),$$

which indicates that the main contribution to the sum is made by the first $N \approx 1/p$ terms.

8.7 Periodic Dirac deltas

In this section we will consider the infinite series of Dirac deltas and the functional series

$$U(x, y) = 1 + 2 \sum_{m=1}^{\infty} e^{-my} \cos(mx) = 2\mathrm{Re} \sum_{m=0}^{\infty} e^{imz} - 1, \quad z = x + iy, \quad (1)$$

which will play a key role in our analysis. For $y > 0$, the latter converges, and following Section 8.6, one can find its sum to be

$$U(x, y) = \frac{\sinh y}{\cosh y - \cos x}. \quad (2)$$

Note that for $y \to 0+$ the above function gives the Abel generalized sum of the divergent trigonometric series

$$U(x) = U(x, y = 0+) = 1 + 2 \sum_{m=1}^{\infty} \cos(mx). \quad (3)$$

The same equality may be rewritten in the complex form to get

$$U(x) = U(x, 0+) = \sum_{m=-\infty}^{\infty} e^{imx}. \quad (4)$$

Setting $x = \pi$ in (3), we get a numerical series $U(\pi) = 1 - 2 + 2 - 2 + \dots$, and we already know (see (8.2.4)) that its Abel sum is $U(\pi) = 1 - 1 = 0$. The same answer $U(x) \equiv 0$ is obtained for any other x if we put formally $y = 0$ in (2). However, a closer inspection of the right-hand side of equality (4) indicates that for $x = 2n\pi$, $n = 0, \pm 1, \pm 2, \dots$, the limit $U(2n\pi) = +1 + 1 + 1 + \dots = \infty$. The situation is elucidated by Fig. 8.7.1, where the graph of function $U(x, y)$ is shown for small $y > 0$. It hugs the x-axis everywhere except for small neighborhoods of points $x = 2n\pi$, where it has sharp peaks.

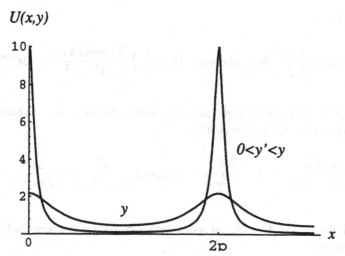

FIGURE 8.7.1
The graph of function $U(x, y)$ for small $y > 0$.

To fully understand the fine structure of the series (4), consider an auxiliary series

$$\Phi(x, y) = \sum_{n-\infty}^{\infty} \frac{2y}{(x - 2\pi n)^2 + y^2}.$$ (5)

For each $y > 0$, by its very structure, $\Phi(x, y)$ is an even, continuous and periodic function of x with period 2π, which can be expanded into a Fourier series

$$\Phi(x, y) = c_0 + \sum_{m=1}^{\infty} c_m \cos(mx).$$ (6)

The zeroth coefficient

$$c_0 = \frac{1}{2\pi} \int_{-\pi}^{\pi} \Phi(x, y)dx = \frac{1}{2\pi} \int_{-\pi}^{\pi} \sum_{n=-\infty}^{\infty} \frac{2y\,dx}{(x - 2\pi n)^2 + y^2}.$$

After changes of variables, and gluing the integrals together, we obtain a formula containing a single integral that can be explicitly evaluated to get

$$c_0 = \frac{1}{2\pi} \int \frac{2y\,dx}{x^2 + y^2} = 1.$$

Similarly,

$$c_m = \frac{1}{\pi} \int_{-\pi}^{\pi} \Phi(x, y) \cos mx \, dx = \frac{1}{\pi} \int \frac{2y \cos(mx) dx}{x^2 + y^2} = 2e^{-my}.$$

Hence, series (6) coincides with series (1), so that, for any $y > 0$, we have equality $\Phi(x, y) \equiv U(x, y)$ or, equivalently,

$$\frac{\sinh y}{\cosh y - \cos x} = 1 + 2 \sum_{m=1}^{\infty} e^{-my} \cos(mx) = \sum_{n=-\infty}^{\infty} \frac{2y}{(x - 2\pi n)^2 + y^2}. \quad (7)$$

Now, if we let $y \to 0+$, the last series weakly converges to a periodic Dirac delta distribution

$$\lim_{y \to 0+} \sum_{n=-\infty}^{\infty} \frac{2y}{(x - 2\pi n)^2 + y^2} = 2\pi \sum_{n=-\infty}^{\infty} \delta(x - 2\pi n),$$

and, in view of (4) and (7), we obtain a distributional Abel summation equality

$$\sum_{m=-\infty}^{\infty} e^{imx} = 2\pi \sum_{n=-\infty}^{\infty} \delta(x - 2\pi n), \quad (8)$$

where the left-hand side is the Fourier series representation of the periodic Dirac delta with period 2π. An obvious extension of this formula for the case of an arbitrary period $2\pi/\Delta$ has the form

$$\sum_{m=-\infty}^{\infty} e^{im\Delta x} = \frac{2\pi}{\Delta} \sum_{n=-\infty}^{\infty} \delta\left(x - \frac{2\pi n}{\Delta}\right), \quad (9)$$

which should be compared with the already familiar formula (3.3.3) for the Fourier image of the Dirac delta.

Equality (7), used here as a tool in deriving distributional formulas (8) and (9), can also be used as a summation tool for more ordinary series often appearing in physical applications. For example, setting $x = 0$ and $x = \pi$, we obtain that

$$\sum_{n=1}^{\infty} \frac{4y}{(2\pi n)^2 + y^2} = \coth \frac{y}{2} - \frac{2}{y}, \quad \sum_{n=1}^{\infty} \frac{4y}{\pi^2(2n-1)^2 + y^2} = \tanh \frac{y}{2}.$$

8.8 Poisson summation formula

A limited supply of elementary and special functions leads to a situation in which analytic solutions of many physical and engineering problems can only be written with the help of series of elementary or special functions. For example, the well-known method of separation of variables in partial differential equations leads to solutions representable in the form of functional series, and the situation is similar for solutions obtained by the method of successive approximations. Often it turns out that a series obtained in this way converges poorly or does not converge at all. In such cases it is desirable to find a transformation accelerating convergence of that series, or an outright analytic expression for its sum. Sometimes, this goal can be achieved through the *Poisson summation formula*

$$\Delta \sum_{m=-\infty}^{\infty} \tilde{f}(m\Delta) = \sum_{n=-\infty}^{\infty} f\left(\frac{2\pi n}{\Delta}\right), \tag{1}$$

which immediately follows from (8.7.9) by multiplying both sides by $f(x)$ and integrating them over the whole x-axis.

Observe the main feature of formula (1). The slower the function $f(x)$ on the right-hand side varies, the faster the Fourier image $\tilde{f}(\omega)$ on the left-hand side decays to zero. This means that the more terms one needs on the right-hand side for good approximation of the infinite series, the fewer terms of the series are necessary on the left-hand side for accurate computation of its sum. So the Poisson summation formula is capable of transforming poorly convergent series into rapidly convergent ones, and many of its applications rest on the above phenomenon.

Relying on properties (3.1.3) and (3.2.5) of the Fourier transform and on formula (1) we can rewrite the Poisson formula in the form

$$\frac{\Delta}{2\pi} \sum_{m=-\infty}^{\infty} f(m\Delta) \exp(-im\Delta s) = \sum_{n=-\infty}^{\infty} \tilde{f}\left(s + \frac{2\pi n}{\Delta}\right). \tag{2}$$

The left-hand side of (2) represents the *discrete Fourier transform* of function $f(t)$, and the right-hand side expresses it in terms of the ordinary Fourier image $\tilde{f}(\omega)$. Hence formula (2) is useful for interpreting results of the computer implementation of the discrete Fourier transform.

Example 1. Poisson summation formula for series of rational functions. Consider the series

$$S(\alpha) = \sum_{m=-\infty}^{\infty} \frac{2\delta\alpha m^2}{(\alpha^2 - m^2)^2 + 4\delta^2 m^2 \alpha^2} = \sum_{m=1}^{\infty} \frac{4\delta\alpha m^2}{(\alpha^2 - m^2)^2 + 4\delta^2 m^2 \alpha^2} \tag{3}$$

of rational functions which is often encountered in physical applications. A use of
the Poisson summation formula gives that

$$S(\alpha) = \sum_{m=-\infty}^{\infty} \tilde{f}(m, \alpha) = \sum_{n=-\infty}^{\infty} f(2\pi n, \alpha), \tag{4}$$

where

$$\tilde{f}(\omega, \alpha) = \frac{2\delta\alpha\omega^2}{(\alpha^2 - \omega^2)^2 + 4\delta^2\omega^2\alpha^2},$$

and

$$f(t, \alpha) = \int \tilde{f}(\omega, \alpha) \cos(\omega t)\,d\omega$$

in view of the evenness of $\tilde{f}$. The above integral can be evaluated by the method
of residues or by checking the tables of integrals. The result is

$$f(t, \alpha) = \frac{\pi}{\gamma} e^{-\beta|t|} \Big(\gamma \cos(\gamma t) - \beta \sin(\gamma |t|) \Big), \tag{5}$$

where

$$\beta = \alpha\delta, \qquad \gamma = \alpha\sqrt{1 - \delta^2}, \qquad \delta^2 < 1.$$

A substitution of (5) into (4) gives

$$S(\alpha) = \frac{2\pi}{\gamma} \left[\frac{\gamma}{2} + \gamma \sum_{n=1}^{\infty} e^{-2\pi\beta n} \cos(2\pi\gamma n) - \beta \sum_{n=1}^{\infty} e^{-2\pi\beta n} \sin(2\pi\gamma n) \right],$$

which is recognizable as a familiar series of complex exponentials. The computation of its sums reduces to

$$Q(\lambda, \nu) = \sum_{n=1}^{\infty} \exp[-(\lambda + i\nu)n] = \frac{1}{\exp(\lambda + i\nu) - 1}$$

for the sum of a geometric series. Since, clearly,

$$S(\alpha) = \frac{2\pi}{\gamma} \left[\frac{\gamma}{2} + \gamma \, \mathrm{Re} \, Q(2\pi\beta, 2\pi\gamma) + \beta \mathrm{Im} \, Q(2\pi\beta, 2\pi\gamma) \right],$$

an explicit calculation of the real and the imaginary parts of function $Q(\lambda, \nu)$ gives

$$S(\alpha) = \frac{\pi}{\gamma} \frac{\gamma \sinh(2\pi\beta) - \beta \sin(2\pi\gamma)}{\cosh(2\pi\beta) - \cos(2\pi\gamma)}. \tag{6}$$

This, and related formulas, can also be found by other methods, but the Poisson formula provides, as a rule, the fastest path to the goal. The typical graph of $S(\alpha)$ (6) is pictured in Fig. 8.8.1.

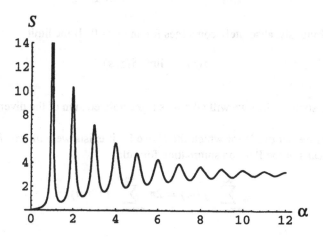

FIGURE 8.8.1
Graph of series (3) in case of $\delta = 0.05$ evaluated with the help of formula (6). The larger α, the more terms in (3) are needed to maintain the validity of the result.

8.9   Summation of divergent geometric series

The Poisson summation formula was used in the Example 1 of the previous Section 8.8 to explicitly sum a nontrivial but absolutely convergent series; later on we will see its applications to accelerate the convergence of already convergent series. In this section, however, we will utilize it to solve a more exotic summability problem for the everywhere divergent series

$$S(z) = 1 + 2 \sum_{m=1}^{\infty} \cos(mz), \tag{1}$$

where z is a complex variable, and where the Abel method and thus even more so the Cesaro method, fail.

Let us form an auxiliary perturbed series

$$S(z, \varepsilon) = 1 + 2 \sum_{m=1}^{\infty} e^{-\varepsilon m^2} \cos(mz) = \sum_{m=-\infty}^{\infty} \exp(-\varepsilon m^2 + imz), \qquad (2)$$

which, obviously, absolutely converges for any $\varepsilon > 0$. If the limit

$$S(z) = \lim_{\varepsilon \to 0+} S(z, \varepsilon)$$

exists for some z, then we will take it as a generalized sum of the divergent series (1).

To find the set of z's for which the above limit exists we will transform series (2) by means of the Poisson summation formula

$$\sum_{m=-\infty}^{\infty} f(m) = 2\pi \sum_{n=-\infty}^{\infty} \tilde{f}(2\pi n), \qquad (3)$$

with $f(t) = \exp(-\varepsilon t^2 + itz)$. The left-hand side of (3) coincides with series (2), and the right-hand side contains function

$$\tilde{f}(\omega) = \frac{1}{2\pi} \int \exp(-\varepsilon t^2 + it(z - \omega))dt = \frac{1}{2\sqrt{\pi\varepsilon}} \exp\left[-\frac{(z-\omega)^2}{4\varepsilon}\right].$$

Hence,

$$S(z, \varepsilon) = \frac{1}{2\sqrt{\pi\varepsilon}} \sum_{n=-\infty}^{\infty} \exp\left[-\frac{(z - 2\pi n)^2}{4\varepsilon}\right]. \qquad (4)$$

Notice that

$$\left|\exp\left[-\frac{(z - 2\pi n)^2}{4\varepsilon}\right]\right| = \exp\left[\frac{y^2 - (x - 2\pi n)^2}{4\varepsilon}\right].$$

It means that for any $z = x + iy$ from the set (blackened out in Fig. 8.9.1)

$$G = \{z \in \mathbb{C} : |y| < |x - 2\pi n|, \quad n = 0, \pm 1, \pm 2, \dots,$$

all the terms of series (4) converge to zero as $\varepsilon \to 0+$. Correspondingly, it is easy to prove that for $z \in G$ we have $S(z) = \lim_{\varepsilon \to 0+} S(z, \varepsilon) = 0$. Thus, we demonstrated that

$$S(z) = 1 + 2 \sum_{m=1}^{\infty} \cos(mz) = 0, \qquad z \in G.$$

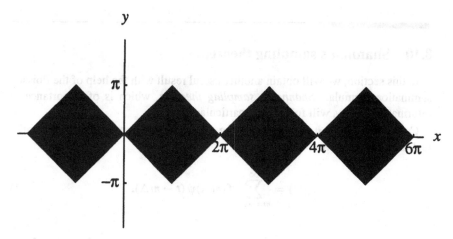

FIGURE 8.9.1
Region G in the complex plane.

In particular, for $x = \pi$, we get that

$$1 + 2 \sum_{m=1}^{\infty} (-1)^m \cosh(my) = 0, \qquad |y| < \pi.$$

Substituting $q = \exp y$, the above equality can be rewritten in the form

$$\sum_{m=0}^{\infty} (-1)^m \left[q^m + (1/q)^m \right] = 1, \qquad e^{-\pi} < q < e^{\pi}. \tag{5}$$

For $q > 1$,

$$\sum_{m=0}^{\infty} (-1)^m (1/q)^m = \frac{q}{1+q}$$

converges absolutely. In view of the linearity and regularity of our summation method, we obtain the following striking extension of the familiar formula for the sum of a geometric series:

$$\sum_{m=0}^{\infty} (-1)^m q^m = \frac{1}{1+q}, \qquad 1 < q < e^{\pi}.$$

8.10 Shannon's sampling theorem

In this section, we will obtain another useful result with the help of the Poisson summation formula. *Shannon's sampling theorem*, which is of importance in information theory, will follow as a particular case.

Consider a function $g(t)$ representing the uniformly convergent on the entire real-axis infinite series

$$g(t) = \sum_{m=-\infty}^{\infty} f(m\Delta)\psi(t - m\Delta). \tag{1}$$

Here $f(t)$ and $\psi(t)$ are known functions with Fourier images $\tilde{f}(\omega)$ and $\tilde{\psi}(\omega)$, respectively. Let us apply the Fourier transform (3.1.1) to both sides of (1). In view of the uniform convergence of the series, the integration and infinite summation operations can be interchanged so that, taking into account the formula (3.1.3a)

$$\frac{1}{2\pi} \int \psi(t - m\Delta)e^{-i\omega t}\, dt = \tilde{\psi}(\omega)e^{-i\omega m\Delta},$$

we have

$$\tilde{g}(\omega) = \tilde{\psi}(\omega) \sum_{m=-\infty}^{\infty} f(m\Delta)e^{-i\omega m\Delta}. \tag{2}$$

Finally, transforming the sum in (2) by means of the Poisson formula (8.8.2), we get

$$\tilde{g}(\omega) = \frac{2\pi}{\Delta} \tilde{\psi}(\omega) \sum_{n=-\infty}^{\infty} \tilde{f}\left(\omega + \frac{2\pi n}{\Delta}\right). \tag{3}$$

Now, let $\tilde{f}(\omega)$ be a function with compact support, identically equal to zero for

$$|\omega| \geq \pi/\Delta. \tag{4}$$

Then the supports of the summands in (3) have empty intersections and, for a given frequency ω, only one term of the infinite series is different from zero. In particular, for $|\omega| \leq \pi/\Delta$, equality (3) is equivalent to the equality

$$\tilde{g}(\omega) = \frac{2\pi}{\Delta} \tilde{\psi}(\omega)\tilde{f}(\omega). \tag{5}$$

Assume additionally that $\tilde{\psi}(\omega)$ is of the form

$$\tilde{\psi}_0(\omega) = \frac{\Delta}{2\pi}\Pi(\Delta\omega/\pi), \tag{6}$$

where the rectangular function

$$\Pi(\nu) = \chi(\nu+1) - \chi(\nu-1). \tag{7}$$

Then, equality (5) is valid for any ω and assumes the form

$$\tilde{g}(\omega) = \tilde{f}(\omega). \tag{8}$$

The equality of the Fourier images implies the equality of the functions themselves. Hence, substituting in (1) the inverse Fourier image

$$\psi_0(t) = \frac{\sin\tau}{\tau}, \qquad \tau = \frac{\pi t}{\Delta}, \tag{9}$$

of the rectangular function (6), we arrive at the equality

$$f(t) = \sum_{m=-\infty}^{\infty} f(m\Delta)\frac{\sin((\pi t/\Delta) - \pi m)}{(\pi t/\Delta) - \pi m}, \tag{10}$$

which expresses the contents of *Shannon's sampling theorem*:

Assume that the Fourier image of function $f(t)$ vanishes outside the interval $|\omega| \leq \Delta$. Then, for any t, the values of $f(t)$ are completely determined via formula (10) by the values of this function at discrete time instants

$$t_m = m\Delta. \tag{11}$$

The theorem has numerous applications in physics and information theory which will not be discussed here. However, we will take a closer look at some of its modifications and extensions which will permit us to grasp the meaning of this result at a deeper level.

First of all, note that if $\tilde{\psi}(\omega)$ is taken to be a one-sided rectangular function (as opposed to the symmetric function (6))

$$\tilde{\psi}(\omega) = \frac{\Delta}{\pi}[\chi(\omega) - \chi(\omega - \pi/\Delta)]. \tag{12}$$

Then the right-hand side of (8) becomes

$$\tilde{g}(\omega) = 2\tilde{f}(\omega)\chi(\omega).$$

As we observed in Section 6.6, the right-hand side is the Fourier image of an analytic signal $F(t)$ corresponding to function $f(t)$. The original function corresponding to the Fourier image (12) is

$$\psi(t) = \frac{\Delta}{\pi i t}[\exp(i\pi t/\Delta) - 1].$$

Substituting this expression into (1), we find an explicit formula for the complex function

$$F(t) = \sum_{m=-\infty}^{\infty} f(m\Delta)\frac{\exp((i\pi t/\Delta) - i\pi m) - 1}{(\pi t/\Delta) - \pi m},$$

representing the analytic signal corresponding to function $f(t)$.

Let us rewrite formula (3), replacing Δ by δ:

$$\tilde{g}(\omega) = \frac{2\pi}{\delta}\tilde{\psi}(\omega) \sum_{m=-\infty}^{\infty} \tilde{f}(\omega + 2\pi n/\delta), \tag{13}$$

and observe that if

$$\delta < \Delta, \tag{14}$$

and $\tilde{f}(\omega)$, as before, identically vanishes for $|\omega| \geq \Delta$, then the supports of summands in the series (13) are separated by gaps of length $2\pi(1/\delta - 1/\Delta)$. As a result, to pass from (13) to (3), it suffices to select $\tilde{\psi}(\omega)$ from a broad class of functions

$$\tilde{\psi}(\omega) = \begin{cases} \delta/(2\pi), & \text{for } |\omega| \leq \pi/\Delta; \\ 0, & \text{for } |\omega| \geq \pi(2/\delta - 1/\Delta); \\ arbitrary, & \text{for } \pi/\Delta < |\omega| < \pi(2/\delta - 1/\Delta). \end{cases} \tag{15}$$

The schematic plot of one of possible Fourier images $\tilde{\psi}(\omega)$ for which the equality (8) is valid is shown in Fig. 8.10.1.

Taking one of the Fourier images (15) and calculating the corresponding original function $\psi(t)$ we arrive at a formula more general than (10):

$$f(t) = \sum_{m=-\infty}^{\infty} f(m\delta)\psi(t - m\delta). \tag{16}$$

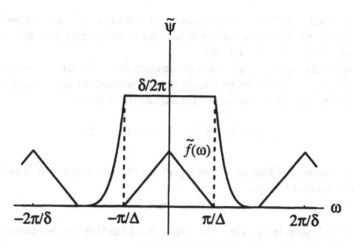

FIGURE 8.10.1

The plot of one of possible Fourier images $\tilde{\psi}(\omega)$ for which the equality (8) is valid. The triangles symbolize here the summands in the series (13).

In particular, killing the Fourier image (15) for $|\omega| \geq \pi/\Delta$ we arrive at the most widely used form of Shannon's formula

$$f(t) = \varepsilon \sum_{m=-\infty}^{\infty} f(m\delta) \frac{\sin((\pi t/\Delta) - \varepsilon\pi m)}{(\pi t/\Delta) - \varepsilon\pi m}, \qquad \varepsilon = \delta/\Delta < 1. \qquad (17)$$

It would seem that formulas (16) and (17) have, in comparison with the simplest formula (10), the shortcoming of requiring the knowledge of values $f(t)$ at densely distributed time instants $t_m = \delta m$. However, the indicated drawback is partly amended by the great flexibility in the selection of function $\psi(t)$. The freedom to impose values of the Fourier image $\tilde{\psi}(\omega)$ in the intervals

$$\frac{\pi}{\Delta} \leq |\omega| \leq \pi \left(\frac{2}{\delta} - \frac{1}{\Delta} \right) \qquad (18)$$

can be utilized to improve the speed of convergence of the series appearing on the right-hand side of the generalized Shannon's theorem. Indeed, the experience gathered by the reader while studying the Fourier transform's asymptotics in Chapter 4 suggests that to achieve that goal one should choose $\tilde{\psi}(\omega)$ decaying to zero inside the intervals (18) as smoothly as possible. An infinitely differentiable (in the classical sense) $\tilde{\psi}(\omega)$ would be ideal. In this case, the corresponding original function $\psi(t)$ would decay for $|t| \to \infty$ faster than any power function $1/t^n$. As a result, it may turn out that in computing values of $f(t)$ with a given accuracy fewer terms are needed in the series (16) than in (10). An example of a damped Fourier image $\tilde{\psi}(\omega)$ and the corresponding rapidly decaying function $\psi(t)$ is provided in Exercises at the end of this chapter.

In electrical engineering applications one often deals with narrow-band $f(t)$, that is, with functions whose Fourier images are concentrated in a narrow neighborhood of specific frequencies $\pm\omega_0$ and vanish outside the intervals

$$|\omega - \omega_0| < \Omega, \qquad |\omega + \omega_0| < \Omega.$$

Using the simplest Shannon formula (10) we should impose values of function $f(t)$ in intervals of length

$$\Delta = \pi/(\omega_0 + \Omega), \tag{19}$$

because the Fourier image $\tilde{f}(\omega)$ of a narrow-band function $f(t)$ is identically equal to zero only for frequencies satisfying condition

$$|\omega| \geq \omega_0 + \Omega,$$

analogous to (4).

At the same time it is intuitively clear that for a narrow-band signal for which $\Omega \ll \omega_0$, one should have a more adequate version of formula (16), with sufficiently large intervals δ between the readings: $\delta \approx 2\pi/\Omega \gg \pi/\omega_0$. Let us derive it.

To begin with, note that the compactly supported Fourier image of a narrow-band signal can be represented in the form of two components

$$\tilde{f}(\omega) = \tilde{f}_+(\omega) + \tilde{f}_-(\omega), \tag{20}$$

concentrated in the neighborhoods of the central frequencies $+\omega_0$ and $-\omega_0$, respectively. Choose δ so that the central frequencies of components $\tilde{f}_+$ in the sum (13) would be located halfway inbetween adjacent central frequencies of the components $\tilde{f}_-$. To accomplish this it is necessary that for some positive integer l we have the equality

$$-\omega_0 + \frac{2\pi}{\delta}l = \omega_0 - \frac{\pi}{\delta}.$$

Consequently,

$$\delta = \pi(l + 1/2)/\omega_0. \tag{21}$$

Select the value of l in such a way that, for any ω, only one of the summands in (13) is different from zero. To achieve this it is sufficient to demand that δ satisfies the inequality $\pi/2\delta > \Omega$. Hence by (21) it follows that

$$l \leq \frac{1}{2}(\omega_0/\Omega - 1). \tag{22}$$

It remains to select as $\tilde{\psi}(\omega)$ a function that would turn equality (13) into

$$\tilde{g}(\omega) = \tilde{f}_+(\omega) + \tilde{f}_-(\omega).$$

By analogy with (15), it is clear that here it is sufficient to let

$$\tilde{\psi}(\omega) = \begin{cases} \delta/(2\pi), & \text{for } |\omega \pm \omega_0| \leq \omega; \\ 0, & \text{for } |\omega \pm \omega_0| \geq \pi/\delta - \Omega; \\ arbitrary, & \text{for } \Omega < |\omega \pm \omega_0| < \pi/\delta - \Omega. \end{cases} \tag{23}$$

In particular, setting $\tilde{\psi}(\omega)$ to be identically zero for frequencies satisfying the two inequalities $|\omega \pm \omega_0| > \Omega$, that is selecting

$$\tilde{\psi}_0(\omega) = \frac{\delta}{2\pi} \left[\Pi\left(\frac{\omega - \omega_0}{\Omega}\right) + \Pi\left(\frac{\omega + \omega_0}{\Omega}\right) \right], \tag{24}$$

where $\Pi(\nu)$ is given by (7), we can calculate the corresponding original function to be

$$\psi_0(t) = \frac{2\delta}{\pi t} \sin(\Omega t) \cos(\omega_0 t). \tag{25}$$

Substituting it into (16), we arrive at the simplest variant of the narrow-band Shannon's theorem:

$$f(t) = \frac{2\delta}{\pi} \sum_{m=-\infty}^{\infty} f(m\delta) \frac{\sin(\Omega t - m\delta\Omega)}{t - m\delta} \cos(\omega_0 t - m\omega_0\delta). \tag{26}$$

Note that, if $f(t)$ is a real narrow-band process, that is if $\Omega \ll \omega_0$ then, without violating inequality (22), one can select $l \gg 1$, and make the distance δ (21) between readings t_m much larger than Δ (19), as in the case of the standard Shannon's theorem.

8.11 Divergent integrals

Summability problem for divergent integrals is close in spirit to that for divergent infinite series. Let us demonstrate this using the Fourier transform

$$\tilde{\chi}(\omega) = \frac{1}{2\pi} \int_0^\infty e^{-i\omega t} dt$$

of the Heaviside function $\chi(t)$ as an example. It is a typical divergent integral. Evaluating it by the infinitesimal relaxation method gives

$$\int_0^\infty e^{-i\omega t} dt = \lim_{\alpha \to 0+} \int_0^\infty e^{-i\omega t - \alpha t} dt = \frac{1}{i\omega + 0} \tag{1}$$

and illustrates an application of the Abel summation method to divergent integrals.

The generalized sum (1) of the above divergent integral is quite stable with respect to a wide class of regularizing functions. Indeed, let us replace the above exponential regularizing function $\exp(-\alpha t)$ by an arbitrary, absolutely integrable and continuously differentiable function $f(\alpha t)$, and consider the integral

$$\int_0^\infty e^{-i\omega t} f(\alpha t) dt. \tag{2}$$

The change of variable $x = \alpha t$ transforms (2) into the integral

$$\frac{1}{\alpha} \int_0^\infty e^{-ipx} f(x) dx,$$

where we also introduced a new parameter $p = \omega/\alpha$. If $\alpha \to 0$ then $p \to \infty$. Hence, to evaluate the above integral, we can employ results of Chapter 4 on the asymptotic behavior of Fourier transforms of discontinuous functions. In particular, by analogy with (4.3.3), we have that

$$\frac{1}{\alpha} \int_0^\infty e^{-ipx} f(x) dx \sim \frac{1}{i\alpha p} f(0) = \frac{1}{i\omega} f(0), \qquad \alpha \to 0, \quad \omega \neq 0. \tag{3}$$

The regularity assumption for our summation method requires that $f(0) = 1$. So, for $\omega \neq 0$, the summation result is the same as in (1).

The above summation method obviously satisfies the separation of scales condition (as discussed in the context of series summability methods). However, one can easily produce examples of methods which violate it and thus can potentially give nonunique answers.

Example 1. Divergent integral with nonunique generalized sum. Consider the divergent integral

$$\int_0^\infty \sin x \, dx. \tag{4}$$

Summation of this integral following the prescription given in (2) gives value 1, which can be also found by formally writing the integral as a series $\sum a_n$ with terms

$$a_n = \int_{\pi n}^{\pi(n+1)} \sin x \, dx = 2(-1)^n,$$

so that

$$\int_0^\infty \sin x \, dx = 2 - 2 + 2 - 2 + \ldots = 1.$$

However, if one considers a method of summation based on the integral

$$I(\alpha) = \int_0^\infty e^{-\alpha x} \sin x \, [1 + 2\alpha q \sin x] \, dx, \qquad (5)$$

one obtains a result different from 1. Indeed, the integral (5) can be easily evaluated to give

$$I(\alpha) = \frac{1}{\alpha^2 + 1} + \frac{4q}{\alpha^2 + 4}.$$

As $\alpha \to 0$, the integrand in (5) converges uniformly on bounded sets of x's to $\sin x$—the integrand in (4). However, for any $q \neq 0$,

$$\lim_{\alpha \to 0} I(\alpha) = 1 + q \neq 1,$$

and for different values of parameter q, one gets different summation results. This is related to the fact that the relaxation function $e^{-\alpha x}[1 + 2\alpha q \sin x]$ is not of the form $f(\alpha x)$ and violates the principle of separation of scales.

8.12 Exercises

1. Find the sum of the series

$$S(x, y) = \sum_{m=1}^\infty e^{-my} \sin(mx), \qquad y > 0.$$

2. Find the sum of the divergent series

$$S = \sum_{m=1}^\infty m \sin(mx).$$

3. Using the Poisson summation formula find the functional action of the distribution

$$E = \sum_{m=1}^\infty \sin(mx).$$

4. The analysis of waves propagating in resonators and waveguides leads to series of the following type:

$$S(\omega, \Delta) = \sum_{m=-\infty}^{\infty} f(m\Delta) \exp(-img(\omega)),$$

where $f(x)$ is an absolutely integrable function such that $f(0) = 1$. Transform the above series by means of the Poisson summation formula and find its weak limit for $\Delta \to 0$.

5. Using formula (8.7.8), transform the series

$$P(x) = \sum_{m} \exp\left(i(m \cdot a(x))\right),$$

where the summation is extended to all vectors m with integer components (m_1, m_2, m_3), and $y = a(x)$ is an infinitely differentiable vector field which provides a one-to-one mapping of the x-space into the y-space such that the Jacobian $J(x) = |\partial a/\partial x|$ of the transformation is everywhere positive and continuous. The inner product appearing under the summation is $(a(x) \cdot m) = a_1(x)m_1 + a_2(x)m_2 + a_3(x)m_3$.

6. The 3-D Poisson summation formula turns out to be useful in solid state physics and, in particular, in the study of crystal properties. Using the result from Exercise 5, derive the right-hand side of the formula if the left-hand side is

$$F(x) = \sum_{m_1, m_2, m_3 = -\infty}^{\infty} f(x + m_1 a_1 + m_2 a_2 + m_3 a_3),$$

where a_1, a_2, a_3 are three not coplanar vectors, $x = (x_1, x_2, x_3)$ in a certain Cartesian coordinate system, and $f(x)$ is an absolutely integrable function with 3-D Fourier image $\tilde{f}(k)$.

7. Calculate the discrete Fourier transform of the rectangular function

$$\Pi(t) = \chi(t+1) - \chi(t-1). \tag{1}$$

8. Find the discrete Fourier transform of the function

$$f(t) = \Pi(t) \cos^4(\pi t/2)$$

and, then, compare it with its usual Fourier image (4.3.22).

9. Find the discrete Fourier image of the periodic function $f(t)$, with period $T = M\Delta$, where M is a positive integer.

10. Find the sum of the functional series

$$S(\Delta) = \sum_{m=1}^{\infty} \frac{\sin(m\Delta)}{m}$$

with the help of the Poisson summation formula.

11. The functions

$$f_N(t) = \sum_{m=1}^{N} \frac{\sin(mt)}{m}$$

are partial sums of the Fourier series of the periodic function $f(t) = (\pi - t)/2 + \pi \lfloor t/(2\pi) \rfloor$ (see solution to Exercise 10). Compare functions $f_N(t)$ and $f(t)$, and investigate the asymptotic behavior (for $N \to \infty$) of $f_N(t)$ in the vicinity of the discontinuities $t = 2\pi n$ of $f(t)$.

12. The Gibbs phenomenon discovered in Exercise 11 is undesirable in many physical and engineering applications. Find the method of summation of the first N terms of the series

$$f_N(t) = \sum_{m=1}^{N} \tilde{h}(m, N) \frac{\sin(mt)}{m}$$

which, at the continuity points of $f(t)$ from Exercise 11, would guarantee convergence of $f_N(t)$ to $f(t)$, and would avoid the Gibbs phenomenon at the discontinuity points of $f(t)$.

13. Derive a formula analogous to the formula (8.9.26) for the analytic signal $F(t)$ corresponding to the narrow-band signal $f(t)$, whose Fourier image is identically zero for $|\omega \pm \omega_0| \geq \Omega$.

14. Construct a function $\psi(t)$ entering into the Shannon series (8.9.16), and possessing a Fourier image of the type (8.9.15). Use common sense.

15. Find the maximal distance δ between readings in the narrow-band Shannon's formula (8.9.25) for $\Omega = \omega_0/10$.

16. Suppose that $f(t)$ is a narrow-band function with the Fourier image vanishing for $|\omega \pm \omega_0| \geq \Omega = \omega_0/10$. Find function $\psi(t)$ entering into the generalized Shannon formula (8.9.16), which decays sufficiently rapidly as $|t| \to \infty$. Utilize results of the Exercises 14 and 15.

Appendix A

Answers and Solutions

A.1   Chapter 1. Definitions and operations

1. (a) $(\sqrt{\pi}/2)\delta'(x)$;

(b) $((-\pi/2)\delta'(x)$;

(c) The "zero distribution" 0;

(d) $\chi(x)$.

2. $\int f(x)dx = 0$, $\int F(x)dx = 1$, where $F(x) = \int_{-\infty}^{x} f(y)dy$.

3. (a) $\chi'(ax) = \delta(x)$ and is independent of a.

(b)

$$\chi'(e^{\lambda x} \sin ax) = \sum_{n=-\infty}^{\infty} (-1)^n \delta\left(x - \frac{\pi n}{a}\right)$$

and is independent of λ.

4. $f'(x) = \delta(x - 1) - \delta(x + 1)$, and

$$\lim_{\varepsilon \to 0_+} f'(x/\varepsilon)/\varepsilon^2 = -2\delta'(x).$$

5. $y(x) = A\delta(x - 1) + B\delta'(x - 1) + C\delta(x)$, where A, B, C are arbitrary constants.

6. Taking into account the multiplier probing property, we have

$$\left(e^{\lambda x}\chi(x)\right)^{(n)} = \lambda^n e^{\lambda x}\chi(x) + \sum_{k=1}^{n} \lambda^{k-1}\delta^{(n-k)}(x).$$

8. $\delta_\sigma = \delta(g(x) - a)|\nabla g(x)|$. The action of this distribution as a functional on an arbitrary test function $\phi \in \mathcal{D}(\mathbf{R}^3)$ is

$$\delta_\sigma[\phi] = \int \delta(g(x) - a)|\nabla g(x)|\phi(x)d^3x = \int_\sigma \phi\, d\sigma.$$

© Springer Nature Switzerland AG 2018
A. I. Saichev and W. Woyczynski, *Distributions in the Physical and Engineering Sciences, Volume 1*, Applied and Numerical Harmonic Analysis, https://doi.org/10.1007/978-3-319-97958-8

9.

$$\delta_\ell = \left| \left[\nabla g_1(x) \times \nabla g_2(x) \right] \right| \prod_{k=1}^{2} \delta(g_k(x) - a_k).$$

10. $\int |\nabla \psi(x)| \delta(\psi(x) - c) d^2 x.$

11. $P[\phi]$ is equal to the flux of the vector field ϕ inside the region bounded by a level surface $\psi(x) = c$:

$$P[\phi] = \int_\sigma \phi \cdot n d\sigma,$$

where n is the interior unit normal vector to the level surface σ. The interior points x are those for which $\psi(x) > c$.

A.2   Chapter 2. Basic applications

Ordinary differential equations

1. $\ddot{y} + \gamma \dot{y} + \omega^2 y = (b + \gamma a)\delta(t) + a\dot{\delta}(t).$

2. $y(0) = 1, \dot{y}(0) = -\gamma.$

3. $y(t) = (1/2) \sin |t|.$

4. $y(t) = \chi(t)\left(a + (b - a) \cos t\right).$

5. $\ddot{y} + y = \delta(t).$

6. $y = Ae^{ht} + B\chi(t)e^{ht}.$

Wave equation

7. The Green function $G(x, t)$ is a solution of the following initial value problem:

$$\frac{\partial^2 G}{\partial t^2} = c^2 \frac{\partial^2 G}{\partial x^2}, \quad G(x, t = 0) = 0, \quad \frac{\partial}{\partial t} G(x, t = 0) = \delta(x).$$

Using the D'Alembert formula we obtain that

$$G(x, t) = \frac{1}{2c} \chi(t)\left(\chi(x + ct) - \chi(x - ct) \right).$$

Hence the solution of a non-homogeneous wave equation is of the form

$$u(x, t) = \frac{1}{2c} \int_{-\infty}^{t} d\tau \int_{x - c(t-\tau)}^{x + c(t-\tau)} dy \, f(y, \tau).$$

8. $u(x, t) = g(x) * \partial G(x, t)/\partial t + h(x) * G(x, t)$, where $*$ denotes the convolution in the spatial variable.

Continuity equation

9.

$$\rho(x, t) = \int \rho_0(y)\delta(x - ye^{gt})dy = e^{-gt}\rho_0(xe^{-gt}),$$

$$C(x, t) = \int C_0(y)\delta(y - xe^{-gt})dy = C_0(xe^{-gt}).$$

For $g > 0$, as time increases, the density at each point x decreases to 0, and the concentration asymptotically becomes homogeneous and converges everywhere to a constant. If the total mass of the passive traces is equal to m, then for $g < 0$ and $t \to \infty$ the density weakly converges to the distribution $m\delta(x)$ and the concentration to the zero distribution.

10. $\rho(x, t) = \int \rho_0(y)\delta(x - y - gyt)dy = (1/1 + gt)\rho_0((x/1 + gt))$. For $g > 0$ the density converges everywhere to zero but much slower than in the previous problem. However, for $g < 0$ the density becomes singular in finite time $t^* = -1/g$.

11. $\rho(x, t) = \rho_0 v \int \delta(x - v\tau - g\tau^2/2)d\tau = \rho_0(1 + 2xg/v^2)^{-1/2}$. One can see a similar phenomenon watching the stream of water flowing out of a faucet. The further it is from the faucet, the thinner it gets.

12. The equation in question has the integral $\rho(x)v(x) = const = \rho_0 v$. Eliminating the time t from the equations of motion

$$x = vt + gt^2/2, \qquad v(t) = v + gt,$$

we get $v(x) = (v^2 + 2gx)^{1/2}$. Thus, $\rho(x) = \rho_0 v/(v^2 + 2gx)^{1/2}$.

13. In this case the x axis is simply reversed upwards and we are not solving the "rain" problem but a "fountain" problem where the droplets move upwards. In this case the density of droplets is

$$\rho(x) = 2\rho_0\left(1 - 2xg/v^2\right)^{-1/2}.$$

if $x < v^2/2g$, and it is $= 0$ if $x > v^2/2g$. The infinite singularity of the density as $x \to v^2/2g - 0$ describes the effect of concentration of droplets at the point where rising droplets reach their highest elevation. By the way, explain why an "extra" coefficient 2 appeared in the numerator of the above formula.

14. The continuity equation

$$\frac{\partial \rho}{\partial t} + \frac{\partial \Psi}{\partial x_2}\frac{\partial \rho}{\partial x_1} - \frac{\partial \Psi}{\partial x_1}\frac{\partial \rho}{\partial x_2} = 0$$

has the integral of motion $\rho = g(\Psi(x_1, x_2))$, where $g(\Psi)$ is an arbitrary function. If we additionally impose a physical requirement that $g \geq 0$ then we obtain a class of densities of the passive tracer that will be independent of time.

15. Since in this example the dimension of the Dirac delta is $[\delta] = [t]/[x^2]$, we get that $[a] = [m]/[t]$. Also, $m = a \oint dl/|v|$, where the integration is performed on the contour (or contours) given by equation $\Psi(x) = b$.

16. *The Liouville equation* for the particle density is

$$\frac{\partial f}{\partial t} + (v \cdot \nabla_x)f + (\nabla_v \cdot g(x, v)f) = 0.$$

17. In this case the Liouville equation is

$$\frac{\partial f}{\partial t} + (v \cdot \nabla_x)f = h(\nabla_v \cdot vf).$$

Its Green function is

$$G(x, v, y, u, t) = \delta(x - y - (u/h)(1 - e^{-\tau}))\delta(v - ue^{-\tau}).$$

$\tau = ht$ introduced above is dimensionless time. Using the Green function we can write the solution in the form

$$f(x, v, t) = \int f_0(y, u)\delta(x - y - (u/h)(1 - e^{-\tau}))\delta(v - ue^{-\tau})d^3y \, d^3u$$

$$= e^{3\tau} f_0(x - (v/h)(e^{\tau} - 1), ve^{\tau}).$$

18. $\rho(x, t) = \int f(x, v, t)d^3v = m\gamma^3(t)w(x\gamma(t))$, where $\gamma(t) = he^{\tau}/(e^{\tau} - 1)$.

19. $\lim_{t\to\infty} f(x, v, t) = mh^3w(xh)\delta(v)$.

20. Density $\rho(x, t)$ satisfies the continuity equation and velocity $v(x, t)$ satisfies the (nonlinear) Riemann equation

$$\frac{\partial v}{\partial t} + (v \cdot \nabla)v = 0.$$

21. Replacing the substantial derivative by its expression in Eulerian coordinates we obtain the Riemann equation with external forcing acting on particles in the hydrodynamic flow

$$\frac{\partial v}{\partial t} + (v \cdot \nabla)v = g(x, v, t).$$

22. $v(x, t) = \int vf(x, v, t)d^3v/\rho(x, t)$.

23. The analogue of formula (2.4.11) in the 1-D case is

$$\bar{\rho}_0(x) = \frac{\rho}{l} \sum_{i=-\infty}^{\infty} \Delta s g\left(\frac{x - \beta(i\Delta s)}{l}\right).$$

For $\Delta \to 0$ and smooth functions $g(z)$ and $\beta(y)$, the sum converges to the integral

$$\bar{\rho}_0(x) = \frac{\rho}{l} \int g\left(\frac{x - \beta(y)}{l}\right) dy.$$

Now, observe that for $l \to 0$, the function $g(x/l)/l$ weakly converges to the Dirac delta $\delta(x)$. In this fashion the sought limit function is

$$\bar{\rho}_0(x) = \rho \int \delta(x - \beta(y)) \, dy = \rho \alpha'(x),$$

where $y = \alpha(x)$ is the function inverse to the function $x = \beta(y)$. In our case, for $x > 0$ and $\alpha(x) = (x - 1 + \sqrt{x^2 + s^2})/2$,

$$\bar{\rho}_0(x) = \frac{\rho}{2} \left(1 + \frac{|x|}{\sqrt{x^2 + s^2}} \right).$$

Remark. The condition $\Delta = o(l)$ for $l \to 0$ permitted us to correctly choose the order of the limit passages: first let $\Delta \to 0$ for a positive l, and then $l \to 0$.

Pragmatic approach

24. $\mu = (|\alpha| + |\beta|)/|2\alpha\beta|$, if one assumes that Dirac delta is even.

25. $f = -(\alpha + \beta)/2$.

26. $y(0_-) = y(0_+)$, $\dot{y}(0_-) = e^{\gamma} \dot{y}(0_+)$.

27. $y(0_-) = e^{\gamma} y(0_+)$, $y'(0_-) = y'(0_+)$.

28. (a) $R(y) = \delta(y) \int_{-\infty}^{0} \phi(x) \, dx + \chi(y)\phi(y)$;
(b) $R(y) = \chi(|y| - 1)\phi((|y| - 1)$;
(c) $R(y) = \delta(y) \int_{-1}^{1} \phi(x) \, dx + \phi(|y| + 1)$.

29. The generalized particle density is

$$\rho(x, t) = m(t)\delta(x) + \bar{\rho}(x, t),$$

where $m(t) = \rho_0 \sqrt{wt}$ is the mass of the cluster of particles sticking at the origin, and

$$\bar{\rho}(x, t) = \rho_0 \frac{1}{2} \left(1 + \frac{|x|}{\sqrt{x^2 + wt}} \right)$$

is the continuous component of the density. It has a minimum $\bar{\rho}(0, t) = \rho_0/2$ at $x = 0$, and increases to its original value ρ_0 as $|x| \to \infty$ where the particles do not move.

The mass conservation law is here reduced to the requirement that the mass of the cluster is equal to the deficit of the mass of unglued particles:

$$m(t) = \int [\rho_0 - \bar{\rho}(x, t)] \, dx.$$

A direct substitution leads to the equality

$$\int_0^{\infty} \frac{dx}{\sqrt{x^2 + 1}(x + \sqrt{x^2 + 1})} = 1,$$

one of the standard definite integral that can be found in the tables of integrals or via *Mathematica*. Here it was derived via a "physical" argument.

A.3   Chapter 3. Fourier transform

1. We have

$$\tilde{f}_+(\omega) = \frac{1}{2\pi} \int_0^\infty e^{-(\gamma+i\omega)t} dt = \frac{1}{2\pi} \frac{1}{i\omega + \gamma}.$$

2. In view of the properties of the Fourier transform the answer can be obtained from the answer to the previous exercise:

$$\tilde{f}(\omega) = 2 \operatorname{Re} \tilde{f}_+(\omega) = \frac{1}{\pi} \frac{1}{\omega^2 + \gamma^2}.$$

3. $\tilde{f}(\omega) = (1/2\gamma)e^{-\gamma|\omega|}.$

4. $\tilde{f}(\omega) = \delta(\omega) - (\gamma/2)e^{-\gamma|\omega|}.$

5. The answer is

$$\tilde{f}(\omega) = \frac{1}{2\pi} \sum_{n=0}^\infty e^{(i\omega - \gamma)n} = \frac{1}{2\pi} \frac{1}{1 - \exp(i\omega - \gamma)}.$$

6. We have

$$\tilde{f}_o(\omega) = \operatorname{Im} \tilde{f}(\omega) = \frac{1}{4\pi} \frac{\sin \omega}{\cosh \gamma - \cos \omega},$$

$$\tilde{f}_e(\omega) = \operatorname{Re} \tilde{f}(\omega) = \frac{1}{4\pi} \left[\frac{\sinh \gamma}{\cosh \gamma - \cos \omega} + 1 \right].$$

7. $r(\gamma) = \coth(\gamma/2) \to \infty$ for $\gamma \to 0$.

8. The Fourier image of the derivative is $i\omega \tilde{f}(\omega) = (1/2)(e^{i\omega} - e^{-i\omega})$. Hence, by the shift theorem, $f'(t) = \pi[\delta(t-1) - \delta(t+1)]$. Integrating t, and assuming that $f(-\infty) = 0$ we get that $f(t) = \pi \Pi(t)$, where $\Pi(t) = 1$ for $|t| \le 1$, and $= 0$ for $|t| > 1$.

9. The second derivative $f''(t) = 2(\delta(t+1) + \delta(t-1) - \Pi(t))$. Using the answer to the previous problem we get that its Fourier transform is $(2/\pi)(\cos \omega - \sin \omega/\omega)$. Therefore $\tilde{f}(\omega) = (1/\omega^2)(2/\pi)(\sin \omega/\omega - \cos \omega)$.

10. The third derivative

$$f'''(t) = \delta(t+2) - 2\delta(t+1) + 2\delta(t-1) - \delta(t-2).$$

The corresponding Fourier image

$$\widetilde{f'''}(\omega) = \frac{i}{\pi} \Big(\sin(2\omega) - 2\sin(\omega) \Big).$$

Multiplying this expression by $-i\omega^3$ we finally obtain

$$\tilde{f}(\omega) = \frac{2\sin \omega(1 - \cos \omega)}{\pi \omega^3}.$$

11. The second derivative

$$f_l''(t) = \sum_{n=-\infty}^{\infty} \Delta_2 f(n\Delta)\delta(t - n\Delta),$$

where we used the standard notation

$$\Delta_2 f(n\Delta) = f((n+1)\Delta) - 2f(n\Delta) + f((n-1)\Delta)$$

for the second-order difference of function f. The corresponding Fourier image

$$\tilde{f}_l(\omega) = -\frac{1}{2\pi \Delta \omega^2} \sum_{n=-\infty}^{\infty} \Delta_2 f(n\Delta)e^{i\omega\Delta n}.$$

Regrouping terms of the above series, we arrive at

$$\tilde{f}_l(\omega) = \frac{2\sin^2(\omega\Delta/2)}{\pi \Delta \omega^2} \sum_{n=-\infty}^{\infty} f(n\Delta)e^{i\omega\Delta n},$$

which is more convenient for calculations.

12. First, observe that the sought Fourier image is related to the Fourier image of function $h(t)$ via the formula $\tilde{f}(\omega) = 2\pi|\tilde{h}(\omega)|^2$. Calculate first the Fourier image of function $h(t)$. Its second derivative is

$$h'' = \delta(t) - \delta(t - \theta) - \theta\delta'(t - \theta).$$

Thus the Fourier image of function $h(t)$ is

$$\tilde{h}(\omega) = \frac{\theta^2}{2\pi\Omega^2}\left[e^{-i\Omega}(1 + i\Omega) - 1\right],$$

where $\Omega = \omega\theta$ is a new dimensionless argument. In this way,

$$\tilde{f}(\omega) = \frac{\theta^4}{2\pi\Omega^4}\left[e^{-i\Omega}(1 + i\Omega) - 1\right]\left[e^{i\Omega}(1 - i\Omega) - 1\right],$$

or, finally

$$\tilde{f}(\omega) = \frac{\theta^4}{2\pi\Omega^4}\left[\Omega^2 + 2(1 - \cos\Omega - \Omega\sin\Omega)\right].$$

13. $f^{(4)}(t) = \delta(t) - \delta(|t| - \theta) - \theta\delta'(|t| - \theta) - \theta^2\delta''(t).$

14. Apply the Fourier transform to both sides of the recurrence relation to obtain

$$\tilde{f}(\omega; p) = \frac{p(p-1)}{p^2 - \omega^2}\tilde{f}(\omega; p - 2).$$

In the case of even $p = 2n$, the previous identity implies that

$$\tilde{f}(\omega; 2n) = 2n!\tilde{f}(\omega; 0)\prod_{m=1}^{n}\frac{1}{4m^2 - \omega^2}.$$

Substituting

$$\tilde{f}(\omega; 0) = \frac{1}{\pi} \int_0^{\pi/2} \cos(\omega t)\, dt = \frac{\sin(\pi\omega/2)}{\pi\omega},$$

we obtain

$$\tilde{f}(\omega; 2n) = 2n! \frac{\sin(\pi\omega/2)}{\pi\omega} \prod_{m=1}^{n} \frac{1}{4m^2 - \omega^2}.$$

Similarly, for odd $p = 2n + 1$,

$$\tilde{f}(\omega; 2n + 1) = (2n + 1)! \tilde{f}(\omega; 1) \prod_{m=1}^{n} \frac{1}{(2m + 1)^2 - \omega^2}.$$

A substitution

$$\tilde{f}(\omega; 1) = \frac{1}{\pi} \int_0^{\pi/2} \cos t \cos(\omega t)\, dt = \frac{\cos(\pi\omega/2)}{\pi(1 - \omega^2)},$$

gives, finally,

$$\tilde{f}(\omega; 2n + 1) = \frac{(2n + 1)!}{\pi} \cos(\pi\omega/2) \prod_{m=0}^{n} \frac{1}{(2m + 1)^2 - \omega^2}.$$

A.4   Chapter 4. Asymptotics of Fourier transforms

1. $f(x) \sim x^2$ $(x \to 0)$.

2. $f(x) \sim x/3$ $(x \to 0)$, $f(x) \sim 1/x$ $(x \to \infty)$.

3. $f(x) \sim x/3$ $(x \to 0)$.

4. It is clear that the root $x(a) \to 0$ as $a \to \infty$. Hence we can use the asymptotic formula $\cot x = 1/x - x/3 + O(x^3)$ and replace the original equation by a simpler quadratic equation

$$x^2 - \frac{3}{2}ax + \frac{3}{2} = 0 \tag{1}$$

which gives the following asymptotic behavior of the root:

$$x(a) \sim \frac{3}{4}\left(a - \sqrt{a^2 - \frac{8}{3}}\right) \qquad (a \to \infty). \tag{2}$$

Remark. A related question of finding positive roots of the transcendental equation

$$x = \tan x \tag{3}$$

also arises in several mathematical physics problems.

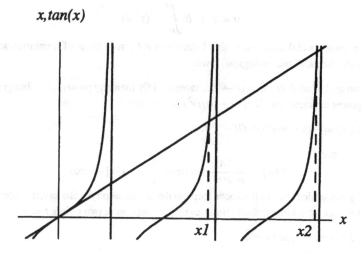

FIGURE A.4.1
Graphs of the functions x and $\tan x$. The first two roots, x_1, x_2, of equation (3) are marked on the x-axis.

The roots x_n (see Fig. A.4.1) can be expressed via the solution $x(a)$ of the equation (1) as follows:
$$x_n = a_n - x(a_n),$$
where
$$a_n = \frac{\pi(2n+1)}{2}.$$
Hence, in view of (2), the corresponding approximate values of the roots of (3) are
$$x_n \approx \frac{a_n}{4} + \frac{3}{4}\sqrt{a_n^2 - \frac{8}{3}}. \tag{4}$$

The larger n the more accurate the approximation is. However, even for $n = 1$, the approximate value $x_1 \approx 4.493397$ given by (4) differs from the true solution by less than $1.2 \cdot 10^{-5}$. So in many practical situations expressions (4) are as good as exact analytic solutions of the equation (3).

5. Taking logarithms of both sides to replace the product by a sum, we obtain that $\ln f(N) = \sum_{n=1}^{N} \ln(1+\alpha_n/N)$. The boundedness of sequence $\{\alpha_n\}$ implies that $\alpha_n/N \to 0$ as $N \to \infty$, uniformly in n. This permits use of asymptotics $\ln(1 + x) \sim x$, for each term separately, and adding them up. As a result, we get $\ln f(N) \sim (1/N) \sum_{n=1}^{N} \alpha_n$. Hence, $f(N) \sim \exp[(1/N) \sum_{n=1}^{N} \alpha_n]$.

6. The answer is
$$\prod_{n=1}^{N}(1 + \Delta\varphi(n\Delta)) = \exp\left(\int_0^t \varphi(\tau)\,d\tau\right)(1 - R),$$

$$0 < R < \Delta \int_0^t \varphi^2(\tau)\, d\tau.$$

We turn your special attention to this formula as it forms a basis of numerous scientific (physical, chemical and biological) laws.

7. Integral $S(\omega)$ decays, as $\omega \to \infty$, more rapidly than any power of ω. Integral $C(\omega)$ has a power asymptotics $C(\omega) \sim -(\alpha/\omega^2)$ $(\omega \to \infty)$.

8. $\tilde{f}(\omega) = -(6\alpha/\pi\omega^4) + O(\omega^{-10})$.

9. We have

$$\tilde{f}(\omega) \sim \frac{n!A}{\pi\omega^{n+1}} \sin(\omega\tau - \frac{\pi}{2}n) \quad (|\omega| \to \infty).$$

The trigonometric factor in the above formula describes, as physicists say, the interference contributions from hidden discontinuities of the function at two points $t = \pm\tau$.

10. $J(\omega) \sim -(3/\omega^2)\cos\omega$.

11. $\tilde{f}(\omega) \sim \sin\omega/(i\pi\omega^2)$, $(\omega \to \infty)$.

12. The answer is

$$\tilde{f}(\omega) \sim \frac{\Gamma(\beta+1)}{2\pi|\omega|^{\beta+1}} \exp\left(i\frac{\pi}{2}(1-\beta)\,\text{sign}\,\omega\right), \quad (\omega \to \infty).$$

13. One obtains

$$\tilde{f}(\omega) \sim \frac{\sin|\omega| - \cos|\omega|}{2\sqrt{\pi|\omega|}|\omega|}, \quad (|\omega| \to \infty).$$

A.5 Chapter 5. Stationary phase and related methods

1. Apply the general asymptotic formula (5.2.3). In our case, take $x = k\rho \to \infty$, $p(t) = \cosh(t)$, $f(t) = -1/2\pi$. There exists only one stationary point $\tau = 0$ where $p''(0) = 1$. Consequently, there remains only one term in the formula (5.2.3), and that term has to be multiplied by 1/2 because the stationary point coincides with the lower limit of integration. Hence,

$$G(\rho) \sim -\frac{1}{\sqrt{8i\pi k\rho}} \exp[-ik\rho], \quad (k\rho \to \infty),$$

an expression equivalent with (9.4.3).

2. Initially, it is easier to find asymptotics of the complex function $D_\omega(z)$ described in Chapter 4. To make use of it notice that $D_\omega(z)$ (5.6.1) is the Fourier transform of function

$$f(\phi) = 2e^{iz\sin\phi}[\chi(\phi) - \chi(\phi - \pi)],$$

which has two jumps: $\lfloor f(0) \rceil = 2$ and $\lfloor f(\pi) \rceil = -2$. So, in view of the asymptotic relation (4.3.3), we get

$$D_\omega(z) \sim \frac{1}{\pi \omega i}(1 - e^{i\omega \pi}), \quad (\omega \to \infty).$$

Now, separate the real and imaginary parts of this expression to obtain the sought asymptotics of Anger and Weber functions:

$$J_\omega \sim \frac{\sin(\omega \pi)}{\omega \pi}, \quad E_\omega \sim \frac{1 - \cos \omega \pi}{\omega \pi}, \quad (\omega \to \infty).$$

3. Again it is simpler to find initially the asymptotics of $D_\omega(z)$. Separation of it real and imaginary parts is straightforward. The asymptotics is obtained by the stationary phase method. There is only one stationary point $\phi^* = \pi/2$. Therefore, with help of formula (5.3.1), we get

$$D_\omega(z) \sim \sqrt{\frac{2}{\pi z}} \exp\left[i\left(z - \frac{\pi}{4}(2\omega + 1)\right)\right], \quad (z \to \infty).$$

4. In our case, calculation of integral (5.6.1) in the Fresnel approximation is reduced to an application of approximate equality

$$\sin \phi \approx 1 - (\phi - \phi^*)^2/2,$$

where, as in Exercise 3, the stationary point $\phi^* = \pi/2$, and to the extension of the domain of integration to the entire line. Thus, in the Fresnel approximation, the exact integral (5.6.1) is replaced by the approximate expression

$$D_\omega(z) \approx \frac{1}{\pi} e^{-iz} \int \exp\left[i\omega\phi - iz(\phi - \phi^*)^2/2\right] d\phi.$$

The last integral can be evaluated in closed form using the standard formula (3.2.3) which gives

$$D_\omega(z) \approx \sqrt{\frac{2}{\pi z i}} \exp\left[i\left(z - \frac{\omega\pi}{2}\right) + \frac{i\omega^2}{2z}\right], \quad (z \gg 1).$$

Notice that as $z \to \infty$ the above expression tends to the expression obtained in Exercise 3.

5. Let us write integral (5.6.1) in the form

$$D_\omega(\rho\omega) = \frac{1}{\pi} \int_0^\pi e^{-i\omega p(\phi)} d\phi, \tag{2}$$

where

$$p(\phi) = \phi - \rho \sin \phi, \tag{3}$$

which is more convenient for asymptotic analysis. In our case $(0 < \rho < 1)$ function $p(\phi)$ is strictly monotone so that there exists the unique and strictly monotone and smooth inverse function $\phi = q(s)$. Consequently, it is possible to write the integral (2) in the familiar Fourier transform form:

$$D_\omega(\rho\omega) = \frac{1}{2\pi} \int f(s) e^{-i\omega s} \, ds, \tag{4}$$

where

$$f(s) = q'(s)[\chi(s) - \chi(s - \pi)]. \tag{5}$$

This function has two jumps

$$\lfloor f(0) \rfloor = 2/P'(0) = 2/(1 - \rho), \quad \lfloor f(\pi) \rfloor = 2/P'(\pi) = 2/(1 + \rho).$$

Thus, in view of (4.3.3), we get

$$D_\omega(\rho\omega) \sim \frac{1}{\pi\omega i}\left(\frac{1}{1 - \rho} - \frac{e^{-i\omega\pi}}{1 + \rho}\right), \quad (\omega \to \infty). \tag{6}$$

Notice that this relation is a natural generalization of the answer to Exercise 2 (in the case $z < \omega$ ($\rho < 1$)) which follows from (6) by taking $\rho \to 0$.

6. Observe that in the case $\rho = 1$, the inverse function $\phi = q(s)$ is no longer smooth over the entire interval $\phi \in [0, \pi]$. Indeed, in the vicinity of $\phi = 0$ the original function $P(\phi)$ and its derivative $P'(\phi)$ have the following asymptotic behavior:

$$P(\phi) \sim \phi^3/6, \quad P'(\phi) \sim \phi^2/2, \quad (\phi \to 0).$$

Consequently, $\phi(s) \sim (6s)^{1/3}$ and the function $f(s)$ (5) has at $s = 0$ a singularity of the order

$$f(s) \sim \left(\frac{16}{9s^2}\right)^{1/3} = A s^{\alpha - 1}, \quad (s \to 0+),$$

where

$$A = (16/9)^{1/3}, \quad \alpha = 1/3.$$

Thus, the asymptotics of $D_\omega(\omega)$ is described by (4.6.2), i.e.,

$$D_\omega(\omega) \sim A \frac{\Gamma(\alpha)}{2\pi(i\omega)^\alpha}, \quad (\omega \to \infty).$$

Inserting the numerical values of constants A and α and utilizing the symmetrization formula $\Gamma(\alpha)\Gamma(1 - \alpha) = \pi/\sin(\pi\alpha)$ one gets

$$D_\omega(\omega) \sim \left(\frac{2}{9\omega}\right)^{1/3} \frac{\sqrt{3} - i}{\sqrt{2/3}}, \quad (\omega \to \infty). \tag{7}$$

7. In this case function $P(\phi)$ (3) has a simple stationary point $\phi^* = \arccos(1/\rho)$. For convenience, express ρ via an auxiliary variable θ: $\rho = 1/\cos\theta$, $0 < \theta < \pi/2$. Then $\rho^* = \theta$ and elementary calculations yield

$$P(\phi^*) = \theta - \tan\theta, \quad P''(\phi^*) = \tan\theta.$$

A substitution of these quantities into (5.3.1) gives

$$D_\omega(\rho\omega) \sim \frac{1}{\sqrt{2\pi\omega\tan\theta}} \exp\left[i\left(\omega(\tan\theta - \theta) - \pi/4\right)\right], \quad \rho\cos\theta = 1, \quad (\omega \to \infty). \quad (8)$$

Noticing that

$$\omega\tan\theta = \sqrt{z^2 - \omega^2}, \quad \theta = \arctan\sqrt{z^2/\omega^2 - 1},$$

it is easy to see that (8) is a natural generalization of the answer to Exercise 3 in the case $z > \omega$ ($\rho > 1$).

8. Transform (5.6.3) applying the following "physical" argument: if t represents the time and ω—the frequency, then it is natural to analyze the integral (5.6.3) in the dimensionless variable of integration $u = \omega t$ and rewrite the former as

$$I(\omega) = \frac{1}{\sqrt{2\pi\omega}}\Lambda(\gamma). \quad (9)$$

The deliberately separated factor $1/\sqrt{2\pi\omega}$ has the dimensionality of the original integral and

$$\Lambda(\gamma) = \frac{1}{\sqrt{2\pi}}\int_0^\infty \frac{e^{-iu}du}{\sqrt{u + \gamma^2}} \quad (10)$$

is a dimensionless function of a dimensionless variable $\gamma = \sqrt{\omega\tau}$. The substitution $v = \sqrt{u + \gamma^2}$ leads to

$$\Lambda(\gamma) = e^{i\gamma^2}\Phi(\gamma), \quad (11)$$

where

$$\Phi(\gamma) = \sqrt{\frac{2}{\pi}}\int_\gamma^\infty e^{-iv^2}dv = \frac{1}{\sqrt{2i}} - C(\sqrt{2/\pi}\gamma) + iS(\sqrt{2/\pi}\gamma)$$

is the complex Fresnel integral expressed through the Fresnel sine and cosine integrals discussed in Sections 5.3-4. The discussions of Section 5.4 indicate that

$$\Phi(\gamma) \sim \begin{cases} 1/\sqrt{2i}, & (\gamma \to 0); \\ 1/(i\gamma\sqrt{2\pi})e^{-i\gamma^2}, & (\gamma \to \infty). \end{cases}$$

Consequently, it follows from (9-11) that

$$I(\omega) \sim \begin{cases} 1/(2\sqrt{\pi i\omega}), & (\omega\tau \to 0); \\ 1/(2\pi i\omega\sqrt{\tau}), & (\omega\tau \to \infty). \end{cases} \quad (12)$$

Having arrived this far, we already have noticed the remarkable fact that the above formula contains the main asymptotics of our integral: for $\tau = 0$ (5.6.5) and for $\tau > 0$ (5.6.4). Although, for any $\tau > 0$, the formula (16) implies the asymptotic power law with $\alpha = 1$ ($\omega \to \infty$), for $\omega \ll \tau$, we already witness the appearance of the intermediate asymptotic power law with $\alpha = 1/2$ (5.6.5). For $\tau \to 0$, the region of frequencies where the power law with $\alpha = 1/2$ obtains, expands towards large ω's, and for $\tau = 0$ the power law with $\alpha = 1/2$ is valid everywhere.

9. Introduce a new function

$$x! = \Gamma(x+1) = \int_0^\infty e^{-t} t^x dt, \tag{13}$$

which is well defined for any real $x > -1$. For $x = n$, $n = 1, 2, \ldots$, it coincides with the factorial $n!$. Let us rewrite the integral (13) in the form suitable for asymptotic analysis by introducing a new variable of integration τ such that $t = x\tau$. Then

$$x! = x^{(x+1)} \int_0^\infty \exp[-xP(\tau)]\, d\tau,$$

where

$$P(\tau) = \tau - \ln \tau.$$

This function has a unique minimum at $\tau = 1$, where $P(1) = P''(1) = 1$. Hence, in view of (5.5.3),

$$x! \sim \sqrt{2\pi x}\, x^x e^{-x}, \qquad (x \to \infty),$$

which, in particular, gives the Stirling formula. Note that the asymptotic Stirling formula gives a good approximation of the factorial for any finite n. For $1! = 1$ we get a decent approximation 0.9221, and even for $(1/2)! = \sqrt{\pi}/2 \approx 0.8862$, the Stirling formula gives 0.7602. In a certain sense one can claim that the Stirling's formula is most precise for $x = 1$. The *relative error* decreases with the growth of n but the *absolute error* increases!

10. Introduce in the integral (5.6.9) a new variable of integration

$$y = x - v(x, t)t. \tag{14a}$$

Taking into account equation (5.6.6) satisfied by the solution v, the old variable of integration is expressed in terms of the new variable via the equality

$$x = P(y, t) := y + v_0(y)t, \tag{14b}$$

and the Fourier integral takes the form

$$\tilde{v}(\kappa, t) = \frac{1}{2\pi} \int v_0(y) e^{-i\kappa P(y,t)} dP(y, t).$$

Integrating by parts we arrive at a more convenient for asymptotic analysis expression

$$\tilde{v}(\kappa, t) = \frac{1}{2\pi i\kappa} \int v_0'(y) e^{-i\kappa P(y,t)} dy. \tag{15}$$

Here, and below, the prime denotes differentiation with respect to y. As long as $0 \le t < -1/u$, the function $P(y, t)$ is a strictly monotone smooth function, with $P'(y, t) > 0$ for any y, and the Fourier image (15) decays to zero, as $\kappa \to \infty$, more rapidly than any power function. However, at the time $t = -1/u$, there appears on the y-axis a point $y = z$ where $P'(z, -1/u) = 0$. It means that the behavior of $P(y, -1/u)$ in the vicinity of this point might have the power asymptotics of the integral (15) as $\kappa \to \infty$. To find this asymptotics expand $P(y, -1/u)$ in the Taylor series in powers of $(y - z)$:

$$P(y, -1/u) = P(z, -1/u) + \frac{1}{2}P''(z, -1/u)(y - z)^2 + \frac{1}{6}P'''(z, -1/u)(y - z)^3 + \ldots$$

The term of order 1 has disappeared since, in view of the assumption of this exercise, $v_0'(z) = -u$ and $P'(z, -1/u) = 0$. Moreover, since z is the minimum of function $v_0(z)$, we also have $P''(z, -1/u) = 0$. Suppose that

$$P = \frac{1}{6}P'''(z, -1/u) \neq 0.$$

Then the above Taylor expansion gives the asymptotic relation

$$P(y, -1/u) - P(z, -1/u) \sim P \cdot (y - z)^3, \qquad (y \to z).$$

Since $(y - z)^3$ is an odd function of $(y - z)$ we have to use the asymptotic formula (5.1.9) and obtain that

$$\tilde{v}(\kappa, -1/u) \sim -\frac{u\Gamma(4/3)}{2\pi i} \exp\left[-i\kappa P(z, -1/u)\right] \mathrm{Re}\left(Pi\kappa^4\right)^{-1/3}, \qquad (\kappa \to \infty). \quad (16)$$

In the concrete case (5.6.10), where

$$z = 0, \quad u = -1, \quad P(z, 1) = 0, \quad P = 1,$$

we get

$$\tilde{v}(\kappa, 1) \sim \frac{\Gamma(4/3)}{2\pi i} \mathrm{Re}\left(i\kappa^4\right)^{-1/3}, \qquad (\kappa \to \infty).$$

Remark. Physicists and engineers usually do not work with the complex Fourier image of the signal but with the real and nonnegative spectral density of the signal's energy

$$E_v(\kappa) = 2\pi |\tilde{v}(\kappa, t)|^2.$$

In this fashion, its integral over the entire κ-axis gives the "total energy" $\int v^2(x, t)\,dx$ of the signal. The corresponding asymptotics of the energy density at the time $t = -1/u$ is then

$$E_v(\kappa) \sim \frac{3u^2\Gamma^2(4/3)}{8\pi}|\kappa|^{-8/3}, \qquad (\kappa \to \infty).$$

A.6   Chapter 6. Singular integrals and fractal calculus

Principal value

1. Transforming the original partial differential equation by Fourier transform in the space and time coordinates x and t produces the algebraic equation

$$\tilde{u}(k^2 - \omega^2) = \frac{i\omega}{2\pi}\tilde{f}(\omega).$$

Its solution

$$\tilde{u}(k,\omega) = \frac{i}{4\pi}\tilde{f}(\omega)\left[\frac{1}{k-\omega} - \frac{1}{k+\omega}\right].$$

First let us calculate the inverse Fourier image in x

$$\tilde{u}(x,\omega) = \int \tilde{u}(k,\omega)e^{ikx}\,dk$$

of each of the two terms in the brackets and write

$$\tilde{u}(x,\omega) = \tilde{u}_-(x,\omega) - \tilde{u}_+(x,\omega),$$

where

$$\tilde{u}_\pm(x,\omega) = \frac{i}{4\pi}\tilde{f}(\omega)\int \frac{e^{ikx}\,dk}{k\pm\omega}.$$

To obtain $\tilde{u}_-(x,\omega)$ observe that in view of the causality principle the frequency ω appearing inside the integral should be replaced by $\omega - i0$. The resulting integral is then evaluated with the help of (6.2.4) which gives

$$\int \frac{e^{ikx}\,dk}{k-\omega+i0} = \mathcal{PV}\int \frac{e^{ikx}\,dk}{k-\omega} - i\pi e^{iax}.$$

To calculate the above principal value integral note that

$$\mathcal{PV}\int \frac{e^{ikx}dk}{k-\omega} = e^{iax}\,\mathcal{PV}\int \frac{e^{isx}}{s}\,ds = e^{iax}i\int \frac{sin(sx)}{s}\,ds = i\pi e^{iax}\,\text{sign}\,(x).$$

Substituting the right-hand side of this equation into the preceding expression we obtain

$$\int \frac{e^{ikx}dk}{k-\omega+i0} = -2\pi i e^{iax}\chi(-x).$$

So

$$\tilde{u}_-(x,\omega) = \frac{1}{2}\tilde{f}(\omega)e^{iax}\chi(-x).$$

Similar calculations give

$$\tilde{u}_+(x,\omega) = \frac{i}{4\pi}\tilde{f}(\omega)\int \frac{e^{ikx}\,dk}{k+\omega-i0} = -\frac{1}{2}\tilde{f}(\omega)e^{-iax}\chi(x).$$

Therefore

$$\tilde{u}(x, \omega) = \frac{1}{2}\tilde{f}(\omega) \exp(-i\omega|x|).$$

Finally, taking the inverse Fourier transform in ω we obtain

$$u(x, t) = \frac{1}{2}f(t - |x|).$$

As expected, the obtained solution satisfies the radiation condition which, in this case, means that the waves generated by a point source at the origin should run away from the origin.

Hilbert transform

2. $\psi(t) = (1/\pi) \ln |(\tau - t)/t|.$

3. The problem can be solved by passing to the Fourier images of the corresponding functions. We know that $\tilde{\varphi}(\omega) = \exp(-|\omega|\tau)/2\tau$. According to (6.5.2),

$$\psi(t) = -2 \, \text{Im} \int_0^\infty \tilde{\varphi}(\omega)e^{i\omega t} \, d\omega. \tag{1}$$

Hence,

$$\psi(t) = \frac{t}{\tau(t^2 + \tau^2)}.$$

4. $\psi(t) = ie^{i\Omega t} \, \text{sign} \, (\Omega), \quad \Omega \neq 0.$ In particular, if $\varphi(t) = \cos \Omega t$, then $\psi(t) = -\sin|\Omega|t.$ If $\varphi(t) = \sin|\Omega|t$, then $\psi(t) = \cos \Omega t.$

5. If one remembers that the Fourier image

$$\tilde{\varphi}(\omega) = \frac{1}{2\nu}[\chi(\omega + \nu) - \chi(\omega - \nu)],$$

then the simplest way to find $\psi(t)$ is to utilize the formula (1) which holds true for any real function $\varphi(t)$. Thus,

$$\psi(t) = -\frac{1}{\nu} \int_0^\nu \sin \omega t \, d\omega = \frac{\cos \nu t - 1}{\nu t}.$$

Analytic signals

7. Recall that the Fourier image of the original function is

$$\tilde{\xi}(\omega) = \frac{1}{4\tau} \left(e^{-|\omega - \Omega|\tau} + e^{-|\omega + \Omega|\tau} \right).$$

The corresponding Fourier image of the analytic signal (6.6.1) is equal to

$$\tilde{\zeta}(\omega) = 2\tilde{\xi}(\omega)\chi(\omega).$$

Utilizing the inverse Fourier transform

$$\zeta(t) = 2 \int_0^\infty \tilde{\xi}(\omega)e^{i\omega t} \, d\omega$$

yields

$$\zeta(t) = \frac{1}{t^2 + \tau^2} \left(e^{i\Omega t} + i\frac{t}{\tau} e^{-\Omega\tau} \right).$$

8. Two cases should be considered separately:

$$\zeta(t) = \frac{\sin \nu t}{\nu t} e^{i\Omega t}, \qquad \Omega > \nu,$$

and

$$\zeta(t) = \frac{1}{i\nu t} \left(e^{i\nu t} \cos \Omega t - 1 \right), \qquad \Omega < \nu.$$

Notice that in the first case the analytic signal $\zeta(t)$ is obtained from the original function $\xi(t)$ just by replacing $\cos \Omega t$ by $e^{i\Omega t}$. This is a particular case of a more general result that is the subject of Exercise 8. In the case $\Omega = \nu$ both expressions coincide.

9. *Proof:* It is evident that

$$\xi(t) = \frac{1}{2}[f(t) - ig(t)]e^{i\Omega t} + \frac{1}{2}[f(t) + ig(t)]e^{-i\Omega t}.$$

Now it follows from the assumptions that the Fourier image of the second summand is equal to zero for $\omega > 0$, hence the Fourier image of the first summand coincides with one-half of the Fourier image of the analytic signal. As a consequence, the analytic signal is twice the first summand.

Remark 1. Recall that the imaginary part of the analytic signal (6.6.2) is equal to minus the Hilbert transform (6.5.1) of $\xi(t)$. This implies the following corollary to the above result: If $\varphi = f \cos \Omega t + g \sin \Omega t$ then its Hilbert transform is $\psi = g \cos \Omega t - f \sin \Omega t$.

Remark 2. Signals with finite-support Fourier image seldom appear in electrical engineering applications. However, for narrow-band signals the replacement of the actual analytic signal by the expression (6.9.3) gives a rather good approximation. For example, the signal from Exercise 6 has an unbounded Fourier image but is narrow-band if $\Omega\tau \gg 1$. In this case, it is easy to see that the approximate expression

$$\zeta_a(t) = \frac{1}{t^2 + \tau^2} e^{i\Omega}$$

is very close to $\zeta(t)$.

10. Function $g(t)$ has the Fourier image $\tilde{g}(\omega) = i\delta'(\omega)$ supported by the single point $\omega = 0$. Hence, according to Exercise 8. $\zeta(t) = -ite^{i\Omega t}$. Generally, if in formula (6.11.1) $f(t) = P_n(t)$, $g(t) = Q_m(t)$, are polynomials of degrees n and m, respectively, then

$$\zeta(t) = [P_n(t) - iQ_n(t)]e^{i\Omega t}$$

because the Fourier transforms of these polynomials have a one-point support $\omega = 0$.

11. The corresponding analytic signal is

$$\zeta(t) = A(t)e^{i\Phi(t)} = A_1 e^{i\Omega_1 t} - iA_2 e^{i\Omega_2 t}.$$

Consequently,

$$A(t) = \sqrt{A_1^2 + A_2^2 + 2A_1A_2 \sin(\Omega_2 - \Omega_1)t}.$$

Assuming, for the sake of concretness, that $A_1 > A_2$ ($p = A_2/A_1 < 1$), we get

$$\Phi(t) = \Omega_1 t - \arctan\left(\frac{p\cos(\Omega_2 - \Omega_1)t}{1 + p\sin(\Omega_2 - \Omega_1)t}\right),$$

and

$$\Omega(t) = \frac{d\Phi(t)}{dt} = \frac{\Omega_1(1 + p\sin(\Omega_2 - \Omega_1)t) + \Omega_2(p^2 + p\sin(\Omega_2 - \Omega_1)t)}{1 + p^2 + 2p\sin(\Omega_2 - \Omega_1)t}.$$

12. For convenience, let us replace the original function $\xi(t)$ by its complex twin $\xi_c = \chi(t)e^{i\Omega t}$ and recall that the sought function $\eta(t)$ is equal to minus Hilbert transform (6.6.2) of the original function. So the corresponding complex twin

$$\eta_c(t) = -\frac{1}{\pi}\mathcal{PV}\int \frac{\xi_c(s)}{s - t}\,ds = -e^{i\Omega t}\frac{1}{\pi}\mathcal{PV}\int_{-\Omega t}^{\infty}\frac{e^{i\theta}}{\theta}\,d\theta.$$

Here we have introduced the new integration variable $\theta = \Omega(s - t)$. After splitting the integral into real and imaginary parts one gets

$$\eta_c(t) = e^{i\Omega t}\left[\frac{1}{\pi}\operatorname{Ci}(|\Omega t|) - i\left(\frac{1}{2} + \operatorname{Si}(\Omega t)\right)\right],$$

where

$$\operatorname{Ci}(x) = -\int_x^{\infty}\frac{\cos\theta}{\theta}\,d\theta, \qquad x > 0, \qquad \operatorname{Si}(x) = \int_0^x\frac{\sin\theta}{\theta}\,d\theta,$$

are, respectively, the *integral cosine and sine functions* introduced before. Clearly, the imaginary part of $\eta_c(t)$ coincides with the sought function $\eta(t)$ so that finally

$$\eta(t) = \frac{1}{\pi}\operatorname{Ci}(|\Omega t|)\sin\Omega t - \left(\frac{1}{2} + \frac{1}{\pi}\operatorname{Si}(\Omega t)\right)\cos\Omega t.$$

Notice that, in a sense, the corresponding analytic signal violates the causality principle. Indeed, at $t < 0$, when the original signal $\xi(t)$ is identically equal to zero, the analytic signal $\zeta(t)$, along with $\eta(t)$, is already nonzero.

13. The Fourier image (4.4.10) of $\xi(t)$ (4.4.9) has already been found. So, the corresponding analytic signal is equal to

$$\zeta(t) = \frac{\Gamma(\alpha)}{\pi i^{\alpha}}\int_0^{\infty}\omega^{-\alpha}e^{i\omega t}\,d\omega. \tag{2}$$

The last integral obviously reduces to integral (4.4.8) which we will rewrite in another, more suitable for our purposes, form ($0 < \beta < 1$):

$$\int_0^{\infty}\omega^{\beta-1}e^{i\omega t}\,d\omega = \Gamma(\beta)|t|^{-\beta} \times \begin{cases} i^{\beta}, & \text{if } t > 0 \\ i^{-\beta}, & \text{if } t < 0. \end{cases}$$

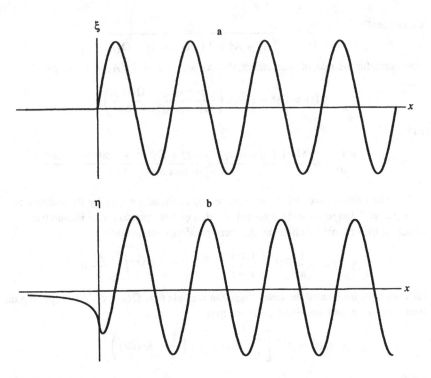

FIGURE A.6.1
Graphs of (a) ξ and (b) η as functions of the dimensionless variable $x = \Omega t$.

Taking $\beta = 1 - \alpha$ and substituting the result into (2) we obtain

$$\zeta(t) = \frac{|t|^{\alpha-1}}{\sin \pi \alpha} \times \begin{cases} \sin \pi \alpha + i \cos \pi \alpha, & \text{if } t > 0; \\ -i, & \text{if } t < 0. \end{cases}$$

14. It follows from the Hilbert transform property (c) of Exercise 5 that

$$\zeta_e(t) = \zeta(t) + \zeta^*(-t),$$

where the asterisk means the complex conjugate and $\zeta(t)$ is the analytic signal from the previous exercise. Thus

$$\zeta_e(t) = |t|^{\alpha-1} \left(1 + i \cot\left(\frac{\pi \alpha}{2}\right) \operatorname{sign}(t)\right).$$

15. Similar to $\xi(t)$, the sought function is a singular distribution. In order to find it, let us multiply the formal equality

$$\eta(t) = -\frac{1}{\pi} \mathcal{PV} \int \frac{\delta'(s)}{s - t} \, ds$$

by a test function $\phi(t)$ and integrate it over all $t1$'s. After simple distribution-theoretical transformations we get that

$$\int \phi(t)\eta(t)\,dt = \frac{1}{\pi}\mathcal{PV}\int \frac{\phi'(s)}{s-t}\,ds.$$

The last functional has been described in detail in Section 6.7 and the results of that section imply that

$$\eta(t) = \frac{1}{\pi}\mathcal{PV}\left(\frac{1}{t^2}\right).$$

16. Applying the Fourier transform to the last equality we get an algebraic equation

$$(i - e^{i\omega T})\tilde{\phi}(\omega) = 0, \qquad \omega > 0. \tag{3}$$

It has no ordinary continuous non-zero solutions $\tilde{\phi}(\omega)$, but there are distributional singular solutions like solution (1.5.5) of the equation (1.5.6). Indeed, if ω_n is a root of the "dispersion equation" $i = e^{i\omega T}$, then $\delta(\omega - \omega_n)$ is a solution of equation (3). Clearly,

$$\omega_n = \frac{1}{T}\left(\frac{\pi}{2} + 2\pi n\right), \qquad n = 0, 1, 2, \ldots,$$

so that we obtain

$$\phi(t) = \exp\left(i\frac{\pi}{2}\frac{t}{T}\right)\sum_{n=0}^{\infty} a_n \exp\left(i\frac{2\pi n}{T}t\right), \tag{4}$$

where a_n are arbitrary constants. In other words, we have found solutions

$$\phi(t) = \exp\left(i\frac{\pi}{2}\frac{t}{T}\right)\zeta(t), \tag{5}$$

where $\zeta(t)$ is an arbitrary periodical analytic signal with period T. Furthermore, since $\hat{H}$ is a real operator, both $\operatorname{Re}\phi(t)$ and $\operatorname{Im}\phi(t)$ are also solutions to this exercise. The simplest example here is obtained for $\zeta(t) \equiv 1$. Then $\phi(t) = e^{i\Omega t}$, $\operatorname{Re}\phi = \cos\Omega t$, $\operatorname{Im}\phi = \sin\Omega t$, $\Omega = \pi/2T$. A little more complicated example is obtained setting $a_n = a^n$, $|a| < 1$. Then, for instance,

$$\operatorname{Re}\phi(t) = \frac{\cos(\Omega t) - a\cos(3\Omega t)}{1 + a^2 - 2a\cos(4\Omega t)}.$$

Fractal calculus

17. As it often happens, it is simpler to derive a more general formula involving the integral

$$\int_{-\infty}^{t} dt_1 \int_{-\infty}^{t_1} dt_2 \ldots \int_{-\infty}^{t_{n-1}} dt_n a(t_1, t_2, \ldots, t_n)g(t_n). \tag{6}$$

This is a linear functional of $g(t)$ satisfying the causality principle and as such it has an integral representation

$$x(t) = \int_{-\infty}^{t} h(t, s)g(s)\,ds$$

the kernel thereof can be found by substituting $g(t) = \delta(t - s)$ in (6):

$$h(t, s) = \int_s^t dt_1 \int_s^{t_1} dt_2 \ldots \int_s^{t_{n-2}} dt_{n-1} a(t_1, t_2, \ldots, s).$$

In particular

$$h(t, s) = a_n(s) \int_s^t dt_1 a_1(t_1) \int_s^{t_1} dt_2 a_2(t_2) \ldots \int_s^{t_{n-2}} dt_{n-1} a_{n-1}(t_{n-1}).$$

18. We shall utilize the identity

$$g(b) = \int g(u) \frac{\partial}{\partial u} \chi(u - b) \, du.$$

Substituting it in the original integral and changing the order of integration we get

$$I = \int g(u) \frac{\partial}{\partial u} \left[\int \overset{n}{\ldots} \int_A a(x_1, \ldots, x_n) \chi\left(u - b(x_1, \ldots, x_n)\right) dx_1 \ldots dx_n \right] du.$$

Note that the inner n-tuple integral is an integral of function a over (not necessarily connected) domain $C_u = A \cap B_u$ created by intersection of domains A and B_u, where the latter is the set of points satisfying the inequality

$$b(x_1, \ldots, x_n) \leq u.$$

Denote that integral by

$$\psi(u) = \int \overset{n}{\ldots} \int_{C_u} a(x_1, \ldots, x_n) dx_1 \ldots dx_n.$$

So, the desired formula (also called the *Catalan formula*) has the form

$$I = \int g(u) \frac{d}{du} \psi(u) \, du.$$

If $b(x_1, \ldots, x_n)$ is a bounded continuous function with minimum m and maximum M then the Catalan formula simplifies to

$$I = \int_m^M g(u) \, d\psi(u).$$

A.7   Chapter 7. Uncertainty principle and wavelet transform

Function spaces:

1. Consider an auxiliary function $I(x) = \int (|f| + x|g|)^2 \, dt \geq 0$. Clearly,

$$I(x) = \|g\|^2 x^2 + 2Px + \|f\|^2,$$

so that the discriminant of this quadratic polynomial must be nonpositive, i.e.,

$$P^2 - \|f\|^2\|g\|^2 \le 0.$$

2. The norm's definition gives

$$\|f + g\|^2 = \int |f + g|^2 \, dt = \|f\|^2 + \|g\|^2 + 2P,$$

where P was defined in Exercise 1. Applying Schwartz inequality to the last term on the right-hand side we get

$$\|f + g\|^2 \le (\|f\| + \|g\|)^2$$

which implies the triangle inequality.

Windowed Fourier transform.

3.
$$\tilde{f}'(\omega, \tau) = i\omega \tilde{f}(\omega, \tau) + \frac{\partial}{\partial \tau} \tilde{f}(\omega, \tau).$$

4. Applying the windowed Fourier transform to both sides of the differential equation we obtain
$$\frac{d\tilde{x}}{d\tau} + (h + i\omega)\tilde{x} = \tilde{f}(\omega, \tau).$$

Its solution is
$$\tilde{x}(\omega, \tau) = \int_0^\infty \tilde{f}(\omega, \tau - \theta)e^{-(h+i\omega)\theta} \, d\theta.$$

5. (a) $f(t)e^{i\omega_0 t} \mapsto \tilde{f}(\omega - \omega_0, \tau)$;
(b) $f(t + \theta) \mapsto \tilde{f}(\omega, \tau + \theta)e^{i\omega\theta}$.

6. If the time-window $g(t)$ is sufficiently well localized to guarantee the existence of the Fourier image
$$\tilde{\varphi}(\omega) = \frac{1}{2\pi} \int g(t)e^{vt - i\omega t} \, dt$$

of function $\varphi(t) = g(t)e^{vt}$, then
$$\tilde{f}(\omega, \tau) = e^{v\tau}\tilde{\varphi}(\omega)e^{-i\omega\tau}.$$

Notice that in this case the Fourier image of the original function does not exists even in the distributional sense. However, the windowed Fourier image of $f(t)$ exists and can me measured experimentally.

7.
$$\tilde{f}(\omega, \tau) = e^{v\tau} \frac{1}{2} \left[\tilde{\varphi}(\omega - \omega_0)e^{-i(\omega - \omega_0)\tau} + \tilde{\varphi}(\omega + \omega_0)e^{-i(\omega + \omega_0)\tau} \right].$$

If $v\lambda \ll 1$ then approximately,

$$\tilde{f}(\omega, \tau) \approx e^{v\tau} \tilde{f}_0(\omega, \tau),$$

where $\tilde{f}_0(\omega, \tau)$ is the windowed Fourier image of the monochromatic signal $f_0(t) = \cos \omega_0 t$:

$$\tilde{f}_0(\omega, \tau) = \frac{1}{2}\left[\tilde{g}(\omega - \omega_0)e^{-i(\omega-\omega_0)\tau} + \tilde{g}(\omega + \omega_0)e^{-i(\omega+\omega_0)\tau}\right].$$

8. We know all the values of the function

$$\bar{F}(\omega, \tau) = \tilde{f}(\omega, \tau)[\chi(-\tau) + \chi(\tau - T)].$$

The latter, in view of equality (49), is related to the original signal as follows:

$$f(t)g(t - \tau)[\chi(-\tau) + \chi(\tau - T)] = \int \bar{F}(\omega, \tau)e^{i\omega t}\, d\omega.$$

After multiplication of both sides by $g^*(t - \tau)$ and integration over all τ we arrive at the equality

$$f(t)A(t) = \int d\tau \int d\omega\, \bar{F}(\omega, \tau)g^*(t - \tau)e^{i\omega t}.$$

If the coefficient

$$A(t) = \left[\int_{-\infty}^{t-r} + \int_t^\infty\right]|g^2(\theta)|\, d\theta$$

is different from zero everywhere then the value of the signal can be recovered via the formula

$$f(t) = \frac{1}{A(t)} \int d\tau \int d\omega\, \bar{F}(\omega, \tau)g^*(t - \tau)e^{i\omega t}.$$

Note that for the validity of the above formula it suffices that $g(t) \neq 0$ for all points of a certain interval of length $\lambda > T$.

Wavelet transforms

9.

$$D = 2\pi \int_0^\infty |\tilde{\psi}(\kappa)|^2 \frac{d\kappa}{\kappa}.$$

10. Passing in the integral to new variable of integration $\theta = s/\lambda$ we get

$$g_\epsilon(s) = \frac{1}{D|s|} F\left(\frac{|s|}{\epsilon}\right),$$

where

$$F(x) = \int_0^x K(z)\, dz.$$

The fact that the autocorrelation function of a real-valued mother wavelet is even was taken into account here. To write $g_\epsilon(x)$ in the standard form weakly converging to the Dirac delta let us introduce another notation:

$$\Phi(x) = \frac{1}{D|x|} F(|x|).$$

Then

$$g_\epsilon(s) = \frac{1}{\epsilon} \Phi\left(\frac{s}{\epsilon}\right)$$

and the necessary and sufficient condition of the weak convergence of the above function to the Dirac delta is that

$$0 < D = 2 \int_0^\infty \frac{F(s)}{s}\, ds < \infty.$$

Note that, in the process, we have obtained a new expression for the coefficient D in (49).

11. It is clear that in this case

$$K(z) = \frac{d^4}{dz^4} K_0(z),$$

where

$$K_0(z) = \int \varphi(s)\varphi(z-s)\, ds = \sqrt{\pi}\, \exp(-z^2/4).$$

Hence,

$$F(x) = \sqrt{\pi}\frac{d^3}{dx^3} \exp(-x^2/4) = \frac{\sqrt{\pi}}{8}(6x - x^3)e^{-x^2/4}.$$

Dividing this expression by x and norming it properly we get

$$\Phi(x) = \frac{6 - x^2}{8\sqrt{\pi}} \exp(-x^2/4).$$

The graph of this function is displayed on Fig. A.7.1.

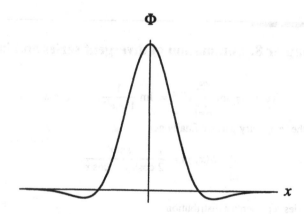

Φ

FIGURE A.7.1
The graph of function $\Phi(x)$ from Exercise 11.

12. First. let us write out the square modulus of the continuous wavelet transform (7.4.1):

$$|\hat{f}(\lambda, \tau)|^2 = |A(\lambda)|^2 \int dt_1 \int dt_2\, f(t_1)f^*(t_2)\psi^*\left(\frac{t_1 - \tau}{\lambda}\right)\psi\left(\frac{t_2 - \tau}{\lambda}\right).$$

Integrating over all τ and utilizing equality (7.4.40) we obtain

$$\int |\hat{f}(\lambda, \tau)|^2 \, d\tau = \int dt_1 \int dt_2 \, f(t_1) f^*(t_2) \lambda |A(\lambda)|^2 K \left(\frac{t_1 - t_2}{\lambda} \right).$$

Now it suffices to integrate both sides of the equality over $\lambda > 0$. To use the distributional relation (1) and the local probing property of the Dirac delta we shall divide both sides by $D\lambda^3 |S(\lambda)|^2$. So, if the continuous wavelet transform is defined by formula (7.4.5) then

$$\|f\|^2 = \frac{1}{D} \int_0^\infty \frac{d\lambda}{\lambda^2} \int_{-\infty}^\infty |\hat{f}(\lambda, \tau)|^2 d\tau.$$

In this case,

$$E_f(\lambda, \tau) = \frac{1}{\lambda^2} |\hat{f}(\lambda, \tau)|^2$$

has a natural interpretation as the energy density of the signal in (λ, τ)-domain.

13.

$$\hat{f}_a(\lambda, \tau) = \frac{1}{\sqrt{a}} \hat{f}(a\lambda, a\tau).$$

14.

$$\hat{f}(\lambda, \tau) = \frac{A(\lambda)}{\lambda^{\alpha+1}} \varphi(\tau/\lambda),$$

where $\varphi(x) = \int |s|^\alpha \psi^*(s - x) \, ds$.

A.8   Chapter 8. Summation of divergent series and integrals

1.

$$S(x, y) = \text{Im} \sum_{m=1}^\infty e^{imz} = \text{Im} \frac{1}{1 - e^{iz}}, \qquad z = x + iy.$$

Separating the imaginary part we finally get

$$S(x, y) = \frac{1}{2} \frac{\sin x}{\cosh y - \cos x}.$$

2. The series represents a distribution

$$S = -\frac{1}{2} \frac{d}{dx} \sum_{m=-\infty}^\infty e^{imx}.$$

Recalling the generalized equality (8.7.8) we finally get

$$S = -\pi \sum_{m=-\infty}^\infty \delta'(x - 2\pi n),$$

The "prime" indicates differentiation with respect to x. If $\phi(x)$ is a test function in $\mathcal{D}$ then the functional corresponding to S acts

$$S[\phi] = \pi \sum_{m=-\infty}^{\infty} \phi'(2\pi n).$$

The compact support of $\phi(x)$ assures that there are only finitely many non-zero terms in the above series.

3. The Poisson summation formula (8.8.2) gives

$$\sum_{m=-\infty}^{\infty} f(m\Delta) \exp(-imx) = 2\pi \sum_{n=-\infty}^{\infty} \tilde{f}(x + 2\pi n).$$

In our case, $f(x) = (1/2) \operatorname{sign}(x)$. Its generalized Fourier image (see Section 4.5) is

$$\tilde{f}(\omega) = \mathcal{PV} \frac{1}{i\omega}.$$

Hence,

$$E[\phi] = -i \, \mathcal{PV} \int \frac{1}{x} \sum_{n=-\infty}^{\infty} \phi(x + 2\pi n)\, dx.$$

4.

$$S(\omega, \Delta) = 2\pi \sum_{n=-\infty}^{\infty} \frac{1}{\Delta} \tilde{f}((g(\omega) - 2\pi n)/\Delta).$$

Since the weak limit ($\Delta \to 0$) of function $\tilde{f}(x/\Delta)/\Delta$ is equal to $\delta(x)$, we have, weakly, that

$$\lim_{\Delta \to 0} S(\omega, \Delta) = \frac{2\pi}{|g'(\omega)|} \sum_{n} \delta(\omega - \omega_n),$$

where the summation is extended over all roots ω_n of the equation $g(\omega) = 2\pi n$.

5. The triple sum under consideration splits into the product of single sums

$$P(x) = \prod_{k=1}^{3} \sum_{m_k=-\infty}^{\infty} \exp(im_k a_k(x)).$$

Aplying formula (8.7.8) to each of these sums we get

$$P(x) = (2\pi)^3 \sum_{n_1,n_2,n_3=-\infty}^{\infty} \delta(a_1 - 2\pi n_1)\, \delta(a_2 - 2\pi n_2)\, \delta(a_3 - 2\pi n_3).$$

In view of the discussion in Section 1.9, each of the products of the 1-D Dirac deltas represents the 3-D Dirac delta

$$\delta(a_1 - 2\pi n_1)\, \delta(a_2 - 2\pi n_2)\, \delta(a_3 - 2\pi n_3) = \delta(a(x) - 2\pi n),$$

where n is a vector with integer components (n_1, n_2, n_3). As a result,

$$P(x) = (2\pi)^3 \sum_n \delta(a(x) - 2\pi n),$$

where summation extends over all vectors n. Recalling relation (1.8.1), we can rewrite the last formula in the form

$$P(x) = \frac{(2\pi)^3}{J(x)} \sum_n \delta(x - b(2\pi n)),$$

where $x = b(y)$ is a vector field inverse to $y = a(x)$.

6. Expressing the components of the above triple integral through the 3-D Fourier image

$$f(x) = \int \int \int \tilde{f}(k) \exp(i(k \cdot x)) \, d^3k,$$

we get

$$F(x) = \int \int \int d^3k \, \tilde{f}(k) \exp(i(k \cdot x)) \times$$

$$\times \sum_{m_1,m_2,m_3=-\infty}^{\infty} \exp\left[i\left(m_1(a_1 \cdot k) + m_2(a_2 \cdot k) + m_3(a_3 \cdot k)\right)\right].$$

Let us write the triple sum inside the integral in the vector form used in the previous exercise:

$$\sum_{m_1,m_2,m_3=-\infty}^{\infty} \exp\left[i\left(m_1(a_1 \cdot k) + m_2(a_2 \cdot k) + m_3(a_3 \cdot k)\right)\right] = \sum_m \exp\left(i(m \cdot a(k))\right),$$

where the vector field $p = a(k)$ has coordinates $a_l(k) = (a_l \cdot k)$, $l = 1, 2, 3$. The corresponding Jacobian of the transformation of k into p is the mixed triple product $J = (a_1 \cdot [a_2 \times a_3])$. In view of the results of Exercise 5, we have

$$\sum_m \exp\left(i(m \cdot a(k))\right) = \frac{(2\pi)^3}{(a_1 \cdot [a_2 \times a_3])} \sum_n \delta\left(k - b(2\pi n)\right),$$

where $k = b(p)$ is the vector field inverse to the vector field $p = a(k)$. Let us find it, noting that by definition $b(p)$ satisfies the identity $p = a(b(p))$. It is clear that $b(p)$ is a linear homogeneous function of p representable in the form

$$b(p) = \frac{1}{2\pi} \sum_{s=1}^{3} b_s p_s.$$

Constant vectors b_1, b_2, b_3 can be determined by substituting the last equality in the previous one. This gives

$$(a_l \cdot b_s) = \frac{1}{2\pi} 2\pi \delta_{ls},$$

where δ_{ls} is the Kronecker symbol, equal to 1 when the indices coincide and zero otherwise. The last equality means, in particular, that vector b_1 is perpendicular to the vectors a_2 and a_3. Consequently, $b_1 = C[a_2 \times a_3]$. The coefficient C can be found from the condition $(a_1 \cdot b_1) = 2\pi$. Proceeding similarly in the case of vectors b_2 and b_3 we get

$$b_1 = 2\pi \frac{[a_2 \times a_3]}{(a_1 \cdot [a_2 \times a_3])}, \quad b_2 = 2\pi \frac{[a_3 \times a_1]}{(a_1 \cdot [a_2 \times a_3])}, \quad b_3 = 2\pi \frac{[a_1 \times a_2]}{(a_1 \cdot [a_2 \times a_3])}.$$

In this fashion,

$$\sum_m \exp\left(i(m \cdot a(k))\right) = \frac{(2\pi)^3}{(a_1 \cdot [a_2 \times a_3])} \sum_n \delta(k - 2\pi b(n)).$$

Substituting the right-hand side of this equality in the original expression for $F(x)$ and using the probing property of the Direc delta we finally obtain

$$F(x) = \frac{(2\pi)^3}{(a_1 \cdot [a_2 \times a_3])} \times$$

$$\sum_{n_1,n_2,n_3=-\infty}^{\infty} \tilde{f}(b_1 n_1 + b_2 n_2 + b_3 n_3) \exp\left[i\left(n_1(b_1 \cdot x) + n_2(b_2 \cdot x) + n_3(b_3 \cdot x)\right)\right].$$

7. Recall that the discrete Fourier transform of function $f(t)$ is given by the series

$$\tilde{f}_\Delta(\omega) = \frac{\Delta}{2\pi} \sum_{m=-\infty}^{\infty} f(m\Delta) \exp(-im\Delta\omega). \tag{1}$$

In our case,

$$\tilde{\Pi}_\Delta(\omega) = \frac{\Delta}{2\pi} \sum_{m=-N}^{N} \exp(-im\Delta\omega),$$

where $N = \lfloor 1/\Delta \rfloor$, is the greatest integer $\leq 1/\Delta$. Clearly,

$$\tilde{\Pi}_\Delta(\omega) = \frac{\Delta}{2\pi}\left[-1 + 2\,\mathrm{Re}\sum_{m=0}^{N} \exp(-im\Delta\omega)\right]$$

$$= \frac{\Delta}{2\pi}\left[-1 + 2\,\mathrm{Re}\frac{1 - \exp(-i\Delta\omega(N+1))}{1 - \exp(-i\Delta\omega)}\right] = \frac{\Delta}{2\pi}\frac{\sin(\omega\Delta(N+1/2))}{\sin(\omega\Delta/2)}, \tag{2}$$

in view of the standard formula $1 + q + \ldots + q^N = (1 - g^{N+1})/(1 - q)$ for geometric sum.

Of course one has to remember that there is a certain indeterminacy embedded in any discrete Fourier transform, and the discrete Fourier image corresponds to infinitely many original functions $f(t)$. In particular, for a given value of N, the discrete Fourier image (2) is the same for all rectangular functions $\Pi(t/\tau)$ as long as τ satisfies inequalities $\Delta N < \tau < \Delta(N+1)$. On the other hand, fixing τ and varying Δ within the interval $\tau/(N+1) < \Delta < \tau/N$, we obtain different discrete Fourier images of the same original function. The selection of a particular value is a matter of taste. Here, we choose $\Delta = 1/(N + 1/2)$, which gives (3) its simplest form

$$\tilde{\Pi}_\Delta(\omega) = \frac{\Delta}{2\pi} \frac{\sin \omega}{\sin(\omega \Delta/2)}.$$

8. Recall that

$$\cos^4(\pi t/2) = 3/8 + (1/2) \cos(\pi t) + (1/8) \cos(2\pi t).$$

Hence, the discrete Fourier image has the representation

$$\tilde{f}_\Delta(\omega) = \frac{3}{8} \tilde{f}_0(\omega) + \frac{1}{2} \tilde{f}_1(\omega) + \frac{1}{8} \tilde{f}_2(\omega) \tag{3},$$

where

$$\tilde{f}_n(\omega) := \frac{\Delta}{2\pi} \sum_{m=-N}^{N} \exp(-im\Delta\omega) \cos(n\pi \Delta m).$$

It is easy to see that

$$\tilde{f}_n(\omega) = \frac{1}{2} \left[\tilde{f}_0(\omega + \pi n) + \tilde{f}_0(\omega - \pi n) \right], \tag{4}$$

where $\tilde{f}_0(\omega)$ is given by (3). Choosing, for the sake of concreteness $\Delta = 1/(N + 1)$, the smallest value for which (3) remains valid, we get from (3) that

$$\tilde{f}_0(\omega) = \frac{\Delta}{2\pi} [\cot(\Delta\omega/2) \sin \omega - \cos \omega]. \tag{5}$$

Substituting (5) into (4), and (4) into (3), we get

$$\tilde{f}_\Delta(\omega) = \frac{\Delta}{2\pi} \sin \omega \times$$

$$\times \left\{ \frac{3}{8} \cot\left(\frac{\Delta}{2}\omega\right) - \frac{1}{4} \left[\cot\left(\frac{\Delta}{2}(\omega + \pi)\right) + \cot\left(\frac{\Delta}{2}(\omega - \pi)\right) \right] \right.$$

$$\left. + \frac{1}{16} \left[\cot\left(\frac{\Delta}{2}(\omega + 2\pi)\right) + \cot\left(\frac{\Delta}{2}(\omega - 2\pi)\right) \right] \right\}.$$

This formula is valid without any additional preconditions. Its drawback is that it is difficult to compare with the ordinary Fourier image (4.3.22)

$$\tilde{f}(\omega) = \frac{3}{2} \pi^3 \frac{\sin \omega}{\omega(\omega^2 - \pi^2)(\omega^2 - 4\pi^2)}. \tag{6}$$

of function $f(t)$. However, after sweating through some heavy-duty trigonometric manipulations, we get the formula

$$\tilde{f}_\Delta(\omega) = \frac{\Delta}{2\pi} \sin^4(\Delta\pi/2) \sin \omega \cot(\Delta\omega/2) \times \tag{7}$$

$$\times \frac{2+3\cos(\Delta\pi)+\cos(\Delta\omega)}{[\cos(\Delta\pi)-\cos(\Delta\omega)][\cos(2\Delta\pi)-\cos(\Delta\omega)]}$$

the structure thereof transparently generalizes the structure of the Fourier image $\tilde{f}(\omega)$ to the case of the periodic function $\tilde{f}_\Delta(\omega)$. In particular, expression (7) permits to easily trace the limiting passage from $\tilde{f}_\Delta(\omega)$ to $\tilde{f}(\omega)$ as $\Delta \to 0$. For specific values of δ, the expression (7) can take an even simpler form. So, for example, for $\Delta = 1/3$ ($N = 2$), we have

$$\tilde{f}_{1/3}(\omega) = \frac{1}{48\pi} \sin\omega \cot\left(\frac{\omega}{6}\right) \frac{7+2\cos(\omega/3)}{1+2\cos(2\omega/3)}. \qquad (8)$$

Now, we can compare the discrete Fourier image (7) with the Fourier image (8) in the interval $0 < \omega < \Omega$, where $\Omega = \pi/\Delta = \pi(N + 1)$ is the halfperiod of the discrete Fourier image (7). To complete the comparison let us denote the ratio

$$R_N(x) = \frac{\tilde{f}_\Delta(\omega)}{\tilde{f}(\omega)}, \qquad x = \frac{\omega}{\Omega} = \frac{\Delta\omega}{\pi} = \frac{\omega}{\pi(N+1)}.$$

The graphs of $R_N(x)$, for $N = 2$ and $N = 10$, are shown of Fig. A.8.1.

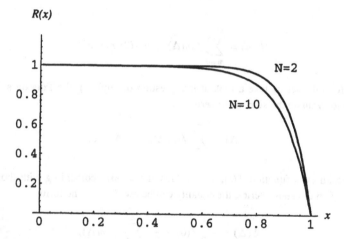

FIGURE A.8.1
The plots, for $N = 2$ and $N = 10$, of discrete and ordinary Fourier transforms ratio for function $f(t)$ from Exercise 8. The function has hidden singularities. The plots demonstrate that the discrete Fourier transform may be smaller than the ordinary one.

The plots indicate an intuitively surprising argument: the "rougher" discrete Fourier image can decay to zero, as ω increases, faster than the ordinary Fourier image of function $f(t)$. There is a simple explanation for this phenomenon which is obvious and convincing to an engineer and can have a heuristic value to a mathematician. The point is that the discrete Fourier image (7) is the Fourier image of not only the original function $f(t)$, the fourth derivative thereof is discontinuous, but also of many other functions, including some that are infinitely differentiable. Those, of course, have Fourier images $\tilde{f}(\omega)$ that decay, as $\omega \to \infty$, faster than any power of ω. As a result, the discrete Fourier image (7)

$$\tilde{f}_{\Delta}(\omega) = \sum_{n=-\infty}^{\infty} \tilde{f}(\omega + 2\pi n/\Delta),$$

in the interval $|\omega| < \Omega = \pi/\Delta$, and can turn out to be smaller than the Fourier image (6).

9. Taking advantage of the periodicity of $f(t)$ we can regroup the terms of the series (1) and rewrite it in the form

$$\tilde{f}_{\Delta}(\omega) = \sum_{m=0}^{M-1} f(m\delta) e^{-im\Delta\omega} \frac{\Delta}{2\pi} \sum_{n=-\infty}^{\infty} e^{-i\omega T n}.$$

Then, in view of the generalized equality (8.7.8),

$$\tilde{f}_{\Delta}(\omega) = \sum_{n=-\infty}^{\infty} \tilde{F}(n)\delta(\omega - 2\pi n/T), \qquad (9)$$

where

$$\tilde{F}(n) = \sum_{m=0}^{M-1} f(m\Delta) \exp(-i2\pi nm/M). \qquad (10)$$

10. Initially, let us take a look at the question of applying the Poisson summation formula to evaluate the sum of the series

$$S(\Delta) = \sum_{m=1}^{\infty} f(m\Delta), \qquad \Delta > 0.$$

Selecting an even function $f(t)$, with values at $t = m\Delta$ coinciding with those of the function f in the above series, the equality can be rewritten in the form

$$S(\Delta) = -\frac{1}{2}f(0) + \frac{1}{2} \sum_{m=-\infty}^{\infty} f(m\Delta),$$

which is more suitable for an application of the Poisson summation formula. The latter gives

$$S(\Delta) = -\frac{1}{2}f(0) + \frac{\pi}{\Delta} \sum_{n=-\infty}^{\infty} \tilde{f}(2\pi n/\Delta).$$

Since the Fourier image $\tilde{f}(\omega)$ of an even function $f(t)$ is also even, the above equality can be rewritten in the form

$$S(\Delta) = \frac{\pi}{\Delta}\tilde{f}(0) - \frac{1}{2}f(0) + \frac{2\pi}{\Delta} \sum_{n=-1}^{\infty} \tilde{f}(2\pi n/\Delta).$$

If $\tilde{f}(\omega)$ has a compact support, then the series on the right has only finitely many terms different from zero. In our case it is convenient to choose

$$f(t) = \Delta \frac{\sin t}{t}, \qquad \tilde{f}(\omega) = \frac{\Delta}{2}[\chi(\omega + 1) - \chi(\omega - 1)].$$

Substituting these function in the preceding equality we get

$$S(\Delta) = \frac{1}{2}(\pi - x) + \pi \sum_{n=1}^{\infty} \chi(1 - 2\pi n/\Delta),$$

so that finally

$$\sum_{m=1}^{\infty} \frac{\sin(m\Delta)}{m} = \frac{1}{2}(\pi - \Delta) + \pi \left\lfloor \frac{\Delta}{2\pi} \right\rfloor,$$

where $\lfloor x \rfloor$ is the largest integer $\leq x$. Note that the expression on the left hand side represent the Fourier series of the periodic function appearing on the righ-hand side.

11. Counting on the curiosity of the reader we take a look at a more general problem of the Fourier integral

$$f(t) = \int \tilde{f}(\omega) e^{i\omega t} d\omega, \tag{11}$$

where, in the case of a 2π-periodic function $f(t)$, the Fourier image appearing inside the integral

$$\tilde{f}(\omega) = \sum_{m=-\infty}^{\infty} \tilde{f}_m \delta(\omega - m). \tag{12}$$

If we replace (11) by the equality

$$f_N(t) = \int \tilde{f}(\omega) \tilde{h}(\omega, N) e^{i\omega t} d\omega \tag{13}$$

where $\tilde{h}(\omega)$ is the rectangular function, for example,

$$\tilde{h}(\omega, N) = \Pi(\omega/N), \qquad \Pi(t) = \chi(t + 1) - \chi(t - 1), \tag{13a}$$

then the series (12) retains only the needed number $2N + 1$ of terms. The corresponding original function is

$$h(t, N) = 2\sin(Nt)/t. \tag{13b}$$

Using formula (3.2.6), we can rewrite the equality (13), with the help of the convolution

$$f_N(t) = \frac{1}{2\pi} f(t) * h(t, N). \tag{14}$$

By (3.3.7), the function $h(t, N)/2\pi$ weakly converges, as $N \to \infty$, to the Dirac delta. So, if $f(t)$ is sufficiently smooth, then $f_N(t)$ converges pointwise (and even uniformly) to $f(t)$. However if, as is the case in this exercise, the original function $f(t)$ is only piecewise continuous, then it is necessary to study in more detail the asymptotic behavior (for $N \gg 1$) of the convolution integral (14) in the vicinity of discontinuities of the first kind.

Assume that the piecewise continuous function $f(t)$ has a jump at $t = \tau$ of size $\lfloor f \rfloor = f(\tau + 0) - f(\tau - 0)$. Hence, asymptotically, as $t \to \tau$,

$$f(t) \sim \frac{1}{2}[f(\tau - 0) + f(\tau + 0)] + \frac{1}{2}\lfloor f \rfloor \operatorname{sign}(t - \tau).$$

Substituting this expression into (13), we find that the behavior of $f_N(t)$ in the vicinity of the discontinuity point is described by the following asymptotics:

$$f_N(t) \sim \frac{1}{2}[f(\tau - 0) + f(\tau + 0)] + \frac{1}{2}\lfloor f \rfloor \operatorname{Si}(x), \qquad x = \frac{t - \tau}{N}, \qquad (15)$$

where the *integral sine function*

$$\operatorname{Si}(z) = \int_0^z \frac{\sin x}{x}\, dx. \qquad (16)$$

Observe certain features of the asymptotic formula (15). First of all, $\operatorname{Si}(0) = 0$, which means that at the discontinuity point $t = \tau$ of function $f(t)$, the truncated Fourier integral (13) (and in our case, the series in Exercise 11) converges to the arithmetic average of the one-sided limits of the function.

Secondly, if we remove the already analized first summand on the right-hand side of (15), place the discontinuity at the origin $t = 0$ ($\tau = 0$), and rewrite (15) in the form

$$f_N(t) \sim \lfloor f \rfloor g(t/N), \qquad g(x) = \frac{1}{\pi} \operatorname{Si}(x)$$

then a new, important in physics and engineering phenomenon can be observed. The odd function $g(x)$ entering in the above formula, normalized by the size $\lfloor f \rfloor$ of the jump, describes $f_N(t)$ behavior in the vicinity of the discontinuity. For $x \to \pm\infty$, we have $g(x) \to \pm 1/2$. However, since the integrand in (16) changes sign, the approach of $g(x)$ to the limit is not monotone. In particular, as is clear from the the graph of function $g(x)$ shown on Fig. A.8.2,
$g(x)$ assumes the maximal value $g(\pi) \approx 0.59$ for $x = \pi$ ($t = \pi/N$). This means that, for arbitrarily large N, at the distance π/N to the left and to the right of the jump point of function $f(t)$, the graph of function $f_N(t)$ has sharp up and down excursions. This anomalous behavior of function $f_N(t)$ in the neighborhood of jump points of function $f(t)$ is called the *Gibbs phenomenon*.

12. As before, we shall rely on formula (14). Observe that the Gibbs phenomenon was caused by the fact that function $h(t, N)$ entering (14) changed signs. So the phenomenon can be avoided if we select a nonnegative function $h(t, N)$ such that its Fourier image has a compact support. The latter is needed so that only finitely terms of the series from Exercise 11 are different from zero. So let us consider function

$$h(t, N) = A\frac{4\sin^2(Nt)}{t^2} \qquad (17)$$

as a candidate. Coefficient A will be selected later from a normalization condition, which, as is clear from (15), takes the form

$$\frac{1}{2\pi} \int h(t, N)\, dt = 1.$$

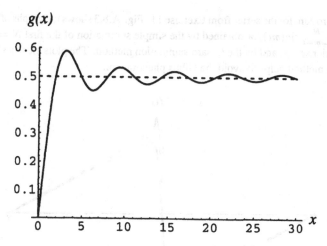

FIGURE A.8.2
Graph of the function $(1/\pi)\,\mathrm{Si}\,(x)$.

First of all, let us check that the Fourier image of h has a compact support. In view of (3.2.10), the Fourier image of the square of the original function is

$$\tilde{h}(\omega, N) = A\Pi(\omega/N) * \Pi(\omega/N).$$

The convolution of two functions with compact support also has compact support. In our case

$$\tilde{h}(\omega, N) = 2AN \begin{cases} 1 - |\omega|/2N, & \text{for } |\omega| \le 2N; \\ 0, & \text{for } |\omega| > 2N. \end{cases} \tag{19}$$

With this information it is easy to find the correct norming A, which has to be such that $\tilde{h}(0, N) = 1$. Thus $A = 1/2N$. It is not difficult to show that the above function $h(t, N)/2\pi$ weakly converges to the Dirac delta and thus guarantees the convergence of $f_N(t)$ to $f(t)$ at the continuity points of the latter. On the other hand, the positivity of function (18) removes the Gibbs phenomenon.

Finally, it should be mentioned that function (19) substituted in (12) preserves in the sum (13) $4N - 3$ terms: $2N - 1$ on the left and on the right of the terms $m = 0$. To keep only N terms of both sides of $m = 0$ one has to replace N by $(N + 1)/2$ in the preceding formulas. As a result we obtain that

$$\tilde{h}(\omega, N) = \begin{cases} 1 - |\omega|/(N + 1), & \text{for } |\omega| \le 2N; \\ 0, & \text{for } |\omega| > 2N; \end{cases}$$

$$h(t, N) = \frac{4}{N + 1} \frac{\sin^2((N + 1)t/2)}{t^2}.$$

The corresponding sum in the formulation of Exercise 12 assumes then the familiar shape

$$f_N(t) = \sum_{m=1}^{N} \left(1 - \frac{m}{N + 1}\right) \frac{\sin(mt)}{m} \tag{20}$$

of the Cesaro sum for the series from Exercise 11. Fig. A.8.3 shows the graphs of functions $f_N(t) = \sum_{m=1}^{N} \sin(mt)/m$ obtained by the simple summation of the first $N = 25$ terms of the Fourier series, and by the Cesaro summation method. The plots clearly shows how the Cesaro method helps to avoid the Gibbs phenomenon.

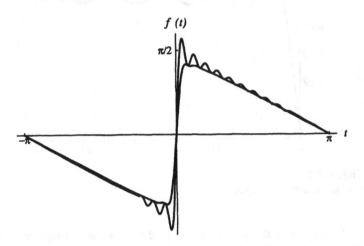

FIGURE A.8.3

The graphs of functions $f_N(t) = \sum_{m=1}^{N} \sin(mt)/m$ obtained by the simple summation of the first $N = 25$ terms of the Fourier series, and by the Cesaro summation method.

13. It suffices to select

$$\tilde{\psi}_0(\omega) = \frac{\delta}{\pi} \Pi((\omega - \omega_0)/\Omega).$$

The inverse Fourier transform

$$\psi_0(t) = e^{i\omega_0 t} \frac{2\delta}{\pi} \sin(\Omega t),$$

substituted into (8.9.16), gives

$$F(t) = \frac{2\delta}{\pi} e^{i\omega_0 t} \sum_{m=-\infty}^{\infty} f(m\delta) \frac{\sin(\Omega t - m\delta\Omega)}{t - m\delta} e^{-im\omega_0\delta}. \qquad (21)$$

Recall that one of the features of the analytic signal that is attractive for the engineers is that it uniquely determines its amplitude $A(t)$ and phase $\phi(t)$ so that

$$f(t) = A(t) \cos(\omega_0 t + \phi(t)).$$

To see this it is sufficient to write the complex signal in the form

$$F(t) = A(t) \exp(i\omega_0 t + \phi(t)),$$

separate the real and imaginary parts

$$F(t) = f_c(t)\cos(\omega_0 t) + i f_s(t)\sin(\omega_0 t),$$

and compare this equality with with the previous one to obtain the slowly evolving in time envelope

$$A(t) = \sqrt{f_c^2(t) + f_s^2(t)}$$

of the narrow-band signal.

14. Initially, let us construct an appropriate class of sufficiently smooth Fourier images $\tilde{\psi}(\omega)$ such that the corresponding origianal function $\psi(t)$ decays to zero (as $|t| \to \infty$) faster than $\psi_0(t)$ (8.9.6), and such that the series (8.9.16) converges faster than the standard Shannon's series (8.9.17). Our experience with the Fourier transform suggests that it is useful to write $\tilde{\psi}(\omega)$ in the form of the convolution

$$\tilde{\psi}(\omega) = \frac{\delta}{\Delta}\tilde{\psi}_0(\omega) * \frac{\mu}{\pi}\tilde{\varphi}(\mu\omega/\pi)$$

of the rectangular function $\delta\tilde{\psi}_0(\omega)/\Delta$, where $\tilde{\psi}_0(\omega)$ is given by the equality (8.9.6), and of the "enveloping" it compact-support function $\mu\tilde{\varphi}(\mu\omega/\pi)/\pi$. The constant μ is determined from the equality $\pi/\mu = \pi(1/\delta - 1/\Delta)$. If $\tilde{\varphi}(\nu)$, $\varphi(\tau)$ being the original function, vanishes identically outside the interval $|\nu| \geq 1$ and satisfies norming condition $\varphi(0) = 1$, then $\tilde{\psi}(\omega)$ is a function of type (8.10.5). With such choice of $\tilde{\psi}(\omega)$, the function of interest

$$\psi(t) = \frac{\delta}{\Delta}\frac{\sin\tau}{\tau}\varphi(\Delta\tau/\mu), \qquad \tau = \pi t/\Delta.$$

As $\tilde{\varphi}(\nu)$ one can take, for example, the sufficiently smooth function

$$\tilde{\varphi}(\Omega) = \begin{cases} (4/3)\cos^4(\pi\Omega/2), & \text{for } |\Omega| < 1; \\ 0, & \text{for } |\Omega| > 1; \end{cases}$$

the graph thereof is shown of Fig. 4.3.3b. Then, in view of (4.3.22),

$$\varphi(\tau) = 4\pi^4\frac{\sin\tau}{\tau(\tau^2 - \pi^2)(\tau^2 - 4\pi^2)}.$$

15. First, let us find the largest value of l which determines the length of the interval δ between readings (8.9.21). In our case, it follows from (8.9.22) that $l < 9/2$. Therefore, we choose $l = 4$. Thus, by (8.9.22), $\delta = 9\pi/2\omega_0$.

16. Taking advantage of the freedom to choose arbitrarily values of the Fourier image $\tilde{\psi}(\omega)$ in the intervals $\Omega < |\omega \pm \omega_0| < \pi/\delta - \Omega$, the widths thereof are $\rho = \omega_0/45$ in our case, let us smooth out the Fourier image $\tilde{\psi}_0(\omega)$ (8.9.24) with the help of convolution with the function

$$\tilde{\psi}(\omega) = \tilde{\psi}_0(\omega) * \tilde{\varphi}(\mu\omega/\pi), \qquad \mu = 90\pi/\omega_0.$$

Calculate the inverse Fourier image with the help of (8.9.25) to obtain

$$\psi(t) = \frac{\mu}{\pi}\varphi(\pi t/\mu)\frac{2\delta}{\pi t}\sin(\Omega t)\cos(\omega_0 t).$$

Recall, that $\Omega = \omega_0/10$, $\delta = 9\pi/2\omega_0$, $\mu = 90\pi/\omega_0$.

Appendix B

Bibliographical Notes

The **history of distribution theory** and its applications in physics and engineering goes back to

> [1] O. HEAVISIDE, On operators in mathematical physics, *Proc. Royal Soc. London*, **52**(1893), 504-529, and **54** (1894), 105-143,

and

> [2] P. DIRAC, The physical interpretation of the quantum dynamics, *Proc. Royal Soc. A, London*, **113**(1926-7), 621-641.

A major step towards the **rigorous theory** and its application to weak solutions of partial differential equations was made in the 1930s by

> [3] J. LERAY, Sur le mouvement d'un liquide visquex emplissant l'espace, *Acta Mathematica* **63** (1934), 193-248.

> [4] R. COURANT R., D. HILBERT, *Methoden der Mathematischen Physik*, Springer, Berlin 1937.

> [5] S. SOBOLEV, Sur une théorème de l'anayse fonctionelle, *Matematiceski Sbornik* **4** (1938), 471-496.

The theory obtained its **final** elegant **mathematical form** (including the locally convex linear topological spaces formalism) in a classic treatise of

> [6] L. SCHWARTZ, *Théorie des distributions*, vol. I (1950), vol II (1951), Publications de l'Institut de Mathématique de L'Université de Strasbourg,

which reads well even today.

In its **modern mathematical depth** and richness the distribution theory and its application to Fourier analysis and differential equations can be studied from many sources starting with massive multivolume works by

> [7] I.M. GELFAND et al. *Generalized functions*, 6 volumes, Moscow, Nauka 1959-1966.

© Springer Nature Switzerland AG 2018
A. I. Saichev and W. Woyczynski, *Distributions in the Physical and Engineering Sciences, Volume 1*, Applied and Numerical Harmonic Analysis, https://doi.org/10.1007/978-3-319-97958-8

and

[8] L. HÖRMANDER, *The Analysis of Linear Partial Differential Operators*, 4 volumes, Springer, Berlin-Heildelberg-New York-Tokyo 1983-1985,

to smaller, one volume monographs from research oriented

[9] E.M. STEIN, G. WEISS, *Introduction to Fourier Analysis on Euclidean Spaces*, Princeton University Press 1971,

[10] L.R. VOLEVICH, S.G. GINDIKIN, *Generalized Functions and Convolution Equations*, Moscow, Nauka 1994,

to, textbook style

[11] R. STRICHARTZ, *A Guide to Distribution Theory and Fourier Transform*, CRC Press, Boca Raton 1994,

[12] V.S. VLADIMIROV, *Equations of Mathematical Physics*, Moscow, Nauka 1981,

An **elementary**, but rigorous, **construction** of distributions based on the notion of equivalent sequences was developed by

[13] J. MIKUSIŃSKI, R. SIKORSKI, *The Elementary Theory of Distributions*, I (1957), II (1961), PWN, Warsaw.

[14] P. ANTOSIK, J. MIKUSIŃSKI, R. SIKORSKI, *Generalized Functions, the Sequential Approach*, Elsevier Scientific, Amsterdam 1973.

The **applications** of distribution theory have appeared in uncountable physical and engineering books and papers. As far as more recent, applied oriented textbooks are concerned, which have some affinity to our book, we would like to quote

[15] F. CONSTANTINESCU, *Distributions and Their Applications in Physics*, Pergamonn Press, Oxford 1980.

[16] T. SCHUCKER, *Distributions, Fourier Transforms and Some of Their Applications to Physics*, World Scientific, Singapore 1991.

which however, have a different spirit and do not cover some of the modern areas covered by our book.

The classics on **Fourier integrals** are

[17] S. BOCHNER, *Vorlesungen über Fouriersche Integrale*, Akademische Verlag, Leipzig 1932,

[18] E.C. TITCHMARSCH, *Introduction to the Theory of Fourier Integrals*, Clarendon Press, Oxford 1937,

with numerous modern books on the subject, including the above mentioned monograph [9] and elegant expositions by

[19] H. BREMERMAN, *Complex Variables and Fourier Transforms*, Addison-Wesley, Reading, Mass. 1965,

[20] H. DYM, H.P. McKEAN, *Fourier Series and Integrals*, Academic Press, New York 1972,

[21] T.W. KÖRNER, *Fourier Analysis*, Cambridge University Press 1988.

The **asymptotic problems** (including the method of stationary phase) discussed in this book are mostly classical. The well known reference is e.g.

[22] N.G. DE BRUIJN, *Asymptotic Methods in Analysis*, North-Holland, Amsterdam 1958,

with the newer reference being

[23] M.B. FEDORYUK, *Asymptotics, Integrals, Series*, Nauka, Moscow 1987.

The **special functions** have a rich literature including

[24] H. BATEMAN, A, ERDÉLYI, *Higher Transcendental Functions*, 2 volumes, McGraw-Hill, New York 1963,

[25] F.W.J. OLVER, *Asymptotics and Special Functions*, New York 1974,

and their connections with harmonic analysis on groups are explained, e.g., in

[26] N. YA. VILENKIN, *Special Functions and Group Representations*, Nauka, Moscow 1965.

As always, it is handy to keep around

[27] JAHNKE-EMDE, *Tables of Higher Functions*, Teubner-Verlag, Leipzig 1960,

[28] I.S. RYZHIK, I.M. GRADSHTEYN, *Tables of Integrals, Sums, Series and Products*, FM, Moscow 1963,

[29] M. ABRAMOWITZ, I.A. STEGUN, *Handbook of Mathematical Functions*, National Bureau of Standards, 1964

although the role of such compendia has been recently diminished with introduction of computer symbolic manipulation software such as *Mathematica* and *Maple*.

A good source on the mathematical theory of **singular integrals** is

[30] E.M. STEIN, *Singular Integrals and Differentiability Property of Functions*, Princeton University Press 1970,

with vast literature spread through mathematical journals. The well known sources on **fractal calculus** are

[31] A.H. ZEMANIAN, *Generalized Integral Transformations*, Interscience, New York 1968,

[32] K.B. OLDHAM, J. SPANIER, *The Fractional Calculus. Theory and applications of Differentiation and Integration to Arbitrary Order,* Academic Press, San Diego 1974,

[33] A.C. McBRIDE, *Fractional Calculus and Integral transforms of Generalized Functions,* Pitman, London 1979,

and more recent advances in the area can be gleaned from the collection of research papers

[34] A.C. McBRIDE, G.F. ROACH, Editors, *Fractional Calculus,* Research Notes in Mathematics, Pitman, Boston 1985.

The **wavelets** have obtained recently several excellent expositions, and the reader can benefit from consulting

[35] Y. MEYER, *Wavelets and Operators,* Cambridge University Press 1992,

[36] I. DAUBECHIES, *Ten Lectures on Wavelets,* SIAM, Philadelphia 1992.

[37] G. KAISER, *A Friendly Guide to Wavelets,* Birkhäuser-Boston 1994.

This is a very active research area and some new results have appeared in the following volumes of articles

[38] I. DAUBECHIES, Editor, *Different Perspectives on Wavelets,* American Mathematical Society, Providence, R.I. 1993,

[39] C.K. CHUI, Editor, *Wavelets: A Tutorial in Theory and Applications,* Academic Press, New York 1992.

The latter contains articles by D. Pollen on construction of Daubechies wavelets and G.G. Walter on wavelets and distributions which were used in Chapter 7. An engineering perspective can be found in

[40] A. COHEN, R.D. RYAN, *Wavelets and Digital Signal Processing,* Chapman and Hall, New York 1995.

The classic text on **divergent series** is

[41] G.H. HARDY, *Divergent Series,* Clarendon Press, Oxford 1949,

but the problem has broader implications and connections with asymptotic expansions and functional analytic questions concerning infinite matrix operators, see e.g.

[42] R.B. DINGLE, *Asymptotic Expansions,* Academic Press, New York 1973,

[43] I.J. MADDOX , *Infinite Matrices of Operators,* Springer-Verlag, Berlin 1973.

The modern viewpoint is presented in

[44] B. SHAWYER, B. WATSON, *Borel's Methods of Summability: Theory and Applications,* Clarendon Press, Oxford 1994.

Finally, an exhaustive discussion of the **Shannon's sampling theorem** and related interpolation problems can be found in

[45] R.J. MARKS II, *Introduction to Shannon's Sampling and Interpolation Theory*, Springer-verlag, Berlin 1991.

Index

© Springer Nature Switzerland AG 2018
A. I. Saichev and W. Woyczynski, *Distributions in the Physical and Engineering Sciences, Volume 1*, Applied and Numerical Harmonic Analysis, https://doi.org/10.1007/978-3-319-97958-8

Printed in the United States
By Bookmasters